Effective Police Administration

a behavioral approach

Effective Police Administration

a behavioral approach

Harry W. More, Jr.
EDITOR

JUSTICE SYSTEMS DEVELOPMENT, INC.
SAN JOSE, CALIFORNIA

© 1975, by Justice Systems Development, Inc.
All rights reserved. Published 1975.
ISBN: 0-914526-01-4
Library of Congress Catalog Card Number: 75-3503
Published by
Justice Systems Development, Inc.
P.O. Box 23884,
San Jose, California 95153

Printed in the United States of America

Administration of Justice Series

CONTEMPORARY CRIMINAL JUSTICE
Harry W. More, Jr. and Richard Chang,
Editors, San Jose State University

EFFECTIVE POLICE ADMINISTRATION
a behavioral approach
Harry W. More, Jr., Editor,
San Jose State University

Contributing Articles

(Reprinted with permission of the authors and/or the publisher.)

John E. Angell, "Toward an Alternative to the Classic Police Organizational Arrangements: A Democratic Model," *Criminology*, Vol. 9, No. 2 and 3, August-November, 1971, pp 185-206.

John Ashby, James L. LeGrande, and Raymond T. Galvin, "The Nature of the Planning Process," *Law Enforcement Planning*, Washington: U.S. Government Printing Office, 1968, pp. 1-17.

Robert R. Blake, Jane S. Mouton and Alvin C. Bidwell, "Managerial Grid," *Advanced Management – Office Executive*, Vol. 1, No. 9, September, 1962, pp. 12-15 and 36.

Eugene G. Columbus, "Management by Systems," *The Police Chief*, Vol. 37, No. 7, July, 1970, pp. 14-16.

J. Laverne Coppock, "Evolution in Police Management," *The Police Chief*, Vol. 39, No. 5, May, 1972, pp. 18-19, 76-79.

James J. Cribbin, "The Protean Managerial Leader," *Personnel*, Vol. 49, No. 2, March-April, 1972, pp. 8-15.

Richard M. Davis, "Police Management Techniques for the Medium-Sized Community," *The Police Chief*, Vol. 37, No. 7, July, 1970, pp. 44-50.

Wendell L. French and Cecil H. Bell, Jr., *Organization Development* Englewood Cliffs, New Jersey: Prentice-Hall, Inc., 1973, pp. 15-20.

G. Douglas Gourley, et. al., *Effective Police Organization and Management*, Report to the President's Commission on Law Enforcement and the Administration of Justice, Volume 7, Los Angeles: California State College at Los Angeles, October, 1966, pp. 965-978.

John P. Kenney, "Team Policing: A Model for Change," A Paper presented to the National Symposium on Urban Police Practices, Quanico, Virginia, Movember 12, 1972, pp. 1-16.

J.W. Lawrie, "Leadership and Magical Thinking," *Personnel Journal,* Vol. 49, No. 9, September, 1970, pp. 750-756.

Melvin J. LeBaron and Fred D. Thibault, "Organizational Team Building: The Next Generation of Training,"*Journal of California Law Enforcement,* Vol. 9, No. 2, October, 1974, pp. 71-80.

John A. McAllister, "Territorial Imperative," *California Law Enforcement Journal,* Vol. 7, No. 1, July 1972, pp. 13-20.

Dougals M. McGregor, "The Human Side of Enterprise," *The Management Review,* Vol. XLVI, No. 11, Nov. 1967, pp. 22-28, 88-92.

National Advisory Commission on Criminal Justice Standards and Goals, *Police,* Washington: U.S. Government Printing Office, January, 1973, pp. 132-133, 136-141 and 154-161.

National Commission on Productivity, *Opportunities for Improving Productivity in Police Services,* Washington: National Commission on Productivity, 1973, pp. 1-29.

Ernst K. Nillson, "Systems Analysis Applied to Law Enforcement," *Allocations of Resources in the Chicago Police Department,* Washington: Law Enforcement Assistance Administration, March, 1972, pp. 1-12.

Gary Pence, "Program Planning Budgeting System," *The Police Chief,* Vol. 38, No. 7, July, 1971, pp. 52-57.

Lyman K. Randall, "Common Questions and Tentative Answers Regarding Organization Development," *California Management Review,* Vol. XIII, No. 3, Spring, 1971, pp. 45-52.

W.R. Rhodes, "A Behavioral Science Appliction to Police Mangement," *The Police Chief,* Vol. 37, No. 5, May, 1970, pp. 46-50.

Revis O. Robinson, II, "Organization Development: An Action Plan for the Ontario Police Department," *Journal of California Law Enforcement,"* Vol. 8, No. 4, April, 1974, pp. 177-183.

George H. Savord, "Organizing for Team Leadership," *Journal of California Law Enforcement,* Vol. 8, No. 1, January, 1974, pp. 22-27.

Richard L. Shell and David F. Stelzer, "Systems Analysis: Aid to Decision Making," *Business Horizons,* December, 1971, pp. 67-72.

Robert Tannenbaum and Warren H. Schmidt, "How to Choose a Leadership Pattern," *Harvard Business Review,* May-June, 1973, pp. 162-180.

Peter C. Unsinger, "PERT and a Planning Problem," Paper presented at the Northern California Police Planners Association, May, 1974.

Preface

It is widely recognized by contemporary police administrators that the scope and complexity of police management is increasing daily. As society changes, it impacts on the organization and the manager must become increasingly sensitive to organizational problems.

Important challenges face the police manager:

- the increasing demand for controlling police discretion
- the court decisions which enforce selective hiring criteria
- the increasing politicization of the police
- the expansion and growth of police unions
- the need for providing adequate career opportunities for employees
- the changing needs of the community
- the changing nature of police employees

There are, of course, many other challenges, but each of these listed above will particularly demand greater sensitivity and flexibility. The successful police manager must be able to understand the changes taking place within the police enterprise and in our society. The police executive must view change as a challenge and not a problem. This book addresses itself to proven techniques of dealing with organizational change.

EFFECTIVE POLICE ADMINISTRATION embodies a wealth of new and important material focusing upon the behavioral aspects of police administration. It provides the major theoretical views and conceptual approaches to management as well as crossing the barrier between theory and practice by including examples to illustrate the application of the concepts.

The history of police management thought is effectively traced and a review of the major theories of management is achieved to include: grid management, theory X — theory Y, force field analysis and systems analysis.

This volume introduces the student and practitioner to the managerial functions of planning, organizing, fiscal management and budgeting. It identifies the major elements of organizational development and the implications of change in police organizations.

This comprehensive anthology of writings by outstanding authorities and renowned specialists is carefully woven together by dynamic introductions preceding each chapter. Special teaching aids include chapter objectives, discussion questions, annotated references and numerous examples of management techniques.

Harry W. More Jr.

Contents

1 HISTORY OF POLICE MANAGEMENT THOUGHT ... 3
    Introduction 4
    The Fielding's 4
    Robert Peel 6
    Leonhard F. Fuld 12
    Raymond B. Fosdick 16
    Elmer D. Graper 20
    Bruce Smith 25
    International City Management Association 28
    O.W.Wilson 31
    V.A. Leonard 34
    Conclusion 35

2 BEHAVIORAL SCIENCES 45
    Introduction 45
        Types of People in an Organization 46
        Forces on the Individual in an Organization ... 47
        A Metamorphosis of Management Theory 48
        Some Implications for Police Management 51
        Toward the Future 52
    Police Management in Transition 54
        Status Quo of Administration 55
        Force Field Analysis Theory 60
        Seven Step Plan for Problem Solving 66
    A Behavioral Science Application to Police Management 68
        Man is a Wanting Animal 68
        The Self-fulfillment Need 69
        A Comparison of Supervisory Roles 71
        Variations in Supervision 72
        The Key is Involvement in Work 73
        Does it Really Work? 73
        Where and How do you Start? 75
        Summary 75
    Theory X – Theory Y 77
        Management's Task: The Conventional View .. 77
        Is the Conventional View Correct? 79
        Physiological Needs 79
        Safety Needs 80

Social Needs	80
Ego Needs	81
Self-fulfillment Needs	82
Management and Motivation	82
The Carrot-and-stick Approach	83
A New Theory of Management	84
Some Difficulties	85
Steps in the Right Direction	86
Applying the Ideas	87
The Human Side of Enterprise	87
Managerial Grid	88
Production Incidental to Satisfaction	90
Aware of the Need to be Nice	91
Output Pressures are Lessened	92
Tries to Tighten up on Lower Levels	92
Accept Conflict as Inevitable	94
Morale Achieved is Task Related	95
Production Improvement	95
Sheds Light on Controversies	96
Territorial Imperative	97
Why Reorganize?	98
Which Way?	99
Valley Operations Bureau	99
Does it Work?	101
Who Gains?	102
Skeptics?	103
How Far?	104

3 SYSTEMS ANALYSIS 113

Introduction	113
Survey	115
Analysis	116
Synthesis	116
Documentation	117
Implementation	117
Planning and Research	117
Additional Benefits	118
Use of Data Processing	118
Systems Analysis Applied to Law Enforcement	119
Identification of Objectives	120
Using Systems Analysis	122

            Role of the Alternatives 124
            An Art in Infancy 125
            Describing a Police System 125
            Criminal Justice System 127
            Criminal Justice Systems Response 131
            Community Response 132
            Model of a Police System 132
            Police System Outputs 136
            Police System Objectives 139
        Systems Analysis: Aid to Decision Making 140
            What is Systems Analysis? 141
            The Basic Acts 141
            The Tenth Step 143
            Why is Systems Analysis Needed? 145
        Systems: Application to an Agency 149
            A Case Study 151
            Model Uses 153
            The Model 154
            Experiments 155
            Model Results 157
            Conclusion 158

4 MANAGERIAL FUNCTIONS 163
   Introduction 163
       Characteristics of the Existing Police Structures   164
       Classic Bureaucracies in General 165
       Problems Related to Police Bureaucracies . . . 166
   Organizing: A Democratic Model 171
       Overview of the Structure 171
       Expected Advantages of This Model 176
       Related Changes 177
       Conclusions 179
   Planning 180
       A Definition of Planning 181
       Departmental Goals and Planning 183
       The Process of Planning 185
           – Discovery of the Problem 185
           – Isolation and Clarification of the Problem 187
           – Collecting and Analyzing Pertinent Data 188
               – Review of the Literature 188
               – Records, Reports, and Documents   188

xvii

- Interviewing 189
- Questionnaire 190
- Observation 190
- Experimentation 190
- Identification and Evaluation
  of Alternatives 191
- Selection between Alternatives:
  The Final Decision 191

PERT and a Planning Problem 192
   The Concept 193
   The Problem 194
Fiscal Management 198
   Budget Planning and Preparation 200
   Fiscal Control 200
   Total Agency Involvement 204
Program Planning Budgeting System 205
   PPBS Requirements 206
   Methodology 207
   Priorities 209
   Budget Guidance 210
   Budget Structure 211
   Police Program Budget Report 212
   Purpose of Program Budget Report 212
   Criteria for Evaluation 213
   Police Planning 213
   Problems Related to the Implementation of PPBS 215
   Summation 217

5 ORGANIZATIONAL DEVELOPMENT 225
Introduction 225
Parameters of OD 230
   Briefly What is OD? 230
   Is OD Simply the Human Relations Movement
    in a New Format? 231
   Is OD Primarily Concerned with Restructuring
    Organizations? 231
   On what Concepts is OD Based? 232
   OD Concepts Regarding Work Groups
    and Organizations 232
   OD Concepts Regarding Individuals 234
   How Do Many OD Efforts Get Initiated? . . . 237

Is it True that OD Must Always Begin at
　　　the Top of the Organization? 237
　　Where Can an Individual Find an OD Program
　　　to Fit His Needs? 238
　　Does OD Have a Beginning, Middle, and End
　　　Like Most Programs? 238
　　Must a Manager Attend a T-Group or Management
　　　Training Laboratory Before He Can Undertake
　　　OD in His Group? 239
　　What if a Manager Cannot Afford the Time or
　　　Expense Demanded by an OD Effort? 239
　　If OD is so Promising, Why Aren't More Groups or
　　　Organizations Practicing It? 240
　　Why is it Often Difficult to get the Top Executives
　　　of an Organization Involved in OD? 240
　　If OD Encourages People to Express Their Feelings,
　　　Doesn't it Lead Toward Subjective Anarchy and
　　　Away from a Rational and Scientific Approach
　　　to Managing? 241
　　What Are the Most Common Obstacles to OD?　241
　　How Can You Evaluate the Pay-off of
　　　an OD Effort? 242
　　Aren't the Values on Which OD is Based Somewhat
　　　at Odds With Practical Organization Realities?　243
　　Is OD an Attempt to Build an Industrial Utopia?　243
Conclusion 244
Organizational Team Building 244
　　The Nickel Auction 244
　　Organizational Development 245
　　Team Building 246
　　Nature of Team Building Workshops 247
　　Results of Team Building 250
　　Readiness for Team Building 257
　　Conclusion 258
OD – An Action Plan 259
　　Purpose of this Study 260
　　Need for the Study 261
　　Scope of the Study 261
　　Organizational Diagnosis 262
　　Action-Planning 263
　　Evaluating the OD Program 267

       Measures to Increase Program Success 268
       Future OD Program Components 269
       Postscript 271

6 MANAGING FOR RESULTS 277
       Introduction 277
       Management by Objectives/Results 280
           Audit Check List 282
           Design 282
           Goals 288
           Writing Objectives 300
       MBO: The Human Side 304
           Appraisal 304
           Key Tasks 309
           Performance Standards 310
           Data Used in Measuring Performance 312
           Comments 312
           Key Results Analysis and Other Aspects of
             of the Personnel Function 313
           Conclusions 315

7 LEADERSHIP STYLES 321
       Introduction 321
           Consequences of the Myth: Followers 324
           Consequences of the Myth: The Leader 327
           Consequences of the Myth: The Organization . 329
           Diagnostic Leadership: An Alternative 330
       Managerial Leadership 334
           The Manager's Concept of the Leadership Process 335
           The Manager's Personality 336
           The Personality of the Work Group 337
           The Situation in Which Leadership is Exercised 340
           The Organizational "Givens" 341
       How to Choose a Leadership Pattern 343
           New Problem 343
           Range of Behavior 345
           Key Questions 348
           Deciding How to Lead 350
           Long-run Strategy 355
           Conclusions 356

8 TEAM POLICING 361
Introduction 361
Team Policing Models 372
    Problems 373
    Team Policing Systems 374
    Personnel Participation 379
    Training for Team Policing 379
    Community Involvement 380
Team Leadership 381
    Organizing for Team Leadership 381
    Compromise 383
    Principles 384
    Job Enlargement 386

9 POLICE PRODUCTIVITY 393
Introduction 393
    Problems with Current Police Measures 393
    Improving Measurement of Police Services . . . 397
        The Realm of Police Management . . . 397
        Measurement to Assist Management . . . 398
Productivity and Police Services 401
    What is Productivity? 401
    Productivity and Effectiveness 404
    The Process of Productivity Improvement . . . 404
    Establishment of Objectives 405
    Systematic Assessment of Progress 405
    Search for Improved Operating Methods 406
    Experimentation 406
    Implementation 407
Productivity in Police Patrol 407
    Measuring Patrol Activity 409
    Making More Patrolmen Available for
      Street Assignment 409
    Increasing the Real Patrol Time of
      Those Assigned 410
    Maximizing the Impact of Patrol 412
        Deterrence of Crime 412
        Apprehension of Criminal Offenders . . . 418
        Provision of Non-crime Services 424

Index 434

● **The study of this chapter will enable you to:**

1. Identify the contributions of the Fielding's to police management thought.
2. Write a short essay that delineates the impact that Robert Peel had on police organization.
3. List the primary functions of the British police.
4. Identify the three principles of detective administration discussed by Leonhard F. Fuld.
5. List the three causes of faulty police organization identified by Raymond B. Fosdick.
6. Write a short essay on Elmer D. Grapers concern for the excessive use of the "police detail."
7. Identify the eight types of police activities proposed by Bruce Smith.
8. Trace the shift in the management perspective in the numerous editions of Municipal Police Administration.
9. Compare the contributions of O.W. Wilson and V.A. Leonard.

1

History of Police Management Thought

In response to society's needs, organizations have evolved from simple and uni-dimensional entities to complex and multi-representations of reality; and management thought has been a product of that historical development. The identification of management as a unique function is of relatively recent origin.

Within this century, management has risen from obscurity to a dominant position in American society. Ours is a society of organizations shaped by the management process. In this short time, management has projected a number of philosophies that have communicated meaning and value to organization.

Management permeates every aspect of our society and has proven effective in meeting the demands of a highly industrialized society. It serves as a positive vehicle in arranging material and human resources to accomplish desired goals. The character of our nation is conditioned by our organizations and conversely our organizations condition the nation. The viability of organizations are shaped by management and the expertise being created has contributed to the nations pre-eminence.[1]

Never in the history of mankind has management played such an important part in the destiny of man. It exerts a dynamic influence on every facet of human existence through organizations such as business, education and government.

Introduction

Police organization has its roots in the political and social struggle that, through the centuries, has embroiled nations, resulting in divergent efforts of government to control the conduct of individuals.

The American police systems heritage is directly traceable to the police service of England and Wales.[2] The management of the Anglo-Saxon police system differed markedly from that found on the European continent. In the early part of the eighteenth century, even the word "police" was unacceptable. When considering this topic, Englishmen immediately turned their thoughts to the highly developed French police system that had proven a sinister and menacing force.

In France, the police were not a part of society but were superimposed upon it. They had their own code of conduct. Although their task was to regulate the people, it was never held accountable. The police depended directly upon the regime; and the regime was in desperate need of its police. Consequently, there was very little demand for impartial, disinterested control. The determination to preserve the status-quo made them mutually indispensable; it was a supreme objective that transcended any other consideration. Police corruption and police injustice were trifling things as long as the machinery of the police was strong, ruthless, and determined to provide for the continuing existence of the regime.[3]

The Fieldings

The image projected by the French police apparently mediated against police implementation in England and it wasn't until John Fielding began to publish a number of pamphlets, that the word "police" began to have some degree of respectability in English society.[4]

Henry Fielding was appointed to the office of magistrate at Bow Street in London in 1748. He must be given credit for having been the first one to seriously question the effectiveness of the parochial police forces existing at that time. He and his brother John believed that the position of magistrate was something more than a judicial appointment. In fact, their principal efforts were directed toward transforming the metropolitan magistrate into a more efficient police administrator.[5] Their goal was swift and fair justice. Typical of some of their innovations was to provide reliable information of offenses and offenders to justices of the peace, constables, and the general public.[6] It is interesting to note that John Fielding was positively convinced that offenders should be required to register with the magistrate. This, combined with a network of communications and advertisements by which the offenders crimes, movements, and personal appearance could be known throughout the country, was certain to lead to the ultimate eradication of crime.[7]

When Henry Fielding took office he found the parish police totally inadequate, so he selected the most reliable individuals from amongst the parish constables and organized them under his leadership and that of a second-in-command, Saunders Welch. This force proved to be real "Thief-catchers," and when compared to their predecessors, they were quite capable.[8]

After Henry Fielding's death in 1754 his brother John continued to work as a magistrate and for many years kept the Bow Street runners together in order to combat crime. John Fielding maintained a number of runners and a horse patrol that responded in a moments notice for the purpose of pursuing offenders. In addition, he later created a foot patrol consisting of unpaid constables, led by one of his men, who guarded parks and apprehended offenders. Unfortunately, through his many years of faithful service, the Bow Street runners were always in jeopardy due to the lack of financial support from the Treasury. It is also unfortunate that it was not until after his death that he and his brother were acknowledged as being true pioneers in police administration. The Fielding's perceived the police function as the apprehension of the offender. They created a fledging police organization based upon careful selection of personnel, centralization of authority, dissemination of information and strong leadership, but society was not ready to accept, support or expand this improved mode of police organization. In retrospect, the Fielding's were instrumental in laying the foundation for a civilian law enforcement agency whereas those that followed generated new concepts, techniques and procedures as a means of providing improved police services.

Robert Peel

In order to fully understand the difficulties in establishing a modern police force and the many barriers mediating against police reform, one must seriously consider the political and social events leading to the Metropolitan Police Act of 1829. The insipidiency of the industrial revolution and its social and economic consequences are of particular importance. Crime, hunger, poverty, and disease were commonplace. The local governments were in general disarray. Rioting and civil disorder were commonplace and the parocial police forces were unable to contain the numerous acts of civil disobedience.

The totally inadequate mechanisms for the maintenance of law and order and a constantly increasing crime rate had for many years defied attempts at reformation. The central issues at stake were political and social in origin and included:

1. A general fear that the police would be used as spies.
2. Concern that mercenary forces would be at the disposal of the central power.
3. Local autonomy would be undermined.
4. Local self-determination was a heritage too precious to be sacrificed.
5. Difficulty in identifying the proper role of local and central government.
6. A general displeasure with any body resembling the military.[9]

The opposition to police reform was overcome with the passage of Robert Peel's: "An act for Improving the Police In and Near the Metropolis."[10] Peel challenged those who saw his proposals as a direct threat to freedom and stated that it was in vain to talk of liberty of the subject when, "a large proportion of the inhabitants reside in the neighborhood of Twickenham and Brentford were under constant apprehension that their life and properties would be attacked; and such fears were entirely inconsistent with the free enjoyment of liberty and peace. It was the duty, as Peel viewed it, of the legislature to afford them protection against the causes which gave rise to such apprehension."[11]

The Police Act prescribed that a new police office be established at Westminister and that two justices be appointed to carry out the purpose of the act.[12] It provided that the justices control a single police force for Westminister and other parishes named in the act and that the new force replace the present parish watches.[13] The two justices were given the task of organizing, manning and disciplining the service and making regulations for it subject to the approval of the Secretary of State.[14]

The two men chosen to implement this new policing concept were Charles Rowan, who had been a military officer, and Richard Mayne, who had been a successful barrister.

According to British historian Charles Reith, Commissioner Rowan was instrumental in applying military concepts of organization to the new police service. Several models were in existence at that time; the French Gendarmerie Police, the County Police Forces of Scotland, the Irish Constabulary, or Fielding's Bow Street runners. But with Peel's new concept of a consolidated police system, Rowan patterned the new police force after the concepts developed by Sir John Moore. In 1803, Charles Rowan was a young lieutenant in the 52nd Regiment commanded by Sir John Moore. At that time Moore developed a single system, the purpose of which was to create a new flexible drill. He also envisioned a new system of discipline, predicated upon an improved relationship between officers and men making each group more effective.[15]

Moore's plan for reform resulted in a two part manual, one stressing Military Training and the other, Moral Training. The former emphasized in detail the new light infantry drill and movements; the latter provided the methodology for raising the status of the soldier.[16]

Rowan was apparently strongly influenced by Moore's plan and although no comprehensive description of the drill was recorded, it is clear that Rowan called upon his experience with the 52nd Regiment in adapting it to the new police system. The historian, Charles Reith, points out that Rowan's plan conceived the police force as a regiment and that he cleverly adapted the regimental organization to the demilitarized requirements of the civil police. This is particularly noteworthy in the General Instructions which were prepared for Peel by Commissioner Rowan:

Instructions

The following General Instructions for the different ranks of the Police Force are not to be understood as containing rules of conduct applicable to every variety of circumstances that may occur in the performance of their duty; something must necessarily be left to the intelligence and discretion of individuals; and according to the degree in which they show themselves possessed to these qualities and to their zeal, activity, and judgment, on all occasions, will be their claims to future promotion and reward.

It should be understood, at the outset, that the principal object to be attained is the Prevention of Crime.

To this, great and every effort of the Police is to be directed. The security of person and property, the preservation of the public tranquility, and all the other objects of a Police Establishment, will thus be better effected, than by the detection and punishment of the offender, after he has succeeded in committing the crime. This should constantly be kept in mind by every member of the Police Force, as the guide for his own conduct. Officers and Police Constables should endeavour to distinguish themselves by such vigilance and activity, as may render it extremely difficult for any one to commit a crime within that portion of the town under their charge.

When in any Division offences are frequently committed there must be reason to suspect, that the Police is not in that Division properly conducted. The absence of crime will be considered the best proof of the complete efficiency of the Police. In Divisions, where this security and good order have been effected, the Officers and Men belonging to it may feel assured that such good conduct will be noticed by rewards and promotion.

Local Divisions

1. The Metropolitan Police District will be formed into Police Divisions.
2. That part of the District which is at present under the charge of the Police Force, in the first instance, comprising

a large proportion of the town, is formed into five divisions.

3. The number of Men and Officers, and the constitution of the Force, are the same in each Division; but in laying out the Division, attention has been paid to local and other circumstances determining the number of men required; the superficial extent varies in the several Divisions, and consequently that portion of each which is committed to the care of one man.
4. Each division is divided into eight Sections.
5. Each Section into eight Beats.
6. The limits of each of these is clearly defined; each is numbered, and the number entered in a book kept for the purpose.
7. Each Division has an appropriate local name, and is also designated by a letter of the alphabet.
8. There is in every Division a Station or Watch-house, placed as conveniently for the whole, as may be, according to circumstances. From this point all the Duty of the Division is carried on.
9. The men belonging to each Section shall, as far as may be found practicable, lodge together near to the place of the Duty, in order to render them speedily efficient, in case the services of such as are off Duty should be required for any sudden emergency.

Police Force

1. The Police Force consists of as many Companies as there are Divisions, one Company being allocated to each Division.
2. Each Company is formed as follows:
Superintendent,
Inspectors,
Sergeants,
Police Constables.
3. The Company is divided into sixteen Parties, each consisting of one Sergeant and nine men.
Four Sergeant's parties, or one-fourth part of the Company, form an Inspector's party.

4. The whole Company is under the command of the Superintendent.
5. Each Man is conspicuously marked with the letter of his Division, and a number, corresponding with his name in the books, so that he can at all times be known to the public.
6. The first sixteen numbers in each Division denote the Sergeants.
7. A police Company is attached to the Office of the Commissioners, for the duty in the immediate neighborhood, and is also applicable to general purposes.[17]

In addition to the above, the General Instructions set out the basic conditions of the police service, as well as an outline of general duty. In the section of the police constable, Rowan extended Moore's idea of the ideal relationship between officers and men when he established the criteria for relationship between policemen and the public. The constable was told:

> He will be civil and attentive to all persons, of every rank and class; insolence and incivility will not be passed over . . .
>
> He must be particularly cautious not to interfere idly and unnecessarily; when required to act he will do so with decision and boldness . . . .
>
> He must remember that there is no qualification more indispensable to a Police Officer than a perfect command of temper, never suffering himself to be moved in the slightest degree, by any language or threats that may be used; if he does his duty in a quiet and determined manner, such conduct will probably induce well-disposed by-standers to assist him should he require it.[18]

The primary functions of the police were to be:

1. Preservation of the peace.
2. The prevention of crimes.
3. The detection and committal of offenders.

The two new justices moved swiftly, and within short order they had approval for a force of eight superintendents, twenty inspectors, eight-eight sergeants and nearly nine hundred constables, to serve the divisions selected for the first application of the act. Within a year all of the parishes were brought under the control of the single police force which resulted in seventeen divisions with some three thousand men.[19]

The general guidelines established by the two justices under the supervision of Robert Peel, provided:

1. The principal objective of the police is crime prevention.
2. The absence of crime is the best proof of police efficiency.
3. Daily reporting (presentment) of incidents to the Commissioners.
4. Serious crimes circularized in the "Police Gazette."
5. Separation of the supervision of the police machinery from the judiciary.
6. Creation of police divisions, sections and beats.
7. Establishment of various ranks with assigned duties:
    a. Superintendent
    b. Inspector
    c. Sergeant
    d. Constable
8. Establishment of centrally located police stations.
9. Creation of a divisional reserve.
10. Issuance of a "Beat-Card", to each constable, which listed the names of the streets on the beat.
11. Modification of the system to meet local needs.
12. Rotation of shifts; two-thirds of the force employed by night.[20]

The military background of Charles Rowan was obviously evident in the organization and structuring of the metropolitan police, nevertheless numerous precautions were taken to insure that this new body not resemble the military.[21] Blue uniforms were adopted and each member of the force was required to be "conspicuously marked" with the letters of their division and their own identification number. From the outset a general rule was adopted that the police should not carry sidearms but be equipped with a staff, a truncheon, and a rattle.

Probably the most important distinction in the function of the police was to emphasize crime prevention rather than detection.

In view of the tremendous opposition to a police force in England it is remarkable that within five years of its founding the force ceased to be regarded, except in the most radical circles, as a threatening innovation.[22]

Prior to Peel's time, the managment of the police enterprise in England was apparently conducted on a trial and error basis. There was no philosophical foundation to serve as a guide for police procedures or techniques. Rowan undoubtedly recognized the concept of managerial responsibility and provided an excellent foundation for centralized police services.

Rowan acknowledged the functions of management, particularly emphasizing organization, staffing and directing.

The military model provided a structural framework for the New Police based upon hierarchy, span of control, unity of command, and unity of direction.

A unique contribution to the art of police administration was the positively defined goal of crime prevention as the principal objective of the police service.

The Pellian concepts contributed to the initiation and development of police management. At the time of its inception it was utopian and served as a bridge between antiquated police services and the renaissance of law, order, and justice. Police management, as a unique process, has progressed from an undifferentiated process to an organized, sophisticated state utilizing the unifying force of military principles of organization and the establishment of an ideal relationship between the police and the public.

Leonhard F. Fuld

In 1909, Dr. Fuld released his critical study of police organization in the United States. This outstanding treatise seems to be forgotten by many of the more recent critics; however, it is decidedly a monumental work. He was concerned with the individual police officer as well as the police organization. He considered topics in a broad spectrum ranging from the selection of the police officer, to police organization, and the enforcement of the law. Fuld stressed that in all problems of administration the two factors of efficiency and humanity must be borne constantly in mind.[23]

He believed that before a police organization could be effective, the police commissioner or chief of police had to be a strong executive. He emphasized the fundamental principal in administrative organization which demands that the operatives be under the personal directions of a superior.[24]

Fuld was critical of the non-professional heads of police departments with particular reference to boards and commissions. He recognized the great administrative inefficiency in American municipal government and especially police departments where there was the greatest opportunity for corruption. He pointed out that in order to become a chief of police at that time it was necessary for an individual to have political influence, a good physical condition, the ability to pass the civil service examination, seniority, and in many cities, sufficient money to lubricate the appointing machinery.[25]

Fuld expressed the belief that in order to be efficient, the police must act quickly and decisively, particularly in times of emergency and he strongly suggested that this could only be accomplished by a single head of the organization.[26]

In discussing the investigative process he pointed out three principles of detective administration:

1. Appointment of a permanent Chief of Detectives.
2. Tenure for the individual detectives.
3. The squad of the central office detectives should be under the command of a Chief of Detectives.[27]

Regarding the individual police officer, Fuld stressed the need for them to live up to a higher code of conduct and morality than that demanded from other citizens and he believed the most important official duty of the police officer was the duty of knowledge. By this he meant that the officer must be thoroughly familiar with the occupation and character of every individual who resided within his jurisdiction and it was necessary for him to become acquainted with everything that took place in order to be a successful police officer.[28]

Furthermore, in Fuld's opinion, police business differed little from other great organizations pertaining to the need for personal direction by a superior because he found patrolmen to be almost invariably men of limited intelligence and education. He pointed out that patrolmen have a primary need for supervision because:

> The authority with which they are invested, and the respect shown them by the citizens, create in them inordinate desire to shirk their work or, as they themselves express it, to take it easy.[29]

In discussing the qualifications required for candidates of the police service, Fuld reviewed the need for civil service reform with special reference to the selection process; the written, mental, and physical examination; and personal history but he argues that the police officer must possess intellectual aptitude as well as physical power in order to succcessfully apply his analytical skills to the proper actions.

> It is this element of individual discretion which distinguishes a police officer from the soldier. The soldier is merely a part of a great military machine; it is his duty to obey the orders of a superior without individual reflection. The policeman, on the other hand, does not always, nor even generally act under the immediate supervision of the superior officer and, accordingly, he must determine by the exercise of a sound discretion whether he shall act or not, and if he decides to act, what he shall do and how he shall do it.[30]

Fuld was particularly concerned with the selection and the training of the police officer and he stated emphatically that all political considerations should be eliminated during the selection process. He stressed training by pointing out that the ideal police officer should be professionally trained by having completed a good secondary education followed by a special course of study in sociology and the problems of police duty.[31]

When considering the first line supervisor, Fuld saw the need for the ideal sergeant to possess four qualifications:
1. Ability to write and prepare written reports.
2. A thorough knowledge of police business.
3. Capable of being discrete and intelligent.
4. A rudimentary knowledge of criminal law.[32]

When considering other positions within the American police system Fuld expressed the belief that the captain of police is undoubtedly the most important police officer. In an average city of sufficient size to permit division into districts or police precincts he found that a captain was placed in charge. He felt the position had two broad duties—police and administrative. The police duties held the captain responsible for the preservation of the public peace and the protection of life and property in his precinct. Fuld defined the captains administrative duties as being of three kinds—clerical, janitorial and supervisory—but he expressed the belief that the latter was the most important and that in order to maintain good order in the precinct the captain had the power to post men under his command and assign them duties which would accomplish that goal.[33]

In discussing the staff of a police department, Fuld pointed out that the functions of the inspector was of prime importance in every large administrative system. He found that the lack of inspectors was undoubtedly one of the causes of the inefficiency of some of the American police departments. He suggested that inspection work was of the greatest value when it consisted of affirmative betterment by means of criticism of methods and not of individuals, and by the aid of general suggestions for improvements. He found this to be the principal function of the true inspector although he noted they should also report cases of serious delinquency uncorrectable by this method.[34]

Fuld viewed the numerous extraneous duties being acquired by police departments as inappropriate, describing this as one of the most serious defects in the administrative organization of American police departments. He stated there was absolutely no justification for any kind of special detail indirectly related to the police function. Police officers should not perform clerical duties, act as the major-domo for the head of another city department nor assist a private corporation protect their money.[35]

He also counseled against using police officers as probation officers or detailing them to higher courts, but he did feel that police officers should be detached to the fire department in order to conduct fire inspections, because the firemen were of low calibre and incapable of performing that duty.

Fuld's problem approach to police administration in the United states presented a logical exposition of the principles of police administration. The issues he discussed can be summarized as:

1. The elimination of politics from police administration.
2. Specialization of duties.
3. Duties clearly defined.
4. Constant supervision by supervisors.
5. Strong chief executive leadership.
6. Constant audit by inspectors.
7. Maintenance of discipline.
8. Comprehensive training of patrolmen.
9. Selection of personnel.
10. Elimination of non-police duties.

Fuld's principal focus was on the management process of controlling, emphasizing that man would only work effectively when closely supervised.

The problem approach to police administration, as he conceived it, could be rectified through strong leadership and comprehensive control mechanisms.

Raymond B. Fosdick

In 1915, Raymond Fosdick completed a comprehensive study of seventy-two cities in the United States with populations in excess of 100,000. He identified numerous police problems such as political control of the police, the disgraceful administration of criminal law by district attorneys and the judiciary, inadequate police leadership, limited tenure of office for chiefs of police, inadequate police techniques, organizational rigidity, and lack of supervision in investigative work. Fosdick points out that the advances made in police administration have been discouragingly irregular and places much of the blame on the political environment engulfing the American police system. He stated:

> Contrasted with other countries in this regard we stand ashamed. With all allowances for the peculiar conditions which make our task so difficult, we have made a poor job of it. Our progress has fallen far behind our needs. Successful in the

organization of business and commerce, pre-eminent in many lines of activity, we must confess failure in the elementary responsibility laid on all people who call themselves civilized, of preserving order in their communities.[36]

In his study, Fosdick found that the American police departments had the same basic organization. The structure of the department generally included at least two principal branches: the uniformed force, and the detective bureau. Most communities were divided into precincts and were in turn sub-divided into beats for patrol purposes. In some cities he found independent traffic squads and it was his conclusion that in order to have sound mechanical organization, it was necessary to fulfill three conditions:

1. The relationship between supervision and work must be well balanced.
2. The different parts of the mechanism must be adjusted to each other.
3. The whole machine must be adapted to its task.[37]

These three important aspects of management must be considered a significant contribution to police management.

Regarding the first principal, he found that the relationship between supervision and the police force was poorly balanced in most cities in the United States. He attributed this in part to too much supervision rather than too little.

He stressed the need for a single executive and decried the conflict apparent in many American cities between commissions, city managers, directors of public safety, and the chief of police. He found that when there was more than one executive the duties were poorly defined, if at all, and consequently many departments were top heavy with administrators.

When Fosdick analyzed the second condition of good organization demanding the harmonious adjustment of its parts, he found that its application was distinctly discouraging. He seldom found any considerable degree of orderly or systematic relationship between the various bureaus or divisions of the police department and he blamed this on a haphazard growth and a lack of deliberate organizational planning. Police departments were not organized at all and there was no concept of the organic relationship which one function or division logically bears to another. His analysis reflected that many police departments consisted of a number of small and relatively unimportant branches of the service reporting directly to the office of chief of police thus resulting in an excessive span of control. It was his opinion that faulty adjustment of the parts or branches of the police organization terminated in an inadequate distribution of the administrative burden and placed the head of the force in the position of being unable to control his organization.

He found that the harmonious relationship essential to good management practice was discouragingly inevident. In many cities he found the detective bureau managed as an independent organization consequently leading to a great deal of internal friction and misunderstanding. He felt this was obviously fatal to effective police service as was the general lack of understanding of the meaning of organization.

The third characteristic of good organization called for the adaptability of the police machinery to its work; this was, in his opinion, one of the greatest weaknesses in the American police department. He found the departments primitive and crude, having developed without design and whose purposes were never accurately determined or for the most part, only vaguely conceived. Fosdick stated that little exact study had been given to the relationship between patrol duties and crime conditions and he found that patrol arrangement had not been a subject of careful analysis. The lack of careful investigative study of the tasks the police were called upon to do made it impossible to determine the number of men needed to staff the police force. In brief, Fosdick stated that American cities have police organizations that are merely a conventional arrangement, sanctioned by tradition. The adoption of means-to-end, and of machinery-to-purposes, was conspicuously lacking.[38]

In discussing the faulty organization of American police departments, Fosdick pointed out many causes but emphasized three in particular. First he found that many cities had been saddled by legislatures with stereotype forms of organization ill-fitted to local conditions, and ordinances which spelled out the methods of operation which in turn were unadaptable to local needs. He found that legislative prescriptions had rendered immobile the internal arrangements of municipal police agencies.[39]

The second causal factor identified was the willingness of legislatures to complicate the police organization by the addition of extraneous and unrelated functions such as issuing licenses for saloons, restaurants, taverns and lodging houses; inspecting steam boilers; supervising the dog pound; censorship; collecting taxes; and many other unrelated tasks. The inclusion of the numerous miscellaneous activities detracted from the primary functions of the police—the protection of life and property and the maintenance of order.[40]

Fosdick felt, however, that the third and most important factor hindering additional development of the American police organization was inadequate leadership. He stated that unskilled, unfit, and unprofessional executives had wrought almost irremediable injury to the forms of police organizations.[41]

In combating issues confronting the police, Fosdick stated that the uniformed or patrol force was the first line of defense, but as he pointed out, they were incapable of handling the whole police task, necessitating a second stronghold identified as a corp of trained investigators or detectives. The third defense in a police department is the crime prevention service consisting of miscellaneous squads or units engaged in identifying conditions that created crime and suppressing offenses against the public morals. He explained that the functions of these three main branches of police organization were not mutually exclusive in that they must be interrelated, but that in one form or another the task of every police force involved these three approaches.[42]

Fosdick was especially concerned with personnel management and stated:

> In the final analysis, police business is but the aggregate of personal enterprise, and the quality of a department's work depends on the observation, knowledge, discretion, courage and judgment of the men, acting as individuals, who compose the various units or branches of the organization. The operation of mere organization machinery may achieve a high degree of perfection, but if the members of the force, acting in their individual capacities, go about their tasks without intelligence, tact or public spirit, the "system" counts for little and the whole administration is a failure. Only as the training of the

policeman is deliberate and thorough, with emphasis upon the social implications and human aspects of his task, can real success in police work be achieved.[43]

Fosdick found that police officers had a prevailing disposition to view their department as a mechanism created to operate automatically. More thought had been given to the perfection of the organization as a machine of smoothly working parts than to the task of adjusting the individual to his work. He found that this emphasis upon the mechanical side of organization could, in part, be attributed to numerous rules and regulations prescribing in detail the duties of each member of the department. Fosdick did not underestimate the value of an orderly machine but felt that it could be overemphasized to the point of rigidity and consequently the organization could cease to be a vital living thing. He found that the analogy between the army and the police service was largely fallacious and the insistence upon system and procedure making the military successful, had become the cause of defeat in the police force. He found that in many instances the organization stood between the police and the accomplishment of their objectives and the organizational machinery became an end in itself. He stressed that the work of a police force depended upon the capacity of the individual police officer working on his beat. He stressed that the work of a police force depended upon the initiative and imagination of the individual officer and that an overtly rigid organization would stifle good police work.[44]

Elmer D. Graper

In 1921, Elmer D. Graper published a hand-book on American police administration. He emphasized that administration relates to the organization, supervision and control of the processes of police work with a particular emphasis on police practice and procedures relating to the actual doing of daily routine work. In this work he made an effort to avoid dogmatic conclusions in discussing forms of organization and methods of work which vary in effectiveness according to the spirit, intelligence and ability of the police force. He identified certain forms of organization and systematic methods of directing the operations of police personnel that were observed to be best adapted to the most successful utilization of the personnel under varying conditions in many police departments throughout the United States.

One of his first concerns was the issue as to whether or not a police department should be headed by a single individual or by the board system. He found that the arguments advanced in behalf of the board form of organization for administrative heads of police departments was far from being conclusive and that those for the single commissioner type of organization were much stronger. Graper pointed out that the supervisor of a police department must make decisions and carry them out promptly and that his work was almost wholly of an administrative character.[45] The single commissioner or director who gives his whole time to the work, can supervise a police department better than can a board of commissioners.

Graper stated that, "centralization of authority in the hands of a single person—and that person held responsible for the way in which that authority is exercised—is the most promising principle to follow in police departments."

In the organization of a police force, he pointed out that there is a hierarchy of officers, as in a military organization, for the purpose of supervising patrolmen; and in both there are territorial sub-divisions such as districts and precincts for the distribution and control of the force by these officers. The object is to provide for each policeman, a superior officer charged with the duty of enforcing discipline in effective police service. Unsupervised patrols soon degenerate into loafing; hence the efficiency of any police organization depends largely upon the strictness with which each officer from the chief down to the sergeant enforces discipline and the honest performance of duty.

When considering the internal organization of a police department, Graper felt that two matters deserved attention. These were: function and distribution. That is to say, the members of the department should be organized into divisions or groups according to the special kinds of police service they perform, and they should be properly distributed throughout the various sections of the city. The functional organization should depend, therefore, upon the activities undertaken by the police. Dividing the organization into precincts or other divisions should depend to some extent on the particular geographical lay-out of the city in question. He found that in any sizeable city there were at least 5 different kinds of police work that every police department was called upon to perform. These were: (1) patrol service and traffic regulation; (2) detective work; (3) plainclothes service concerned with the suppression of gambling, prostitution, and violations of the liquor laws as well as with crime prevention duties; (4) business and clerical or secretarial services necessary for the control and maintenance of the entire organization; and (5) the medical and surgical care of the force. He felt that logically then, the functional organization of the police force would require at least 5 bureaus or divisions, one to correspond to each of the above-mentioned necessary police services. He felt that such an organization centered responsibility on a few individuals and promoted the most efficient management. Graper expressed concern over the proliferation of separate divisions or bureaus within a police department and pointed out that it was not uncommon to find the bureau of criminal identification a separate division responsible to the chief rather than a sub-division of the detective bureau with which it must obviously work in close cooperation if it is to fulfill the purpose of its existence. It is obviously easier for the head of a police department to supervise his agency through a small number of subordinates than through a large number. He pointed out the similar services should be organically related, allowing for centralized responsibility and promoted efficiency.[46]

With reference to the proper distribution of the force throughout the city, Graper argued that it was customary to divide large cities for the purpose of police service, into precincts or districts, and sometimes both. With reference to the detective function, it was pointed out that a detective function should be centralized and that even when detectives were assigned to precincts for general detective services, they should be under and directly responsible to the head of the detective bureau—not to the commanding officer of the police precinct or district. He emphasized that the wasteful system of having policemen detailed to perform clerical work cannot be too strongly condemned. Clerks, stenographers and statisticians should be employed in the secretarial bureau and policemen who enter the police department should perform only police services.

In his study, Graper expressed grave concern for the common practice of assigning patrolmen to a police detail. A detail was defined as an assignment relieving a policeman from his regular police work and giving him special duties to perform. The system of details, as it exists in many cities, is subversive to discipline because policemen generally look upon details as "easy berths" to be secured through political influence. This harmful influence upon police morals has been amply stated by a leading authority on police administration as follows:

> The baneful effect of political influence in a police force ... lies in the unavoidable conclusion that if a politician can secure for a policeman transfers, assignments, etc., which the policeman wants, that policeman "owes favors" to the politician and the favors are pretty sure to be paid in ways which give privileges to the politician and his friends, which should not be given, and which usually constitute a license to break the law. It is an inevitable circle; the policeman wants a detail—the politician gets it for him; the policeman is bound to do a similar service for the politician. And this does the double harm of breeding crime and breaking police morale.
>
> The proper procedure is quite opposite: the policeman wants a detail; he is granted an interview with the Police Commissioner, states his case, and if his record and ability warrant it, he gets the detail; this time, however, he is under no obligation to any one, except to his own record, and he has got the detail, not as a favor, but because his own work showed that he had rendered good service. The effect is to encourage the men to do ambitious, effective police service,—the diametric opposite of the other procedure.[47]

Graper stressed that the number of police details involving the members of the uniform police force should be kept at a minimum because they tended to weaken the strength of the patrol force and in addition, were certainly uneconomical.[48]

When discussing the organization of a detective bureau it was pointed out that although the detective division differs in its functions and methods from the uniform force, it is still a vital part of the police department and the success or failure of the whole organization depends largely upon the cooperation that exists between the two divisions. When the proper relationship is established, each of the divisions supplement each other, but when petty jealousy exists, the shifting of responsibility for crime conditions invariably follows. Hence, Graper stressed that the organization of the detective bureau should be such that it would tend to promote the proper degree of cooperation within the department.

Three methods of organizing detective divisions in municipal police departments were identified: the centralized system; the decentralized system of having detectives attached to a police district or precinct and subject to the control of that district commander; and in smaller cities, the system of temporary detective details. Graper strongly supported the central office squad form of organizations, where detectives are placed under the control of a chief of detectives who is responsible for every member of that division or bureau. This allowed for control over the men and made it possible to hold the chief of detectives responsible for local crime conditions.

With reference to the organization of special bureaus which had to do with the regulation of vice, gambling, liquor and drugs, Graper found that in the United States, some of the departments decentralized this function, and others centralized it. He also pointed out that in some cities, the two forms of organization were combined whereby precinct officers would be charged with the enforcement of these regulations but there would also be special squads operating throughout the city; the theory being that the latter will act as checks upon the efficiency of the former and will thus enable the head of the department to judge more accurately the efficiency of his men. Graper supported the centralized form of organization including special bureaus necessitated by the unique nature of the work and the requirement of a careful type of supervision.

With reference to the utilization of police women, Graper pointed out that it was a movement of recent years and one that was full of promise for the future. He noted that this work frequently was entitled 'welfare work' and he saw it as a widening of the scope of police activities so as to include measures to prevent or minimize the factors that contribute to vice, gambling, and lawlessness.[49] His study pointed out that the experience of those cities having employed police women showed that crime prevention was one of the most important and useful services a police department could perform and he recommended that the city should make adequate financial provision for such a function in their further development.

He described the numerous kinds of welfare work rapidly

gaining ground in police departments and he supported the position that the older theory of suppression was being replaced by the ideal of crime prevention—generally recognized as the highest and best kind of police work.

Graper emphasized every police department's need for a bureau whose function is the maintenance of permanent records of police activities. These records should deal with matters of personnel operations and the business aspects of the department. The records of the entire department should be under the supervision and control of the chief clerk or secretary responsible directly to the chief of police. He identified two purposes for keeping such records. In the first place, they should bring before the administrative head of the department such information relative to crime conditions and the activities of the police that would enable the chief executive to direct the men at his disposal to the best advantage. In the second place, records should provide the data which can be given to the public in order to allow it to judge intelligently what the needs of the department are, and the efficiency with which it is doing its work.[50]

Graper set forth the best practices prevailing in police administration in 1921 and he perceived the handbook as being most useful to superior officers who are charged with the task of management and to those members of the rank and file who aspired to higher administrative positions. The issues he discussed can be summarized as: (1) single executive leadership; (2) functional organization; (3) hierarchy; (4) coordination of divisions; (5) centralization vs. decentralization; (6) classification of police activities; (7) reduction of police details; (8) selection of detectives; (9) vice; (10) police women; (11) crime prevention; and (12) records management.

Grapers principal focus was on the functional organization of a police department combined with the distribution of personnel according to the needs of various sections of a city.

Bruce Smith

In 1940, Bruce Smith released the first edition of "Police Systems in the United States" and he found that the problems of administrative structure of police departments had often been neglected or ignored. He expressed the belief that the broad principles of organization that had found common acceptance in industrial and military circles had a definite application to the structure of police forces. The failure to apply the techniques of organization to police agencies stemmed from two major problems. The first was the fact that police departments had sprung from small beginnings and had only gradually acquired complex structural features. The second cause of defective organization was that only in the rarest instances had the American police force been led by men with prior experience in large scale management, and consequently they were not even aware of the existence of a problem of organization.[51]

Smith found that the vast majority of American police agencies continued to function according to patterns laid down several generations ago and that the administrative structure had not kept pace with the growth of the departments and the creation of specialized units. Such a pattern of administrative development resulted in inadequate controls and the organizations failed to accomplish the ultimate purpose—unity of action.[52] In tracing the stages of organizational growth, Smith identified the initial subdivision of the organization by the time of the day. With further growth in the size of the force, he then identified the process of subdivision by kinds of activity. This latter process illustrates the increasing move toward specialization of function in law enforcement and Smith points out that in larger departments specialization seems to be essential but that the head of an agency should be careful that he does not weaken the department by introducing too much specialization. The third process, evident in the growth of American police departments, is subdivision by areas of activity—the result of the territorial jurisdiction of a force which services a wide geographic area.

In his study, Smith used the term "span of control"—the concept is that one supervisor can control only a limited number of subordinates. He points out that a determination of the actual span of control must rest upon informed opinion in each specific situation, though no doubt a span of five or six is proper in many instances.[53]

He identifies a broad classification of police activities allowing for a preliminary grouping for the purpose of structural reorganization of a law enforcement agency: (1) patrol; (2) traffic regulation;

(3) criminal investigation; (4) communication and records control; (5) property management; (6) personnel management; (7) crime prevention; and (8) morals regulation.[54] Smith discusses the dangers in conflicting lines of authority between line and staff and the desire to effect some degree of functional supervision over local operations in a large or decentralized agency. He proposes that specialized headquarters units should develop uniform standards and methods for the field and should not exercise actual power to direct local operations under a given set of conditions. Staff operational control weakens the chain of command, destroys administrative responsibility, and impairs morale.

With reference to aides to the administrative head, Smith emphasizes the importance of the complete subordination of the deputy to the chief administrative officer. He expresses the belief that the administrative head should have the power to both designate and remove his deputy or assistant because to do less is to create a situation in which continuity of administrative policy is absent.[55] Smith calls for simplicity in the scheme of the organization because it reduces the number of responsible officers with whom the administrative head must deal and control if his policies are to be effective. Spheres of action should be clearly and boldly outlined in the scheme of the organization—this is best accomplished by grouping related activities allowing for the application of modern administrative processes.[56]

When discussing the difficult problem of controlling commercialized vice, Smith recommends the responsibility for enforcement be laid upon the patrol force but there should be a headquarters squad maintained either for the purpose of reviewing centralized vice enforcement or as a means of conducting raids on vice establishments. He points out that vice is often the natural ally of political machines—making it difficult to control. In the long run, each police administrator, in order to have successful vice control, must work to adjust the structural organization of the force to meet the demands made upon it.[57]

Smith shows a concern for the prospect of overspecialization in law enforcement and describes the continuing shift between generalization and specialization in each of the police functions. And while he acknowledges the increase in complexity in some of the areas of law enforcement and the need for specialization, he also emphasizes the importance of administrators striving to maintain flexible arrangement, particularly when specialized techniques threaten unity of action.[58]

Smith was especially interested in the application of the broad principles of organization to the structure of American police forces. The major issues which he presented can be summarized as: (1) subdivision by time of day (2) subdivision by kinds of activities (3) subdivision by areas of activities (4) the span of control (5) classification of major police activities (6) dangers of conflicting lines of authority (7) aides of the administrative head (8) application of organization principles (9) grouping of related activities (10) control of commercialized vice and (11) specialization.

Smith's principal focus was on the broad principles of organization in terms of their application to the structure of police forces. His extensive experience as a consultant to federal, state and municipal law enforcement agencies provided him with a unique background for a positive and enlightening analysis of management problems.

International City Mangement Association

This book first appeared in 1938 and there have been six subsequent editions. All of the editions have been created by multiple authorship and some of the most distinguished police administrators have served as editors. The 1938 edition was edited by the late Lyman S. Moore and three subsequent editions were edited by the late O.W. Wilson. The 1961 edition was edited by Richard L. Holcomb and the last edition was under the editorship of George Lee Eastman with the assistance of Esther M. Eastman.

The principal focus of each edition has been to provide police chiefs and other commanding officers with a better understanding of contemporary police principles and practices and to assist them in the management of their agencies.

In the 1943 edition of this text the task of the police administrator was divided into four parts: planning, organization, direction, and external relations.[59]

Police organization was defined as "the grouping of related tasks to assure more effective accomplishment, and the establishment of clear-cut channels of communication, authority, and responsibility." Contributors to the text pointed out that it is difficult to generalize about organization and to isolate any principles that are applicable to all organizations at all times. But they felt that there appeared to be at least six "principles" that could be suggested as being generally applicable to law enforcement agencies: (1) the work should be apportioned among the various individuals and units according to some logical plan. (2) The more complex the organization and the more highly specialized the division of work, the greater the need for coordinating authority and machinery at the top. (3) Lines of authority and responsibility would be made as difinitive and direct as possible. (4) Authority should be commensurate with responsibility. (5) There should be "unity of command" throughout the organization. (6) There is a limit to the number of subordinates who can be effectively supervised by any one officer, and care should be taken not to exceed this span of control.[60]

This study points out that in any police department where there is more than one officer, the problem of division of work will arise and in the field of municipal administration there are at least five different factors which determine organization: purpose, process or similarity of method, clientele, area, and time.[61] It was pointed out that if a police department were organized strictly according to the factors mentioned above, certain weaknesses would appear. It is recommended that a combination of the above five factors be used in the organization of the average police department and that inherent in these factors are the dangers of overspecialization.[62]

The study emphasizes that the problem of coordination can best be solved by arranging the personnel of a police department into a pyramid-like structure of responsibility and authority. Each part of the pyramid is responsible for, and has authority over, some portion of the work of the department. In addition to being a simple form of organization, the pyramid has other advantages: lines of authority are clearly defined, each member of the department knows exactly who supervises him, and the principle of unity of command is adhered to in this type of organizational pattern.

The disadvantages of the organizational pyramid are also emphasized in that it can cause excessively formal procedures, it does not allow for the full achievement of the various bases of specialization, and it tends to place too much responsibility on the chief administrator. It is suggested that the dangers inherent in the simple, direct-controlled type of organization can be eliminated, or at least partially offset by the use of auxiliary units and by providing the chief executive officer with managerial aides.

Typical of the auxiliary functions recommended are records, personnel, communications, maintenance and property management.[63] By grouping "housekeeping" functions into a separate unit, it is pointed out that competent and skilled personnel can be employed, duplication of effort is avoided, and auxiliary units can serve as channels of control and coordination for the total department.[64] In addition to auxiliary services or units, the problem of coordination can be solved by increasing the capacity of the chief by providing him with an assistant or a staff to assist him in his tasks. The managerial aide can assume some of the work load of the chief thus allowing the executive administrator to devote more time to the problem of coordination and management of the agency. It is pointed out that the addition of managerial aides to the staff of the chief of police does not contradict the principle of unity of command because the aides have no actual authority.[64]

Each of the subsequent editions have shown a shift in position, best illustrated by the 1971 edition that discusses the principles of organization and identifies them as a set of concepts or propositions believed by many to be the basis for sound organization. It is interesting to note that two additional principles are identified—coordination of effort, and administrative control.[65]

The text also shifts its position in terms of allocation of responsibilities, and identifies the three primary bases for determining appropriate allocation in this vital area as: function or purpose, process or method, and clientele. The secondary bases are identified as time and area, which are additional influential modifiers of police organization.

Acknowledgement is also given to the importance of the informal organization. It is noted that it arises from the interactions of people and though seldom structured, it can develop into somewhat of a loose hierarchical arrangement. It is acknowledged that informal organization is both natural and inevitable and that while it can be disruptive to departmental management, it can also be an invaluable aid.[66] Administrative processes are viewed as those interrelated means employed on a continuing basis whereby a police administrator achieves his organizational objectives and goals. There is no precise way to identify all the kinds of actions that can be included in such a general designation, but for purposes of discussion the text identifies the five principal processes as: planning, organizing, assembling resources, directing and controlling.[67] Each of these management processes is not considered to be mutually exclusive and their interdependence is self evident.

Municipal Police Administration has had considerable impact on the police field by supporting the classical school of management which emphasizes principles and related processes. Each of its editions has undergone considerable refinement as illustrated in the most recent edition with added chapters on "governmental setting for police work" and "federal assistance and law enforcement planning".

O.W. Wilson

In 1950, O.W. Wilson released his monumental work entitled POLICE ADMINISTRATION, that soon became a much used and quoted text. This text was very comprehensive in nature and considered such diverse subjects as organization, control, the juvenile offender, public relations, and leadership. Subsequent editions were issued in 1963 and as recently as 1972; in the latter edition Roy C. McLaren joined him as a co-author.

O.W. Wilson viewed the police function as being the protection of life and property against criminal attack and the preservation of the peace as the primary purposes of police departments in the United States.[68] In order to accomplish this, he viewed police duties as being classified according to their immediate objectives as: (1) the prevention of the development of criminal and anti-social tendencies in individuals (2) the repression of the criminal activities of those so inclined (3) the arrest of criminals, the recovery of stolen property, and the preparation of cases for presentation in court and (4) the regulation of people and their noncriminal activities and the performance of a variety of nonregulatory services.[69]

Wilson suggested that in critically anlayzing the administrative practices in management of a police agency, the following principles of organization should be kept in mind:

1. Tasks similar or related in purpose, process, method, or clientele are grouped together in one or more units under the control of a single person. In order to facilitate their assignment, these tasks are divided according to (a) the time and (b) the place of their performance and (c) the level of authority needed in their accomplishment.
2. Lines of demarcation between the units are clearly drawn by a precise definition of duties which are made known to all members so that responsibility may be placed exactly. Such definition avoids both duplication in execution, and neglect resulting from an unassigned duty.
3. Channels are established through which information flows up and down, and through which authority is delegated. These lines of control permit the delegation of authority, the placing of responsibility, the supervision of operations, and the coordination of effort; the individuals and groups are thus tied together into a unified force susceptible of direction and control. The lines of control must be clearly defined and well understood by all members so that each may know to whom he is responsible and who, in turn, is responsible to him.
4. Each individual, unit, and situation must be under the immediate control of one, and only one, person, thus achieving the principle of unity of command and avoiding the friction that results from duplication of direction and supervision.
5. No more units or persons are placed under the direct control of one man than he is able to manage.
6. Each task is made the unmistakable duty of someone; responsibility for planning, execution and control (implemented by inspection) is definitely placed on designated persons.
7. Supervision is provided of each person at the level of execution regardless of the hour or place.
8. Each assignment of responsibility carries with it commensurate authority to fulfill the responsibility.
9. Persons to whom authority is delegated are invariably held accountable for its use.[70]

Wilson viewed that the purpose of organization was to simplify the direction, coordination, and control of members of the force so the objectives of the department could be gained easily, effectively, and satisfactorily.[71] He suggested the personnel of the department must be directed and they must be given definite tasks and instructed in their performance. Coordination was identified as being essential and, finally, control must be provided. A function of leadership of command was viewed as requiring the exercise of authority and it had to be provided for in the organizational structure. The importance of these was stressed by O.W. Wilson and when reviewing the related problems of command in a police organization he gave considerable attention to the following: (1) delegation of authority; (2) unity of command; (3) span of control and chain of command; (4) organization for administration; (5) direction and coordination; (6) the exercise of authority; (7) timed divisions of tasks of command; and (8) order of rank.[72]

At each level of authority, the head of that unit must plan, direct, and control—from the chief down—each successive level to the officer who performs the task. The rule of control, as proposed by Wilson, stated that authority is delegated by some form of command; responsibility is effectively placed by some form of control. Inspection was enviewed as an implement of control, the purposes of which were fourfold: (1) to ascertain whether the task was being performed as outlined; (2) to learn whether the anticipated results are being attained; (3) to discover whether the resources of the department are being utilized to the best advantage; and (4) to reveal the existence of need, thus accomplishing the first step in planning.[73]

Each of the subsequent editions of "POLICE ADMINISTRATION" extended and expanded its consideration of vital management topics. This is noted in the last edition which gave consideration to police leadership to the community, consolidation, and behavioral aspects of management. Wilson's contribution to police management thought is clearly in the classical school as noted by the emphasis on principles, processes and functions. And when the three editions are compared, control would seem to play a dominant place in this excellent textbook.

V.A. Leonard

In 1951, Dr. V.A. Leonard published the first edition of "POLICE ORGANIZATION AND MANAGEMENT" which was subsequently followed by three other editions. He presented tested principles and procedures for the organization and operation of the police enterprise but he emphasized that the formula for modern police service was not simple. Considerable attention was given to the dynamics of police efficiency emphasizing the importance of organizational structure as modified by external and internal factors which control a police force and condition its performance. This included a consideration of the administrative controls of municipal law enforcement agencies as well as the various types of pressure groups influencing police management.[74]

Leonard stressed that no greater responsibility confronted the appointing power in American cities than the choice of the individual who was to be given control of the police department. Leadership was identified as the most important single factor in the success or failure of police operations and considerable attention was given to the qualifications of the police executive, to include such factors as tenure, salary, and specific qualifications.[75] Unity of command was emphasized as being the indispensible prerequisite to successful organization, however while the rigid adherance to the principle of unity of command may have its absurdities they are however, decidedly unimportant in comparison with the certainty of confusion, inefficiency and irresponsibility which occurs when that principle is violated.[76]

Leonard was a strong exponent of the line and staff concept as adopted from the military and he created the following classifications of staff functions and line operations as follows:

1. Staff functions—planning, inspection, personnel administration, police records system, statistical operations, follow-up control, identification services, property control, communication system, budgetary control, purchasing, transportation, jail administration, supply, crime detection laboratory, and public relations.
2. Line operations—police patrol system, criminal investigation, vice investigation, traffic regulation and control, and crime prevention.[77]

Leonard pointed out that within the structure of the organization, it was necessary to arrange for the distribution of the work load according to some logical plan. Fundamentally, this involves the grouping of related functions into administrative units for economical and effective supervision and control. These were identified as including the major purpose of the activity or function, the process or method to be employed in achieving its immediate objectives, the nature of the clientele with which it must deal, the geographical distribution of operations and the time element.[7,8]

Later editions of this text (the latter two editions were co-authored with Harry W. More) emphasized such factors as the dominance of the idea for the organization, administrative alternatives, informal organization, team policing, performance budgeting, and coordination with other agencies in the criminal justice system.

Conclusion

Management is a relatively new concept and primarily a product of this century. It has attained its imminent position through the contributions of numerous individuals beyond those mentioned in this discussion. But it serves to point out the complexity of the management process, particularly when one considers its application to the thousands of law enforcement agencies in this nation. The contributions of various writers have been reviewed and an effort has been made to examine in some reasonable detail the key issues discussed by each author. It is clear that these authors have been strongly influenced by the classical theories of management as proposed by Fayol, Urwick, Gulick, and other nineteenth and twentieth century writers. The quantitative school of management thought has had little effect, while there has been an obvious trend indicating a merging acknowledgment of behavioral sciences by some of the latter writers.

The writings of August Vollmer, the father of modern police administration, have not been reviewed because he did not produce a specific text in the area of police administration. But unquestionably he has influenced, to a great extent, both O.W. Wilson and Dr. V.A. Leonard. Vollmer, who served first as the marshal and then the chief of police in Berkeley, California from 1905 to 1932 was one of the most innovative police leaders who ever lived. He believed in a strong generalist approach to law enforcement and consequently limited specialization in his agency and he also was a strong exponent of training and education for police officers. He was instrumental in the creation of the motorized police forces as well as the development of communications and records for law enforcement agencies. His philosophies and beliefs are clearly evident in the latter two works reviewed in this chapter. Police managerial history clearly provides a departure point for the consideration of law enforcement management in transition.

Chart 1-1

MANAGEMENT THOUGHT CONTINUUM

Year

1748	Henry Fielding and John Fielding	Fledging police organization. Selection of personnel. Centralization of authority. Information dissemination. Strong leadership.
1829	Robert Peel, Charles Rowan and John Moore	Principle objective of crime prevention. Absence of crime is the best proof of police efficiency. Daily reporting. Creation of divisions, sections and beats. Establishment of ranks. Centrally located police stations. Divisional reserve. Beat card. Shift rotation. Military model.

1909	Leonhard F. Fuld	Elimination of politics. Specialization of duties. Defining duties. Supervision. Leadership. Audit. Discipline. Training. Selection of personnel. Elimination of non-police duties.
1915	Raymond B. Fosdick	Mechanical organization. Single executive. Harmonious relationship. Patrol force the first line of defense. Investigation. Crime prevention. Personnel management. Military analogy false. Importance of the individual officer.
1921	Elmer D. Graper	Single executive. Functional organization. Hierarchy. Coordination. Centralization. Classification of police activities. Reduction of police details. Selection of detectives. Records management. Vice control. Policewomen. Crime prevention.
1938	International City Management Association	Principles. Processes. Police mission.

		Informal organization. Governmental relations. Federal assistance. Allocation of responsibilities.
1940	Bruce Smith	Principles. Unity of action. Organizational growth. Specialization. Span of control. Classification of major police activities. Lines of authority. Aides. Grouping related activities. Vice control.
1950	O.W. Wilson	The police function. Police duties. Principles. Processes. Functions. Control. Planning. Leadership. Span of control. Unity of command. Authority.
1951	V.A. Leonard	Leadership Dominance of patrol. Internal organization. Climate of police organization. Tactical unit. Audit. Unity of command. Line and staff. Organization by purpose. Functional organization.

Topics for Discussion

1. Discuss the influence of Charles Rowan on the British police system.
2. Discuss the importance of the qualifications which Fuld thought the ideal sergeant should possess.
3. Discuss the consequences of the board system of administering police departments.
4. Discuss the application of the principles of management to police organizations.
5. Discuss the impact of O.W. Wilsons text on police administration.
6. Discuss the line and staff concept proposed by V.A. Leonard.

FOOTNOTES

I. HISTORY OF POLICE MANAGEMENT THOUGHT

[1] Felix M. Lopez, *The Making of a Manager* (New York: American Management Association 1970), p. 13.

[2] Leon Radzinowicz, *A History of English Criminal Law and Its Administration From 1750,* Volume 3 (London: Stevens and Sons Ltd., 1956), p. 572.

[3] For detailed historical accounts of this period, see: John Coatman, *Police* (London: Oxford University Press, 1959): James Cramer, *The World's Police* (London: Cassell and Co., Ltd, 1964), and William H. Hewitt, *British Police Administration* (Springfield: Charles C. Thomas, 1965).

[4] *Ibid.,* p. 2.

[5] *Ibid.,* p. 41.

[6] See: Samuel G. Chapman and T. Eric St. Johnston, *The Police Heritage in England and America* (East Lansing, Michigan State University, 1962); Harry W. More, *The New Era of Public Safety* (Springfield: Charles C. Thomas, 1970), and Charles Reith, *The Blind Eye of History* (London: Faber and Faber, 1952).

[7] *Ibid.,* p. 54.

[8] Patrick Pringle, *The Thief-Takers* (London: Museum Press Limited, 1958), pp. 9-19.

[9] Leon Radzinowicz, *A History of English Criminal Law and Its Administration*

From 1750, Volume 4 (London: Stevens and Sons Limited, 1956), pgs. 291-292.
[10] J.L. Lymann, "The Metropolitan Police Act of 1829," *Journal of Criminal Law, Criminology and Police Science*, 1955, p. 53.
[11] Charles Reith, *The Police Idea* (London: Oxford University Press, 1938) p. 251.
[12] Radzinowicz, *op. cit.*, p. 160.
[13] E.H. Glover, *The English Police* (London: Police Chronicle, 1934), p. 60.
[14] *Ibid.*, pgs. 160-161.
[15] Charles Reith, *A New Study of Police History* (London: Oliver and Boyd, 1956), p. 18.
[16] *Ibid.*, p. 23.
[17] *Ibid.*, pgs. 135-137.
[18] *Ibid.*, p. 140.
[19] Radzinowicz, *op. cit.*
[20] W.L. Melville Lee, *A History of Police in England* (London: Methuen and Co., 1901), pp. 234-243.
[21] Leonard Felix Fuld, *Police Administration* (New York: G.P. Putnam's Sons, 1909), p. 21.
[22] Radzinowicz, *op. cit.*, g^{-190-}
[23] Leonhard Felix Fuld, *Police Administration* (New York: G.P. Putnam's Sons, 1909), p. 304.
[24] *Ibid.*, p. 48.
[25] *Ibid.*, p. 41
[26] *Ibid.*, p. 37
[27] *Ibid.*, pgs. 174-181.
[28] *Ibid.*, p. 112.
[29] *Ibid.*, p. 49.
[30] *Ibid.*, p. 152.
[31] *Ibid.*, p. 153.
[32] *Ibid.*, p. 56.
[33] *Ibid.*, pgs. 59, 60.
[34] *Ibid.*, pgs. 68-69.
[35] *Ibid.*, p. 186
[36] Raymond B. Fosdick, *American Police Systems* (Montclair, New Jersey: Patterson Smith, 1969), pgs. 382-383.
[37] *Ibid.*, pgs. 189-190.
[38] *Ibid.*, pgs. 191-201.
[39] *Ibid.*, p. 203.
[40] *Ibid.*, p. 213
[41] *Ibid.*, p. 215.
[42] *Ibid.*, pgs. 268-270.
[43] *Ibid.*, p. 306.

44 *Ibid.*, pgs. 314-315.
45 Elmer D. Graper, *American Police Administration* (New York: The Macmillan Company, 1921), p. 10.
46 *Ibid.*, p. 57.
47 *Ibid.*, pp. 156-157, citing Arthur Woods, *New York City Police Report* 1914-1917, pp. 3-4.
48 *Ibid.*, p. 158.
49 *Ibid.*, pp. 226-233.
50 *Ibid.*, pp. 276-277.
51 Bruce Smith, *Police Systems in the United States* (New Yrok: Harper and Brothers, 1960), p. 208.
52 *Ibid.*, p. 209.
53 *Ibid.*, p. 218.
54 *Ibid.*, pp. 219-220.
55 *Ibid.*, p. 225.
56 *Ibid.*, p. 229.
57 *Ibid.*, p. 238.
58 *Ibid.*, p. 241.
59 International City Management Association, *Municipal Police Administration* (Chicago: International City Managers; Association, 1943), p. 52.
60 *Ibid.*, pp. 69-71.
61 *Ibid.*, p. 74.
62 *Ibid.*, p. 76.
63 *Ibid.*, p. 78.
64 *Ibid.*, p. 81.
65 *Ibid.*, 1971, p. 21.
66 *Ibid.*, pp. 27-29.
67 *Ibid.*, p. 39.
68 O.W. Wilson, *Police Administration* (New York: McGraw-Hill Book Company, 1950), p. 2.
69 *Ibid.*, pp. 2-3.
70 From *POLICE ADMINISTRATION* by O.W. Wilson, copyright, 1950 by McGraw-Hill, Inc. Used with permission of McGraw-Hill Book Co., p. 9. A revised 3rd Ed. was published in 1972.
71 *Ibid.*, p. 36.
72 *Ibid.*, pp. 36-56.
73 *Ibid.*, p. 59.
74 V.A. Leonard, *Police Organization and Management* (Brooklyn: The Foundation Press, 1951), p. 27.
75 *Ibid.*, pp. 45-60.
76 *Ibid.*, p. 64.
77 *Ibid.*
78 *Ibid.*, p. 70.

SELECTED READINGS

Fosdick, Raymond B., *American Police Systems,* Montclair, New Jersey: Patterson Smith, 1969. Based upon a study of seventy-two cities, Fosdick identifies numerous police problems such as political control, inadequate leadership, poor police techniques and a lack of supervision. Oriented toward a mechanistic view of police organization.

Fuld, Leonhard Felix, *Police Administration,* New York: G.P. Putnam's Sons, 1909. A critical study of police organization which is concerned with the individual police officer as well as the police organization. Stresses the factors of efficiency and humanity within the organization.

Graper, Elmer D., *American Police Administration,* New York: The Macmillan Co., 1921. A hand-book on American police administration emphasizing organization, supervision and control of police practices and procedures. Designed for those who are police managers.

International City Management Association, *Municipal Police Administration,* Chicago: International City Management Association, 1943. Focuses on providing police executives with a better understanding of contemporary police principles and practices. Emphasis is placed on traditional management concepts, processes and functions.

Leonard, V.A., *Police Organization and Management,* Brooklyn: The Foundation Press, 1951. Presents tested procedures and principles for the organization and operation of the police enterprise. Reviews the dynamics of police efficiency as well as external and internal factors which control a police force.

Smith, Bruce, *Police Systems in the United States,* New York: Harper and Brothers, 1960. Describes the broad principles of organization that can be applied to the structure of a police force. Includes a consideration of unity of action, span of control and the classification of police activities.

Wilson, O.W., *Police Administration*, New York: McGraw-Hill Book Co., 1950. A comprehensive analysis of police administration to include a consideration of structure, control, public relations and leadership. Reviews the principles of management as a problem of command in a police organization.

 **The study of this chapter will enable you to:**

1. Differentiate between classical and behavioral schools of management.
2. Identify the key elements of Force Field Analysis Theory.
3. Describe the Seven Step Plan for Problem Solving.
4. List Maslow's hierarchy of needs.
5. Identify the assumptions about human behavior which are implicit to "Theory X."
6. Contrast "Theory X" and "Theory Y."
7. Compare country club management with task management.
8. Describe the wide-arc pendulum theory.
9. Write a short essay on the concepts of "Territorial Imperative."

2

Behavioral Sciences

Introduction*

The behavioral sciences include the disciplines of anthropology, psychology, social psychology, and sociology. Research and findings in these areas have focused on systems of belief, attitudes, values, and other facets of the motivational complex, as well as the dynamics of communication, leadership, informal group norms, and the decision-making process at various levels in the organization. "It seems useful to define behavioralism by its essential orientation which is toward the systematic study of human behavior. Behavioralism often asks: How do individuals or groups act, as contrasted with institutionalized expectations and conventional assumptions about their behavior?"[1] There has been an awareness for some time that the formal organizational mechanism (organizational charts, lines of authority and responsibility, rules and regulations, downward communication and various other formal organizational controls) reveals very little about human behavior in the organization. "All

---

*Source: G. Douglas Gourley, et. al., *Effective Police Organization and Management,* Report to the President's Commission on Law Enforcement and the Administration of Justice, Volume 7, Los Angeles: California State College at Los Angeles, October, 1966, pp. 965-978.

organizations have both manifest and latent functions. Manifest or public functions are the official, conventional modes of behavior and mission which partially motivate an organization and legitimate its existence. Latent private values include the drive for power, security, and survival. Behavioral research and theory are often concerned with the latent facet of analysis."[2] The behavioral approach attempts to meet the need for understanding how and why an organization functions as it does. The orientation is, obviously: people. Managerial techniques to be successful must reflect a philosophy of attunement to people within and without the organization, sensitive to their norms, attitudes, values, and motivating forces. Within the organization, management is concerned with a maximization of human and physical resources to effectively fulfill agency goals, as well as providing the climate for individual growth and development. Research findings of the behavioral sciences indicate, for example, that detailed supervision and an excess of detailed rules and regulations to meet every contingency, stifles initiative, individual growth, innovative solutions to problems, and promotes dependency.[3] In external operations, the police department's task would be greatly facilitated if the community has a high compliance-with-law attitude. An agency that enjoys a good measure of public confidence, respect and support, has generally carried out its mission with overtones of sensitivity to community needs and problems.

Types of People in an Organization

To understand the behavior of personnel in an organization requires some insight into the different types of people in the agency. Herbert Simon[4] refers to the "administrative man" who does not wait for the perfect solution or decision. He cannot wait until all possible alternative are in. In any complex problem there is an inherent skills and knowledge limitation. Robert Presthus[5] presents this perspective of types of people in the organization:

1. The upward mobile: he is a conformist and is opportunistic, obeys rules and enjoys working in the hierarchical climate with a view toward moving ahead.
2. The indifferent: he is apathetic and uninvolved, seeks satisfaction and fulfillment outside the job.

3. The ambivalent: the neurotic type, can't make up his mind.

Amitai Etzioni[6] views personnel in an organization as:
1. Alienative: negative involvement (somewhat similar to Presthus' indifferent type).
2. Calculative: emphasis on personal betterment primarily, and the organization secondarily.
3. Normative: this person is deeply committed and dedicated, with attitudes that have become internalized.

Chris Argyris[7] refers to two types of personnel in an organization:
1. Those who are achievement-oriented, who need a balanced tension (to meet challenging problems, solve them and go on to the next unprogrammed situation, forever traveling, never to arrive). In this process, self-esteem and interpersonal competence are developed.
2. Those who prefer a role of dependency. It is obvious, therefore, that individuals in an agency differ and will consequently react differently to forces that have an impact on them.

Forces on the Individual in an Organization

Formal forces: Excessive hierarchical pressures with overly detailed rules and regulations may cause the achievement oriented individual to become dependent and seek satisfaction and fulfillment outside the job, or resign. Formal pressures may result in short-term gains (initial compliance) but at the long-run expense of human resources.

Informal forces: cliques, the informal group, peer groups all create an impact on the individual by shaping norms and attitudes.

Another force has been termed "congruency" by Chris Argyris:[8] Unintended consequences may follow when incongruences develop between individual needs and the organizational climate. For example, if the individual is a mature personality and needs challenging, problem-solving opportunities, a repressive organizational atmosphere will develop an incongruency. Or, if the individual is dependency-oriented, a democratic or permissive organizational climate will also develop an incongruency in this case. In both of these situations, compensatory social defense structures are created and psychological energy is diverted from organizational tasks and goals.

To place the impact of the behavioral sciences on organizational management in another perspective, it is useful to briefly sketch the evolution of three main schools of thought concerning management theory.

A Metamorphosis of Management Theory

Theories of management have gone through three main stages during the past sixty years.[9]

The first stage has been referrred to as the classical school, focusing on: the individual as a passive instrument in the agency, the distribution of authority in a formal way, organizational blueprint and rules, and an impersonal rationality taken for granted. The organization was viewed as a machine, with emphasis on efficiency and economy.

The research and findings of the second stage (social science, behavioral school) reflected the impact of the psychological and sociological disciplines. During the decades (1940-1960) a large amount of evidence was accumulated revealing that people tend to support what they help to create, that democratically run groups develop loyalty and cohesiveness, that strong identification with and commitment to decision-making are generated by honest participation in the formulation of these decisions. Attunement to people in the agency transcended the earlier mechanistic view of organizational behavior (individual satisfaction and fulfillment often assumed more significance than material benefits). Rigid, supervisory controls were seen as an impediment to individual growth and development which, in turn, could jeopardize the health and vitality of the agency.[10] Democratic management with a generalized supervision (trust, confidence, large areas of delegation) became a core rationale of the behavioralists. "For the manager, creative growth is the product of policies and actions that focus on the growth potential of many individuals... too much concentrated power invites rigidity and a slowing down of innovation."[11] In other words, "The function of administration is to provide an environment in which the individual employee, of whatever rank, can function at or near the level of his talents, training and experience."[12]

It has become apparent, however, that some of the behavioral science findings have been distorted. For example, concerning group behavior in the organization:[13]

Statements of Behavioral Scientists	Distortions
1. Social detriments of behavior are important.	1. One's behavior is determined by his membership in social groups.
2. It is helpful to realize that people's aspirations at work have many aspects; they seek membership, self-esteem, economic security, prestige, etc.	2. Even incentive pay and other material benefits aren't of much importance any more.
3. Social behavior and individual satisfaction are related in some significant fashion to productivity.	3. You ought to give workers whatever they want. If they're happy, they'll produce more.
4. Social sub-systems in organizations exert much control over their members. Consequently administrative control is limited.	4. Nobody can really be a boss, only a follower. We should all be soft and tender-hearted.

The third stage (contemporary view of organization theory) has not excluded the classical and social science doctrines, but has added additional dimensions for a more complete analysis: the systems approach, and operations research. The distinctive qualities of modern organization theory are its conception-analytical base, its reliance on empirical research data and, above all, its integrating nature. These qualities are framed in a philosophy which accepts the premise that the only meaningful way to study organization is to study it as a system... It treats organization as a system of mutually dependent variables. As a result, modern organization theory which accepts system analysis shifts the conceptual level of organization study above the classical and neo-classical (social science) theories."[14] William Scott identifies five parts of the organizational system:[15]

1. The individual (the personality structure replete with motives and attitudes).
2. The formal organization (the organizational structure or blueprint with its rules and regulations).
3. The informal organization (its influence on behavior and attitudes of members). The individual has expectations of satisfaction he hopes to derive from association with people on the job.
4. Status and role patterns (a fusion of the foregoing components of the organizational system that acts to keep the organization institutionally fit.
5. The physical setting (the work environment).

The systems approach is therefore useful in analyzing behavior in an organization by focusing on the inter-relationship of the parts of the organizational system. Similarly, the systems method (identification of all inter-reacting components) is brought to bear on solving complex problems.

Organizations exist to solve problems. Due to the information and technological revolution, management has become increasingly dependent upon the advice of teams of experts representing the components of the problem. In problem situations where prior solutions or programs are not available, or have been largely ineffective, the operations research approach provides the optimum answer. Ad hoc teams of experts from all relevant fields of knowledge (inside and outside the agency) are structured to provide the greatest number of communication interchanges to maximize creativity.[16]

It is obvious that systems analysis and the use of ad hoc teams of experts are geared toward "quality" solutions of problems rather than concentrating on democratic management. Herein lies a dilemma. How can broad acceptance and commitment to a program be developed in an agency, when a "quality" solution or policy has been superimposed with limited participation by members of the agency? Program development and implementation requires a blend of quality, and acceptance by those carrying out the program. Yet, an emphasis on participation (to assure acceptance) is not necessarily a criterion of the quality of a program arrived at in such a democratic fashion.[17] Democratic management focuses on the willingness of people to implement a program that they helped to create. However, there are complex situations where a systems approach by an ad hoc team of experts is needed to produce the most effective (quality) program. Quality of the solution in such instances should be the primary consideration.[18] Implementation of the program is then sought by explanation and persuasion. The ideal is obviously the attainment of quality plus acceptance.

Some Implications for Police Management

What are the implications of the foregoing for police management? The police department is obviously only one of the components in the system of administration of criminal justice. If this component attempts to fulfill its mission without regard to the other components (prosecution, courts, probation, parole, corrections, etc.) the over-all result may jeopardize the effective administration of criminal justice. For example, a police department may conduct a "drive" and regard itself as highly effective if it increased the number of arrests by 200% over a specified period. This can create problems for the other components (overload, overcrowded facilities with consequent pressure for not filing complaints by the prosecution, or dismissal, or premature release). Administration of criminal justice can then easily become a mockery. Therefore one "quality" solution to the crime problem would be to form ad hoc teams of experts from the various components of criminal justice. Policy and program recommendations would be transmitted to all components for implementation. Conceivably the members of a police agency might regard their mission as being frustrated. Also, since the program did not originate within the police agency, whole-hearted support might be lacking from a behavioral view. Any attempt to deal with crime from a long range viewpoint cannot ignore a systems approach of the widest possible perspective.[19]

Police departments have a vital interest in reducing crime from a long-range viewpoint as well as putting out the immediate "brush fires." What is required is the task of explanation and persuasion in order to yield maximum compliance with a program or policy arrived at by experts representing all relevant components of the problem.

Another illustration of a "quality" approach to a police problem concerns riots and similar disturbances. A team of experts from relevant disciplines (anthropology, psychology, sociology, legal, judiciary, police), as well as representative community leadership may propose a policy that encompasses control as well as prevention.[20] True, such a program may require a bit of selling to the police agency since (from a behavioral view) an organization tends to support a policy arrived at in its own parochial framework.

In another phase of police management, a team of experts representing the Los Angeles Police Department and System Development Corporation, Santa Monica, conducted a System Design Study for Phase I of the Los Angeles Police Department Information System. The Phase I System was limited at the outset of the study to the following applications: crime and related reports, want and warrant information, and field interview information. The foregoing is an example of a "quality" created program developed by an ad hoc team of experts representing police and data processing experts. Acceptance of the program study was enhanced by the creation within the Police Department of a Review and Concurrence Authority.[21]

Toward the Future

Due to the information and technological revolution, police management will become increasingly dependent upon ad hoc teams of experts (from within and without the agency) to provide "quality" solutions to problems.

Police management will be effective if it can guide its resources into dynamic organizational units which attain their objectives to the satisfaction of management and the community that it serves. The measure of such attainment is heavily dependent upon the degree of morale, and the sense of accomplishment on the part of those rendering the service.

Dr. Robert Tannenbaum, University of California at Los Angeles, Graduate School of Business Administration, in a recent talk, referred to the new frontier with respect to organizational values and methods. "If management conveys trust, it leads to a much greater willingness on the part of both individuals and groups to assume responsibility for their own behavior and decisions ... To meet the needs of the future, trends will be:

1. From a mechanistic, formalistic view of organization to an organic, systems-oriented view.
2. From linkages between individual units by organizational chart to linkages by organizational needs.
3. From motivating individuals from the outside in, to motivating them from the inside out.
4. From the use of formal authority to developing an atmosphere conducive to freedom, creativity and growth.
5. From telling people, to listening to them.
6. From a primary concern with the isolated individual to a concern with teams.
7. From inter-individual and group competition to collaboration.
8. From maskmanship (hiding one's thoughts and feelings) to a more appropriate degree of openness.
9. From a primary concern for the immediate utilization of individuals to a greater concern for their development."[22]

Police management will face increasingly complex problems. These problems will be handled from the wide perspective of the totality of their interacting components (systems approach). Wider exposure to all fields of relevant knowledge (operations research) will afford mind-stretching, innovative programs to deal not only with the immediate problem at hand, but also its long-range implications so that, hopefully, the same "brush-fire" problems do not constantly flare up.

The challenge for management will be to generate enthusiasm within the agency for a "quality" program (often developed by teams of experts from within and without the agency) arrived at by systems analysis and operations research.

Police Management in Transition*

OUR INPUT to create styles of leadership has traditionally been limited to the visibly recognized hierarchy of bureaucrats—principally police chiefs, criminologists and teachers in the police science field. Today's administrators are fortunate to have the outside influence of a host of disciplines, such as schools of government, business administration and members of the behavioral sciences, who advocate flexible management systems adopting the newer concepts of leadership.

To accept these vital changes it takes a manager with intestinal fortitude, since his existing policies will be questioned as well as the objectives of any being proposed in the future. If the hierarchy is weak, it may crumble; but if the manager is courageous and capable, he will not only improve but can surpass expectations.

All of the proposed changes emphasize involvement, the tapping of forgotten resources and the full utilization of facilities. Every police administrator should recognize that his greatest asset lies hidden under a blanket of blue.

Decision-making and management of police are on the threshold of being actively challenged. Inquiries are on the upsurge and will continue to rise with a crescendo until they can no longer be ignored. These actions may indeed be valid for a myriad of reasons. Law enforcement personnel are not of the same caliber today as of yesteryear. The ranks are being permeated with educated, trained, inquisitive and meticulous individuals who are not content with the belief that the administrator's decisions are sacred.

To imply a "revolution" is forthcoming is not a sound judgment when we view that term in its present-day literary meaning. Rather, it would be more germane to suggest an evolutionary movement based upon the experiences of past performances giving way to newer methods of operation.

The principles of police management, as they are universally applied today, are considered by many to be dogma and the applicable concepts taken for granted. These criteria are being

---

*Source: J. Laverne Coppock, "Evolution in Police Management," *The Police Chief,* Vol. 39, No. 5, May 1972, pp. 18-19, 76-79. Reproduced from the Police Chief magazine with permission of the International Association of Chiefs of Police.

challenged, not only by the progressive administrator but by the questioning minds of the law enforcement officers. The validity of these managerial precepts by the police hierarchy may be viewed as suspect in light of present-day, upgraded and more professionalized personnel.

Management and organization are steeped in history and tradition. This tends to create both mental and physical barriers to problem-solving. The pragmatic precepts and philosophical attitues become intellectual roadblocks when a solution is sought. A phenomenon exists within the power structure whereby the feeling of subvention is prevalent if the administration were to allow personnel other than staff to become involved with management decision-making.

Police problems are complexities involving all community aspects. This is not a new and sudden revelation but a problem which has been with us for a long time, and for a long time ignored—if not ignored, then overlooked, or perhaps, even more degrading, indicating the police to be as the proverbial "Three Monkeys." These complexities have recently been brought to our attention by the various task forces which concern themselves with the negative conditions between police and citizens, police and groups and police and society in general. Few administrators were aware of these preexisting conditions and most considered them problematical.

Status Quo of Administration

The true enigma lies with the status quo of administration. Administration is saturated with members who have been unable or unwilling to change. Additionally, these administrators are being left behind in education and are being surpassed by subordinates. Many agencies' managerial teams are stacked with the dregs of longevity. Many of these top positions are currently held by individuals who attained their status merely because they outlasted everyone else—the result being that police and citizen alike suffer.

It is a known maxim that an administrator must make use of all his tools in order to be as effective as possible. The administrator's tools are personnel. The manner in which they are handled is called—for lack of a better term—human relations. Without opening

"Pandora's Box" on the subject, it might be wise to say that whatever direction traveled, goals attained and objectives met will depend upon those same personnel. All the actions and reactions will set the stage for feedback, which is the measuring device for success or failure. Personnel management is a must and no administrator can ignore this all-important function. His progress and the effective progress of the department will depend upon mutual cooperation, mutual goals and understanding—all brought about by effective communication. Through this free flow of communication, the manager will receive the necessary and required data with which to proceed. The executive must receive all available facts bearing upon the subject matter at hand. He determines the superfluous, discarding it; considers bias and unbias, retaining the relevant; and, in general, seeks out every facet of pertinent data for decision-making.

The standard police organizational chart will indicate in graphic form the strata of manpower allocation in the traditional triangle of the administrative hierarchy. The chief of police will be at the apex with his subordinates fanned out below him. The administrative policy-making team will be limited to those same individuals who are listed as lieutenants and above. Obviously, the number of personnel sought to consider major decisions will be limited as indicated by the chart.

As a general and usually accepted rule, most administrators have the preconceived idea that one must be above the position of first-line supervisor in order to be proficient enough and possess the intellect suffciently profound to form executive ideas. Additionally, there exists the status symbol of equating rank to executorship. This should not necessarily be so. There are some interesting parallels drawn if one were to disregard the grade or rank preceding an officer's name and compare their experience, duties, accomplishments, and education. If approached as objectively as possible, one might be unable to determine which officer fits into which slot. Then, too, the elevation of an officer in this day may not make him any more proficient than yesterday under a lesser title.

It was not until the California Peace Officer Standards and Training Commission adopted the philosophy of education and broadening the perspectives of police executives that I was exposed to the newer concepts of organizations. Prior to attending these various seminars and courses, I, like many other administrators, accepted the traditional role and felt any radical changes would be rejected because tradition and custom were the proving grounds. Today, I not only accept change—I advocate it for most agencies.

One of the first changes I noticed from within was the development of a questioning mind. Once this process started, reviewing the organizational functions for effectiveness was a more fruitful task.

I felt I was not receiving my money's worth from my personnel. I believed I should become the agent of change and enlarge upon and enliven a staid program into a more meaningful, vibrant and productive operation. I did so with numerous ideas and reasons. I determined significant input could be achieved by enlarging the policy-making body. This expansion would give experience and depth to the administrative field, providing invaluable knowledge and know-how to the previously inexperienced man.

I first included all sergeants into the realm of management without altering their status of being first-line supervisors. The definitions of their duties made this relatively easy, as their responsibilities are not far afield and, in many instances, parallel those of members of my administrative staff. This immediately tripled the body of decision-makers. The monthly staff meetings are now mandatory sessions to which each presents himself for the express purpose of contributing to a fact-finding, decision- and policy-making group.

One member of the patrol is invited to attend and become equally involved. He has an equal vote without regard to rank and title. He is chosen by drawing and, if for some reason he is unable to attend, an alternate is selected by the same method. This step was taken to alleviate the suspicions of the patrol. It is a well-known fact that most administrators are viewed as being separate from the subordinates, or at least plotting change without indepth evaluations. This patrol input has been well-received by the ranks. The contributions made from these officers have been intelligent and extremely valuable. My officers now anticipate these sessions with enthusiasm.

In the recent past this agency had certain cliques, which tended to offer resistance to administrative decisions. There was a lack of communication and flow of mutual understanding. Closed administrative meetings, directives without stated administrative intent, implied political motives, implied prejudices and suggested favoritism were some of the problem-makers. These problem areas were, in fact, fancied. However, having had longevity within the ranks, I understood the implications. My position of authority and responsibility dictated to me the necessity of dispelling any further internal problems. Those have now been effectively dispensed with.

In preparation for these group meetings, or to supplement problem areas, I have taken the attitude that every sergeant and staff member is responsible for research and constructive input. It is not infrequent for me to assign a specific project for an indepth study. In the event the problem is particularly knotty, the main topic may be divided or even subdivided for fact assignments. When the research is completed and dovetailed together, the entire study is scrutinized by the group in one or more of our problem-solving schematic methods.

I have adopted the position of controller/moderator for the group and maintain firm control. I was apprehensive at first and had certain reservations as to the possibility of character assassination. In every group there are individuals who are more outgoing than others; some may border on the introvert, while others may lack the polish of speech and concerted thought in abstract. This concern has been laid to rest. Aside from solving problems, the group has served to add to and build character, as it forces each contributor to substantiate, with fact, the basis of his verbal effort.

The cohesiveness of this entire group has exceeded my expectations and the morale has risen to new heights throughout the department. The inclusion of the sergeants and the patrolman has added new dimensions of supervisory acceptance on the part of both sergeant and patrolman. They are now able to more readily understand administration and some of the limitations set forth upon us as a department by those outside forces. The sergeants administer with new ideas, thoughts, philosophy and courage, all of which are seconded by the grapevine instigated by the patrolmen. This agency no longer has the hidden concern of "What's going to happen" or "who's doing what to whom."

I find that if the limitations to a given problem are set forth at the onset, the group will submit and arrive at a decision accordingly and within those limitations. I wish to emphasize: This is not an agency run by the democratic process. I administer this office as the chief of police and this has never been questioned. In supplement to this, each member has taken the position of answering to repercussions that may be anticipated should his theory be accepted in total. Each member contributes his most profound consideration based upon his knowledge in light of his own experience, values and projected beliefs.

Staff members per se, lieutenants and above, are for the most part removed from the street and away from the direct stream of police work. This may, in some instances, hamper their decision. The sergeants are in the thick of it and are usually in a better position to accurately judge the pulse of society from the attitude of the beat officers. This has proven true in nearly all instances. Their immediate disadvantage is the lack of experience in administration and legislative know-how. I maintain that the best way to get the job done is to go to the man doing the job and ask him. He may frequently be found near the lower plateau of the hierarchy. "You may find the reason the ditch isn't being dug any faster is because the handle of the shovel is too short." Police management of this type and within this department has insured maximum administrative support and decision development among all members.

With each problem-solving session the entire group gains confidence in themselves and in each other. They have nearly reached the apex of being completely free and able to give candid opinions to their superiors. This free-flow system of communication eliminates the bottlenecks and stopgaps which had previously existed. The understanding by communication with involvement is the catalyst for motivation to do a better job throughout.

Video tape is used, apart from training, to insure all personnel the opportunity to receive, firsthand, the studies in progress, latest information concerning administrative policies and directives, while soliciting any authentic problems for consideration. These broadcasts are important because they insure every officer the same interpretation without fear of misrepresentation. Each commander utilizes this media as a means of dispensing information to a group. The chief of police, during his regular presentation, is interviewed as a matter of practicality from the standpoint of the beat officer and as a method of inquiry concerning items of lesser importance.

As an example of study, the last two sessions of problem-solving attacked the existing policy of officer proficiency evaluation reports and the sticky problem of the uniformed officer not wearing his hat in conformance with the existing policy. Each problem was approached with a different technique.

Force Field Analysis Theory

I chose the "Force Field Analysis Theory"[1] as one standard of problem-solving. The problem of proficiency ratings, which called for a new supervisor's guide to evaluation, is offered as an example for solution by use of this system. We needed a more comprehensive guide and a better form which would be more meaningful and accurate. With this accomplished, it was now a matter of who would do the evaluation. Our previous policy had been for the immediate supervisor to complete the form but this was felt to be inadequate grading. The group felt that since we rotate our shifts every six months on the school semester, all supervisors would sit in judgment of each officer. Since these meetings would be only twice a year, it was determined the overtime involved would be minimal when compared to the importance of the task. The probability is a near certainty that more than one supervisor, generally four, would have occasion to work with the officer. This would eliminate the one-to-one ratio. The sergeants and lieutenants contribute, while those without work contact would remain mute. This method of evaluation has been readily accepted and to date not questioned.

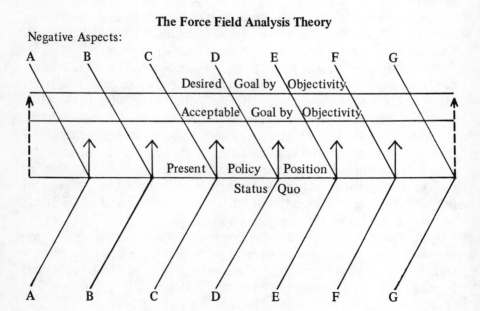

60 *Behavioral Sciences*

AFFIRMATIVE ASPECTS:

The three pressure areas were approximately equal in terms of persuasion in relation to individuals. Each area contained patrolmen, sergeants and staff. An impasse existed which allowed the situation to continue its pre-existing course.

NEGATIVE PRESSURE:

A. The immediate supervisor is the only person qualified to evaluate a subordinate.
B. Not feasible to call all management for evaluation consideration.
C. Too much loss of manpower in cost and compensatory time off.
D. Staff has no direct input.
E. Personalities will not interfere because supervisory personnel are immune to biases.
F. The subordinate has the right to staff appeal if dissatisfied.
G. Proficiency ratings have no bearing on morale or motivation.

STATUS QUO

A. One supervisor is adequate in most instances. In the event scheduling is an interference, the relief sergeant may properly complete the assignment.
B. Immaterial whether evaluation is by an individual within a group or by a group.
C. No cost or manpower loss because the reports shall be completed.
D. Staff will usually have direct knowledge.
E. Personalities could be a factor on an individual basis; however, objectivity should alleviate this problem.
F. The right of appeal still exists to the chief of police.
G. Proficiency ratings may not have much of a bearing, but surely will have some effect.

AFFIRMATIVE ASPECTS:

A. More than one supervisor, generally four, plus two staff commanders will have firsthand knowledge of performance.
B. Each shift sergeant must evaluate the officers at the same time, by calendar, so all can be completed as a group.
C. Minimal loss, if any, in cost and time off, since the reports must be completed according to police regulations.
D. Staff has direct input caused by personnel complaints or lack of complaints Assignments by staff and general information, as well as specific incidents, will be available for constructive additions.
E. Personalities always interfere when evaluation is on a person-to-person basis. Group participation will eliminate a goodly portion. All persons have their prejudices, likes and dislikes, even though one may attempt to be objective.
F. Group sessions eliminate the major portion of appeals by the very fact the opinions are presented as a group.
G. Evaluations are important for morale. It becomes easier for the sergeant to explain deficiencies and excellences in area of concerns and commendations by virtue of consensus.

Any given regulation, rule or policy may contain therein certain real or fancied discrepancies. The interested parties, those governed by the decree, may fall within a broad category of advocate, adversary or conservative middle-of-the-road. In any event, if dissension is prevalent in any area, one of the objects of administration is to manage with smoothness. Consideration, arbitration or counseling is in order, which will ultimatley result in a give and take.

The three pressure areas list examples of both personal and group preferences. Each of the points must be evaluated by the interested parties with mutual acceptance. The moderator, by virtue of his position, moves the present standard up or down the scale according to that acceptance of agreement. Obviously, the ideal situation is one which will erase the negative aspects and obtain for the litigants affirmative aspects. At any rate, this approach will afford involvement and a higher level of satisfaction through participation in the decision process.

The second of the two listed sessions was the uniform hat. Our officers had been exiting their units on police business without their hats. Aside from the safety problem, which was considered, the group was divided in their attitudes. An enthusiastic discussion ensued which called for me to maintain order as the spirit rose. I had my own silent limitations for this policy, yet allowed the group to proceed.

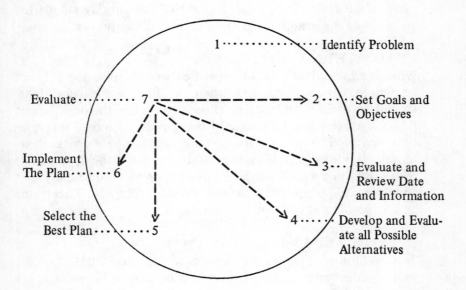

Seven Step Plan for Problem Solving

1 ············ Identify Problem
Evaluate ······ 7
2 ······ Set Goals and Objectives
Implement The Plan ······ 6
3 ···· Evaluate and Review Date and Information
Select the Best Plan ······ 5
4 ····· Develop and Evaluate all Possible Alternatives

ALTERNATIVES:

1. Do nothing.
2. Do something.
3. Do something. Do nothing.
4. Explore (Human tendency is to select #4 just to start a plan of action.) (Steps 1 and 4 are critical areas for consultation.)

Behavioral Sciences 63

I. PROBLEM:
   Not wearing the uniform hat! The uniform patrolmen were ignoring the department policy on wearing the hat or helmet. How do we alleviate the problem?
       A.   Force each officer to wear his hat.
       B.   Change the policy to where it can be enforced.

II. GOAL/OBJECTIVE:
   Perfect an equitable and mutually agreeable procedure on the subject. Arrive at an objective by weighing the input at each point, pro and con, deciding with intelligent objectivity a procedure which can be enforced and one which the officer will obey voluntarily. Determine, in light of present-day standards, an up-to-date method based upon an overall consensus.

III. EVALUATE AND REVIEW DATA:
   What are the existing problems which would cause the officers to ignore the uniform hat when conducting business? Why should an officer deliberately choose to violate the department policy? All possible facts should be considered and all information reviewed. Various factors may come into view, such as morale, lack of motivation or possibly the belief the department is too staid or even—if you will—old-fashioned. Additionally, for any policy to be enforced, the rule must be accepted. Therefore, an acceptable policy must be formulated.

IV. DEVELOP AND EVALUATE PLAN(S):
   Set in motion a series of possible acceptable rules. List all probabilities and consider every eventuality. Formulate, as reasonable as possible, a plan to handle all circumstances, pro and con, the times whenever it becomes an absolute for the wearing of the uniform hat. The group must stipulate to the agreement for seeking alternatives.
       A.   Any existing problem presents a number of "outs" for management. The manager may do nothing at all and just retain the same procedure, albeit not uniform.
       B.   A hastily devised plan may be proffered, which would probably be in truth just another change without

eradicating the problem. Or the administration might begin a preliminary study, but allow the project to fade because of managerial interest.
C. This problem-solving plan eliminates procrastination. The group is faced with a situation and is compelled to explore every possible avenue and consider all alternate plans based upon individual input.

V. PLAN SELECTION:
Which plan best suits the overall philosophy of the department? Will the adoption of the new procedure be generally acceptable and enforceable? Will the patrol volunarily adhere to the plan because they believe their interests were considered and weighed accordingly? After listing all possibilities and considering both the department and individual, the following procedure was unanimously selected:
"Hereinafter, officers have the option of wearing or not wearing headgear, except during those times listed below.
1. Headgear is required whenever an officer attends a funeral in uniform.
2. Headgear is required whenever an officer works a funeral detail.
3. Officers shall be required to wear headgear when ordered to do so by their supervisor or an administrative officer of the department.
Because of the times when wearing of headgear is required for personal safety and special details and assignments, it is an obligation of each officer to have the appropriate headgear available for immediate use."

VI. IMPLEMENTATION:
Any policy effected must be uniform, not only in application but in scope of time. It behooves any administrator to make use of all avenues of communication to insure the receiver sufficient time to comply. Implementation date should be selected to the extent it will allow the last man sufficient time for awareness. Beyond that point, supervision becomes a viable force again. New policies or systems should be in written form, as they will allay those excuses of misinterpretation.

VII. EVALUATION:

Evaluation of any program may be conducted in various ways. Verbal feedback from the patrol is a good indicator of acceptance. Some agencies may prefer any of the varied written questionnaires available. The lack of disciplinary problems is considered a valid scale. Whichever method or whatever manner chosen will give the manager the desired information as to acceptance or rejection.

The Seven Step Plan for Problem Solving

The application of the "Seven Step Plan for Problem Solving"[2] proved to be an interesting challenge. Identifying the root problem without encroaching upon the other solving areas was invigorating. The input from all members began to spread into every category, which necessitated coordination.

The members accepted the policy, which was slightly more strict than I had envisioned for this department. It was accepted by all, and there has been no further discrepancy on the part of the men when on the street.

Management problems generally place administrators in a perplexing state of complicated give and take. The psychology of human involvement dictates to the manager concerted effort toward alleviating the crisis. The doctrine of judgment may take the course of change. However, "change for the sake of change" is an obvious irrational and thoughtless mechanism seized as a vehicle to relieve tension. This relief will be temporary because the root problem has not been aired.

It is not necessary for any administrative regime to be placed in a precarious state of internal dissension. All forces, when welded together, present to the chief an organization which can be wielded effectively to further the mission of the department. The cliche, "an ounce of prevention is worth a pound of cure," is a pragmatic dictum. Open lines of comprehensive communication flowing up and down, coupled with receptive consideration to authentic subject matter of concern, will do much for enduring and preserving solidarity.

I have striven toward this goal with a continuing effort. This problem-solving and policy-making tool has proven my theory. One may hypothesize and speculate continuously without action but the inevitable result shall remain just so much food for thought. There remains one human element necessary to combat and overcome: change. People are creatures of habit and do not enjoy their engrained standards being altered. Problems require solutions which frequently demand alterations, a phenomenal experience resisted by many, if not most. The physical laws of nature are applicable in most situations causing equal pressure, pro and con, which must be defeated in order to overcome inertia.

I do not necessarily advocate either the "Seven Step Plan for Problem Solving" or the "Force Field Analysis Theory" or any of the several others available. I would recommend the administrator select one for his utilization. I belive all chiefs of police should consider their present policy on decision-making and problem-solving. If the present manner and method are not adequate for the times, then change and bring the department up-to-date.

These techniques and many others are being taught in California's Peace Officers' Standards and Training, Executive Development Courses, held in the universities participating in the "P.O.S.T." Program. This exposure, coupled with the experimental application, provides enough insight to make acceptance only a matter of performance. Any administrator can use the same concept and enjoy success if he is willing to study not only the system but to include himself in the analysis. If the atmosphere for change is not conducive, usually the leader has not reflected favorably his willingness or desire to deviate from traditional guidelines. To ignore change is to expose one's lack of self-confidence, but this can be altered by the appropriate training course and personal exposure to a working model.

A Behavioral Science Application to Police Management.*

WHY AREN'T OUR POLICEMEN more productive? We pay good wages, provide good working conditions, have excellent retirement plans, good fringe benefits and unquestioned job security. Yet they don't seem to be willing to put forth more than minimum effort. We are pressured for paid overtime, holiday rights, longevity pay, uniform allotments and escalator clauses—but granting any of these demands seems to have no visible effect on departmental morale. When you and I applied for this job we competed with thirty other men for each vacancy, we worked six days and had one off, we accepted that going to court on our time was part of the job, and although the pay was poor we were glad to have such a good steady job. Perhaps nineteen years have dulled the memory, but it seems that we were a happier bunch than the bulk of the line officers today.

You may find some comfort in the fact that the problem is not unique to law enforcement. It is widespread throughout public and private industry and services. However, both the cause of the malady and the remedy are known. The need is for administrators who will take the lead in the remedial program.

Man Is A Wanting Animal

Behavioral scientists tell us that as soon as one of man's needs is satisfied, another appears in its place. This process is unending. It continues from birth to death. Abraham Maslow and others have listed motivations in a *Hierarcy of Needs*.[1] At the lowest level are *physical needs;* if the gut is empty the man is not concerned about less important problems. But, when he no longer has to worry about his stomach, his thinking will turn to the next order of concern. *A need satisfied no longer motivates.* With concern about his physical needs behind him (hunger, rest, protection from the elements, etc.), he turns his attention to *safety needs*. These are such things as a need for security, or freedom from danger. But when man's physical needs

---

*Source: W.R. Rhodes, "A Behavioral Science Application to Police Management," *The Police Chief,* Vol. 37, No. 5, May, 1970, pp. 46-50. Reproduced from the Police Chief magazine with permission of the International Association of Chiefs of Police.

are satisfied and he is no longer fearful about his welfare, his *social needs* become important—needs for belonging, for association, for acceptance by his fellows, for giving and receiving friendship and love. A few people are so hung up on trying to reach goals of social acceptance that they never become concerned about the next category which is *ego needs.* These relate to one's self-esteem: needs for status, recognition, appreciation, achievement, and self-confidence. Ego needs are rarely completely satisfied, but in this present day there are a lot of men who are content with the attainments they have made in this area—and with no personal goals to strive for we come to face their frustration and your administrative headache. The only category of needs beyond ego is a vague area called *self-fulfillment.* Perhaps this is best described as the need for being creative in the broadest sense of that term, or perhaps it becomes clearer if we call it the need for realizing one's own potentialities.

You and I worked towards goals of safety and social needs: better wages and hours, compensatory time, improved pensions, professional status and public recognition and acceptance. Our first mistake is to hand this package of accomplsihments to a new officer and expect him to be content because he has no problems to solve. An officer, to a greater extent than ever before, is ceasing to be an object to be used. Not clearly understanding his own motivations, he feels the need to assert his individuality. In his own way he is driven by the same restlessness that generates the gropings of minorities, students and religious groups for a more satisfactory existence. The second mistake is to rely on management control that applied to our era but not to the present—control that was based on needs that do not concern the new generation of officers.

The Self-fulfillment Need

Officers, if deprived of the opportunities *at work* to satisfy these ego and self-fulfillment needs, behave as we might expect: passiveness, resistance to change, vague and frequent complaints without recommendations for improvement, and unreasonable demands for economical benefits. The officers are bound to demand more money under these conditions. Unable to attain self-fulfillment needs at work, they try to buy happiness. I disagree with this Utopia concept in which leisure, not work, is the major sphere in life. Man is a problem-solving creature whose real happiness comes in setting and striving towards goals.

The "old" v. the "new" administrative viewpoints on employees are clarified by behavioral scientist Douglas McGregor:

Theory X

1. The average human being has an inherent dislike of work and will avoid it if he can.
2. Because of the human characteristics of dislike of work, most people must be coerced, controlled, directed, threatened with punishment to get them to put forth adequate effort toward the achievement of organizational objectives.
3. The average human being prefers to be directed, wishes to avoid responsibility, has relatively little ambition, wants security above all.

Theory Y

1. The expenditure of physical and mental effort in work is as natural as play or rest.
2. External control and the threat of punishment are not the only means for bringing about effort toward organizational objectives. Man will exercise self-direction and self-control in the service of objectives to which he is committed.
3. Commitment to objectives is related to the rewards associated with their achievement.
4. The average human being learns, under proper conditions, not only to accept but to seek responsibility.
5. The capacity to exercise a relatively high degree of imagination, ingenuity, and creativity in the solution of organizational objectives is widely, not narrowly, distributed in the population.
6. Under the conditions of modern industrial life, the intellectual potentialities of the average human being are only partially utilized.[2]

If Theory X appeals to you, then you should have:
One-way communication (nothing goes up except reports).
Strategy planning by the top leaders only.
Decision-making at the top level only.
Handing down of decisions to be implemented by supervisors.
Clear-cut instructions to direct the workers.

If you see possibilities in Theory Y then you should consider:
Two-way communication.
Line officer involvement in planning and goal-setting.
Decision-making at the most logical level.

A Comparison of Supervisory Roles

Jack Gibb expands on McGregor's Theories by describing two types of supervisory viewpoints:

> People must be led. People perform best under leaders who are creative, imaginative, and aggressive—under leaders who lead. It is the responsibility of the leader to marshall the forces of the organization, to stimulate effort, to capture the imagination, to inspire people, to coordinate efforts, and to serve as a model of sustained effort. The leader should keep an appropriate social distance, show no favorites, control his emotions, command respect, and be objective and fair. He must know what he is doing and where he wants to go. He must set clear goals for himself and for the group or institution, and then communicate these goals well to all members of the organization. He must listen for advice and counsel before making decisions. But it is his responsibility to make decisions and to set up mechanisms for seeing that the decisions are implemented. After weighting the facts and seeking expert counsel, he must make policy and rules, set reasonable boundaries, and see that these are administered with justice and wisdom, even compassion. The leader should regard good performance and be equally ready to give negative criticism where warranted and to appraise performance frequently, fairly, and unequivocally. He must command strong discipline, not only because people respect a strong leader, but because strength and firmness communicate care and concern.[3]

The foregoing is the traditional or defensive view. Compare it to the following participative concept.

> People grow, produce, and learn best when they set their own goals, choose activities that they see as related to these goals, and have a wide range of freedom of choice in all parts of their lives. Under most conditions persons are highly motivated, like to take responsibilities, can be trusted to put out a great deal of effort toward organizational goals, are creative and imaginative, and tend to want to cooperate with others. Leadership is only one of several significant variables in the life of the group or institution. Leaders can be helpful and often are. The most effective leader is one who acts as a catalyst, a consultant, and a resource to the group to grow, to emerge, and to become more free. He serves the group best when he is a whole person, is direct, real, open, spontaneous, permissive, emotional, and highly personal. The leader at his best is an effective member. He acts in such a way as to facilitate group strength, individual responsibility, diversity, nonconformity, and aggressiveness. The leader is thus not necessary to the group and quickly becomes replaceable, dispensable, and independent.[4]

It seems a great tragedy that the main criteria in the selection process for a police supervisor is that the applicant be a "company man." All too many present-day supervisors are dedicated to the task and concerned about their employees only from a veterinary-hygiene standpoint (their only interest is that the working animals not be off sick, thereby complicating the schedule).

Variations in Supervision

It is not suggested that all traditional leadership practices be discarded in favor of the participative style. Different situations call for different types of supervision. Not only must consideration be given to the situation and the personality characteristics of the subordinates, but also the leader's personal preference and his skill in the various styles. There is certainly no clear-cut set of leadership practices that will always yield the best results. If there is any conclusion that can be drawn from this it is that a management style should be custom-designed to fit the particular characteristics of each task, supervisor and subordinate relationship. One researcher who studied this particular problem suggested that perhaps only 50 percent of a worker group will react favorably to a participative style of leadership. His recommendation was that an organization should have both types of leadership, and that employees should be placed where they will function best.

The Key is Involvement in Work

The men in the Patrol Division distrust those in the Detective Division who seem to be more interested in releasing suspects than charging them; the detectives are too busy to care what the patrolmen think; the Traffic Division feels a coolness because its function does not fit into the pattern of traditional cops and robbers; and the staff meet in secrecy and make final decisions on dozens of matters that are better understood by subordinate experts. I sometimes think the only common unifying factor in most departments is their central heating system.

Thus far we have shared some ideas about the attitudes of officers, and about the role of their supervisors. If what has been said can be reduced to its essence it would be this: *The need is to create an atmosphere of openness and trust where participation is encouraged by placing the decision-making process (with authority and responsibility) at the lowest logical level.* This sounds like an ivory-tower daydream. Let's be realistic. The point is not to build happy squads of officers who feel good about the whole thing; what is wanted are effective teams who work well and can rationally resolve their own problems. This does not mean a fat, dumb, happy police department where all problems have been resolved. Problems are productive if they are in the form of accepting new challenges, gambling on your ability, and expanding yourself. The problems that damage are those frustrations over which you have no control.

Does it Really Work?

Not with all people or in all situations. Research at organizations where participative management plans were initiated show that satisfied, happy workers are sometimes more productive—and sometimes merely happy. Some managers and workers were found to be unable to shoulder the responsibility for decision-making. Some administrators were unable to delegate the authority to lower levels—even though they recognized the need to do so. But few, if any, recommended returning to the traditional supervisory model.

Participative management was almost unknown ten years ago. Union Carbide Corporation was one pioneer in this field. In the early 60's they were a disunited complex of organizations involving chemicals, plastics, silicones, metals, mining and auto products. The internal problems were reflected in low profits and a correspondingly poor stock market rating. In spite of a desire by the division heads to solve the problem it seemed that they were deadlocked in an irresolvable conflict. Everyone had ideas—each one different. The six division heads decided to isolate themselves in a backwoods lodge where they worked on the problem sixteen hours a day for five days. The basic scheme that evolved was simple: that decisions can be made by the working groups affected better than by individuals. This moved decision-making back down the chain, including placing the responsibility for making company heads manage their organizations in terms of corporate goals. They coupled the "Scheme" with "Leveling" which did away with the phoniness of what was supposed to be communication between people. The positive results are shown in Cadillac-sized bonuses for each employee. They don't claim that the Scheme is, or is expected, to be the final answer to everything. The conflict between administrators, who think in terms of the big picture, and the staff, who press for their own advantages, will no doubt always exist to some extent.[5]

In the Union Carbide case, the change was initiated by top management. At TRW Systems in California a group of concerned middle-management people went to the company president requesting permission to experiment with new ideas for increasing effectiveness. "Let's try things," he replied. "If they work, continue them. If they don't, modify them, improve them, or drop them."[6] Thus an internal group at TRW started what has since become a model in organizational development.

Where and How Do You Start?

Experts are divided on a best approach. Some insist that top management is the place to start. Others cite examples where successful conversion was gradually built on demonstrated results with units on the line.

I think the ideal start would be to contract with a firm specializing in management development to conduct a live-in laboratory of several days for your staff. Here you would be taught how to communicate effectively and how to create some openness and trust with others. Then you, as a team, would identify your organizational problems and contribute data that could resolve them. The lab leader will make sure that you understand that the data is yours, the solution is yours and that as owners you are free to use it or not.

The 19-question data collector (see next page) designed by Rensis Likert is a valuable tool to diagnose your management climate and to pinpoint problem areas.[7]

In Seattle, we have started by selecting a dozen officers who displayed natural group leadership talents at human relations laboratories. They are being sent through special training to develop the skills needed to help units identify and resolve problems. Thus we will have an in-house team of grass-roots consultants to work toward the long-range goal of an improved department.

Summary

Probably there is an administrator reading this article who thinks that it is a good idea. He will issue a directive ordering that participative management will become effective immediately. The next day he will have no visible change, and so conclude that it didn't work.

Patience, persistence and confidence are essential if significant change is to occur and be maintained. Most important, the officers on the line must undertake the effort and make it their own. In other words, those on whom we must depend for a lasting change must have ownership of the change effort. You may see in participative management a means of higher morale, more productivity and fewer citizen complaints, but try for a minute to look at it from the

Chart 2-1
A Data Collector

			System 1 Exploitive Authoritative	System 2 Benevolent Authoritative	System 3 Consultative	System 4 Participative Group
Leadership	1.	How much confidence is shown in subordinates?	None	Condescending	Substantial	Complete
	2.	How free do they feel to talk to superiors about job?	Not at all	Not very	Rather free	Fully free
	3.	Are subordinates' ideas sought and used, if worthy?	Seldom	Sometimes	Usually	Always
Motivation	4.	Is predominent use made of 1 fear, 2 threats, 3 punishment, 4 rewards, 5 involvement?	1, 2, 3, occasionally 4	4, some 3	4, some 3 and 5	5, 4, based on group set goals
	5.	Where is responsibility felt for achieving organization's goals?	Mostly at top	Top and middle	Fairly general	At all levels
Communication	6.	How much communication is aimed at achieving organization's objectives?	Very little	Little	Quite a bit	A great deal
	7.	What is the direction of information flow?	Downward	Mostly downward	Down and up	Down, up, and sideways
	8.	How is downward communication accepted?	With suspicion	Possibly with suspicion	With caution	With an open mind
	9.	How accurate is upward communication?	Often wrong	Censored for the boss	Limited accuracy	Accurate
	10.	How well do superiors know problems faced by subordinates?	Know little	Some knowledge	Quite well	Very well
Decisions	11.	At what level are decisions formally made?	Mostly at top	Policy at top, some delegation	Broad policy at top, more delegation	Throughout but well integrated
	12.	What is the origin of technical and professional knowledge used in decision making?	Top management	Upper and middle	To a certain extent, throughout	To a great extent, throughout
	13.	Are subordinates involved in decisions related to their work?	Not at all	Occasionally consulted	Generally consulted	Fully involved
	14.	What does decision-making process contribute to motivation?	Nothing often weakens it	Relatively little	Some contribution	Substantial contribution
Goals	15.	How are organizational goals established?	Orders issued	Orders, some comment invited	After discussion, by orders	By group action (except in crisis)
	16.	How much covert resistance to goals is present?	Strong resistance	Moderate resistance	Some resistance at times	Little or none
Control	17.	How concentrated are review and control functions?	Highly at top	Relatively highly at top	Moderate delegation to lower levels	Quite widely shared
	18.	Is there an informal organization resisting the formal one?	Yes	Usually	Sometimes	No—same goals as formal
	19.	What are cost, productivity, and other control data used for?	Policing, punishment	Reward and punishment	Reward, some self-guidance	Self-guidance, problem solving

patrolman's viewpoint. What would happen to your sense of dignity and self-worth if you were a full participant in the decision-making process which affect you? What would be the effect on you if your sergeant abandoned authority as a way of controlling you and instead shared problems openly with mutual trust? How would you feel if you knew that the lieutenant would treat your errors as part of the learning process and not as something to hold over your head? What would be the effect if you knew that your bosses were interested, not in using you, but in your personal growth?

I think it would make a big difference in your attitude towards work and in your homelife; and that the change would be recognized by the citizens you serve.

Theory "X" — Theory "Y"*

IT HAS BECOME TRITE to say that industry has the fundamental know-how to utilize physical science and technology for the material benefit of mankind, and that we must now learn how to utilize the social sciences to make our human organizations truly effective.

To a degree, the social sciences today are in a position like that of the physical sciences with respect to atomic energy in the thirties. We know that past conceptions of the nature of man are inadequate and, in many ways, incorrect. We are becoming quite certain that, under proper conditions, unimagined resources of creative human energy could become available within the organizational setting.

We cannot tell industrial management how to apply this new knowledge in simple, economic ways. We know it will require years of exploration, much costly development research, and a substantial amount of creative imagination on the part of management to discover how to apply this growing knowledge to the organization of human effort in industry.

Management's Task: The Conventional View

The conventional conception of management's task in harnessing human energy to organizational requirements can be stated broadly in terms of three propositions. In order to avoid the complications introduced by a label, let us call this set of propositions "Theory X":

1. Management is responsible for organizing the elements of productive enterprise—money, materials, equipment, people—in the interest of economic ends.
2. With respect to people, this is a process of directing their

---

*Source: Douglas Murray McGregor, "The Human Side of Enterprise," *The Management Review,* Vol. XLVI, No. 11, November, 1957, pp. 22-28, 88-92. Reprinted by permission of the publisher. Copyright 1957 by American Management Association.

efforts, motivating them, controlling their actions, modifying their behavior to fit the needs of the organization.
3. Without this active intervention by management, people would be passive—even resistant— to organizational needs. They must therefore be persuaded, rewarded, punished, controlled—their activities must be directed. This is management's task. We often sum it up by saying that management consists of getting things done through other people.

Behind this conventional theory there are several additional beliefs—less explicit, but widespread:
4. The average man is by nature indolent—he works as little as possible.
5. He lacks ambition, dislikes responsibility, prefers to be led.
6. He is inherently self-centered, indifferent to organizational needs.
7. He is by nature resistant to change.
8. He is gullible, not very bright, the ready dupe of the charlatan and the demagogue.

The human side of economic enterprise today is fashioned from propositions and beliefs such as these. Conventional organization structures and managerial policies, practices, and programs reflect these assumptions.

In accomplishing its task—with these assumptions as guides—management has conceived of a range of possibilities.

At one extreme, management can be "hard" or "strong." The methods for directing behavior involve coercion and threat (usually disguised), close supervision, tight controls over behavior. At the other extreme, management can be "soft" or "weak." The methods for directing behavior involve being permissive, satisfying people's demands, achieving harmony. Then they will be tractable, accept direction.

This range has been fairly completely explored during the past half century, and management has learned some things from the exploration. There are difficulties in the "hard" approach. Force breeds counter-forces: restriction of output, antagonism, militant unionism, subtle but effective sabotage of management objectives.

This "hard" approach is especially difficult during times of full employment.

There are also difficulties in the "soft" approach. It leads frequently to the abdication of management—to harmony, perhaps, but to indifferent performance. People take advantage of the soft approach. They continually expect more, but they give less and less.

Currently, the popular theme is "firm but fair." This is an attempt to gain the advantages of both the hard and the soft approaches. It is reminiscent of Teddy Roosevelt's "speak softly and carry a big stick."

Is the Conventional View Correct?

The findings which are beginning to emerge from the social sciences challenge this whole set of beliefs about man and human nature and about the task of management. The evidence is far from conclusive, certainly, but it is suggestive. It comes from the laboratory, the clinic, the schoolroom, the home, and even to a limited extent from industry itself.

The social scientist does not deny that human behavior in industrial organization today is approximately what management perceives it to be. He has, in fact, observed it and studied it fairly extensively. But he is pretty sure that his behavior is *not* a consequence of man's inherent nature. It is a consequence rather of the nature of industrial organizations, of management philosophy, policy, and practice. The conventional approach of Theory X is based on mistaken notions of what is cause and what is effect.

Perhaps the best way to indicate why the conventional approach of management is inadequate is to consider the subject of motivation.

Physiological Needs

Man is a wanting animal—as soon as one of his needs is satisfied, another appears in its place. This process is unending. It continues from birth to death.

Man's needs are organized in a series of levels—a hierarchy of importance. At the lowest level, but pre-eminent in importance when they are thwarted, are his *physiological needs.* Man lives for bread alone, when there is no bread. Unless the circumstances are unusual, his needs for love, for status, for recognition are inoperative when his stomach has been empty for a while. But when he eats regularly and adequately, hunger ceases to be an important motivation. The same is true of the other physiological needs of man—for rest, exercise, shelter, protection from the elements.

A satisfied need is not a motivator of behavior! This is a fact of profound significance that is regularly ignored in the conventional approach to the management of people. Consider your own need for air: Except as you are deprived of it, it has no appreciable motivating effect upon your behavior.

Safety Needs

When the physiological needs are reasonably satisfied, needs at the next higher level begin to dominate man's behavior—to motivate him. These are called *safety needs.* They are needs for protection against danger, threat, deprivation. Some people mistakenly refer to these as needs for security. However, unless man is in a dependent relationship where he fears arbitrary deprivation, he does not demand security. The need is for the "fairest possible bread." When he is confident of this, he is more than willing to take risks. But when he feels threatened or dependent, his greatest need is for guarantees, for protection, for security.

The fact needs little emphasis that, since every industrial employee is in a dependent relationship, safety needs may assume considerable importance. Arbitrary management action, behavior which arouses uncertainty with respect to continued employment or which reflects favoritism or discrimination, unpredictable administration of policy—these can be powerful motivators of the safety needs in the employment relationship *at every level,* from worker to vice president.

Social Needs

When man's physiological needs are satisfied and he is no longer fearful about his physical welfare, his *social needs* become important motivators of his behavior—needs for belonging, for association, for acceptance by his fellows, for giving and receiving friendship and love.

Management knows today of the existence of these needs, but it often assumes quite wrongly that they represent a threat to the organization. Many studies have demonstrated that the tightly knit, cohesive work group may, under proper conditions, be far more effective than an equal number of separate individuals in achieving organizational goals.

Yet management, fearing group hostility to its own objectives, often goes to considerable lengths to control and direct human efforts in ways that are inimical to the natural "groupiness" of human beings. When man's social needs—and perhaps his safety needs, too—are thus thwarted, he behaves in ways which tend to defeat organizational objectives. He becomes resistant, antagonistic, uncooperative. But this behavior is a consequence, not a cause.

Ego Needs

Above the social needs—in the sense that they do not become motivators until lower needs are reasonably satisfied—are the needs of greatest significance to management and to man himself. They are the *egoistic needs,* and they are of two kinds:

1. Those needs that relate to one's self-esteem—needs for self-confidence, for independence, for achievement, for competence, for knowledge.
2. Those needs that relate to one's reputation—needs for status, for recognition, for appreciation, for the deserved respect of one's fellows.

Unlike the lower needs, these are rarely satisfied; man seeks indefinitely for more satisfaction of these needs once they have become important to him. But they do not appear in any significant way until physiological, safety, and social needs are all reasonably satisfied.

The typical industrial organization offers few opportunities for the satisfaction of these egoistic needs to people at lower levels in the hierarchy. The conventional methods of organizing work, particularly in mass-production industries, give little heed to these aspects of human motivation. If the practices of scientific management were deliberately calculated to thwart these needs, they could hardly accomplish this purpose better than they do.

Self-fulfillment Needs

Finally—a capstone, as it were, on the hierarchy of man's needs—there are what we may call the *needs for self-fulfillment*. These are the needs for realizing one's own potentialities, for continued self-development, for being creative in the broadest sense of that term.

It is clear that the conditions of modern life give only limited opportunity for these relatively weak needs to obtain expression. The deprivation most people experience with respect to other lower-level needs diverts their energies into the struggle to satisfy *those* needs, and the needs for self-fulfillment remain dormant.

Management and Motivation

We recognize readily enough that a man suffering from a severe dietary deficiency is sick. The deprivation of physiological needs has behavioral consequence. The same is true—although less well recognized—of deprivation of higher-level needs. The man whose needs for safety, association, independence, or status are thwarted is sick just as surely as the man who has rickets. And his sickness will have behavioral consequences. We will be mistaken if we attribute his resultant passivity, his hostility, his refusal to accept responsibility to his inherent "human nature." These forms of behavior are *symptoms* of illness—of deprivation of his social and egoistic needs.

The man whose lower-level needs are satisfied is not motivated to satisfy those needs any longer. For practical purposes they exist no longer. Management often asks, "Why aren't people more productive? We pay good wages, provide good working conditions, have excellent fringe benefits and steady employment. Yet people do not seem to be willing to put forth more than minimum effort."

The fact that management has provided for these physiological and safety needs has shifted the motivational emphasis to the social and perhaps to the egoistic needs. Unless there are opportunities *at work* to satisfy these higher-level needs, people will be deprived; and their behavior will reflect this deprivation. Under such conditions, if management continues to focus its attention on physiological needs, its efforts are bound to be ineffective.

People *will* make insistent demands for more money under these conditions. It becomes more important than ever to buy the material goods and services which can provide limited satisfaction of the thwarted needs. Although money has only limited value in satisfying many higher-level needs, it can become the focus of interest if it is the *only* means available.

The Carrot-and-Stick Approach

The carrot-and-stick theory of motivation (like Newtonian physical theory) works reasonably well under certain circumstances. The *means* for satisfying man's physiological and (within limits) his safety needs can be provided or withheld by management. Employment itself is such a means, and so are wages, working conditions, and benefits. By these means the individual can be controlled so long as he is struggling for subsistence.

But the carrot-and-stick theory does not work at all once man has reached an adequate subsistence level and is motivated primarily by higher needs. Management cannot provide a man with self-respect, or with the respect of his fellows, or with the satisfaction of needs for self-fulfillment. It can create such conditions that he is encouraged and enabled to seek such satisfactions for *himself,* or it can thwart him by failing to create those conditions.

But this creation of conditions is not "control." It is not a good device for directing behavior. And so management finds itself in an odd position. The high standard of living created by our modern technological know-how provides quite adequately for the satisfaction of physiological and safety needs. The only significant exception is where management practices have not created confidence in a "fair break"—and thus where safety needs are thwarted. But by making possible the satisfaction of low-level needs, management has deprived itself of the ability to use as motivators the devices on which conventional theory has taught it to rely—rewards, promises, incentives, or threats and other coercive devices.

The philosophy of managment by direction and control—*regardless of whether it is hard or soft*—is inadequate to motivate because the human needs on which this approach relies are today unimportant motivators of behavior. Direction and control are essentially useless in motivating people whose important needs are social and egoistic. Both the hard and the soft approach fail today because they are simply irrelevant to the situation.

People, deprived of opportunities to satisfy at work the needs which are now important to them, behave exactly as we might predict—with indolence, passivity, resistance to change, lack of responsibility, willingness to follow the demagogue, unreasonable demands for economic benefits. It would seem that we are caught in a web of our own weaving.

A New Theory of Management

For these and many other reasons, we require a different theory of the task of managing people based on more adequate assumptions about human nature and human motivation. I am going to be so bold as to suggest the broad dimensions of such a theory. Call it "Theory Y," if you will.

1. Management is responsible for organizing the elements of productive enterprise—money, materials, equipment, people—in the interest of economic ends.
2. People are *not* by nature passive or resistant to organizational needs. They have become so as a result of experience in organizations.
3. The motivation, the potential for development, the capacity for assuming responsibility, the readinss to direct behavior toward organizational goals are all present in people. Management does not put them there. It is a responsibility of management to make it possible for people to recognize and develop these human characteristics for themselves.
4. The essential task of management is to arrange organizational conditions and methods of operation so that people can achieve their own goals *best* by directing *their own* efforts toward organizational objectives.

This is a process primarily of creating opportunities, releasing potential, removing obstacles, encouraging growth, providing guidance. It is what Peter Drucker has called "management by objectives" in contrast to "management by control." It does *not* involve the abdication of management, the absence of leadership, the lowering of standards, or the other characteristics usually associated with the "soft" approach under Theory X.

Some Difficulties

It is no more possible to create an organization today which will be a full, effective application of this theory than it was to build an atomic power plant in 1945. There are many formidable obstacles to overcome.

The conditions imposed by conventional organization theory and by the approach of scientific management for the past half century have tied men to limited jobs which do not utilize their capabilities, have discouraged the acceptance of responsibility, have encouraged passivity, have eliminated meaning from work. Man's habits, attitudes, expectations—his whole conception of membership in an industrial organization—have been conditioned by his experience under these circumstances.

People today are accustomed to being directed, manipulated, controlled in industrial organizations and to finding satisfaction for their social, egoistic, and self-fulfillment needs away from the job. This is true of much of management as well as of workers. Genuine "industrial citizenship"—to borrow again a term from Drucker—is a remote and unrealistic idea, the meaning of which has not even been considered by most members of industrial organizations.

Another way of saying this is that Theory X places exclusive reliance upon external control of human behavior, while Theory Y relies heavily on self-control and self-direction. It is worth noting that this difference is the difference between treating people as children and treating them as mature adults. After generations of the former, we cannot expect to shift to the latter overnight.

Steps in the Right Direction

Before we are overwhelmed by the obstacles, let us remember that the application of theory is always slow. Progress is usually achieved in small steps. Some innovative ideas which are entirely consistent with Theory Y are today being applied with some success.

DECENTRALIZATION AND DELEGATION

These are ways of freeing people from the too-close control of conventional organization, giving them a degree of freedom to direct their own activities, to assume responsibility, and importantly, to satisfy their egoistic needs. In this connection, the flat organization of Sears, Roebuck and Company provides an interesting example. It forces "management by objectives," since it enlarges the number of people reporting to a manager until he cannot direct and control them in the conventional manner.

JOB ENLARGEMENT

This concept, pioneered by I.B.M. and Detroit Edison, is quite consistent with Theory Y. It encourages the acceptance of responsibility at the bottom of the organization; it provides opportunities for satisfying social and egoistic needs. In fact, the reorganization of work at the factory level offers one of the more challenging opportunities for innovation consistent with Theory Y.

PARTICIPATION AND CONSULTATIVE MANAGEMENT

Under proper conditions, participation and consultative management provide encouragement to people to direct their creative energies toward organizational objectives, give them some voice in decisions that affect them, provide significant opportunities for the satisfaction of social and egoistic needs. The Scanlon Plan is the outstanding embodiment of these ideas in practice.

PERFORMANCE APPRAISAL

Even a cursory examination of conventional programs of performance appraisal within the ranks of management will reveal how completely consistent they are with Theory X. In fact, most such programs tend to treat the individual as though he were a product under inspection on the assembly line.

A few companies—among them General Mills, Ansul Chemical, and General Electric—have been experimenting with approaches which involve the individual in setting "targets" or objectives *for himself* and in a *self*-evaluation of performance semiannually or annually. Of course, the superior plays an important leadership role in this process—one, in fact, which demands substantially more competence than the conventional approach. The role is, however, considerably more congenial to many managers than the role of "judge" or "inspector" which is usually forced upon them. Above all, the individual is encouraged to take a greater responsibility for planning and appraising his own contribution to organizational objectives; and the accompanying effects on egoistic and self-fulfillment needs are substantial.

Applying the Ideas

The not infrequent failure of such ideas as these to work as well as expected is often attributable to the fact that a management has "bought the idea" but applied it within the framework of Theory X and its assumptions.

Delegation is not an effective way of exercising management by control. Participation becomes a farce when it is applied as a sales gimmick or a device for kidding people into thinking they are important. Only the management that has confidence in human capacities and is itself directed toward organizational objectives rather than toward the preservation of personal power can grasp the implications of this emerging theory. Such management will find and apply successfully other innovative ideas as we move slowly toward the full implementation of a theory like Y.

The Human Side of Enterprise

It is quite possible for us to realize substantial improvements in the effectiveness of industrial organizations during the next decade or two. The social sciences can contribute much to such development; we are only beginning to grasp the implications of the growing body of knowledge in these fields. But if this conviction is to become a reality instead of a pious hope, we will need to view the process much as we view the process of releasing the energy of the atom for constructive human ends—as a slow, costly, sometimes discouraging approach toward a goal which would seem to many to be quite unrealistic.

The ingenuity and the perseverance of industrial management in the pursuit of economic ends have changed many scientific and technological dreams into commonplace realities. It is now becoming clear that the application of these same talents to the human side of enterprise will not only enhance substantially these materialistic achievements, but will bring us one step closer to "the good society."

Managerial Grid*

The purpose of this article is to compare seven managerial theories in terms of how each deals with (1) organizational needs for production and profit and (2) human needs for mature and healthy relationships.

Five of these seven are shown in the chart, "The Managerial Grid."[1] They are referred to as *country club management, impoverished management, dampened pendulum, team management,* and *task management.* The remaining two are combined forms.

The term *concern for,* as used in the Grid, is a theoretical variable which reflects basic attitudes or styles of control. The term does not necessarily reflect how much production or profit is obtained or the degree to which human needs are met. The horizontal axis of the Grid represents concern for production and profit. The vertical axis represents concern for mature and healthy relations among those engaged in production. Each axis is on a 1 to 9 point scale, with the 1 representing minimum interest or concern and the 9, maximum concern.

In the discussion that follows, the emphasis is on analysis of the corners and mid-point of the Grid, although these extreme positions are rarely found in pure form in the working situation.

---

*Source: Robert R. Blake, Jane S. Mouton, and Alvin C. Bidwell, "Managerial Grid," *Advanced Management–Office Executive,* Vol. 1, No. 9, September, 1962, pp. 12-15 and 36. Reprinted by permission of the publisher. Copyright 1962 by the American Management Association.

Figure 2-1
The Managerial Grid

Concern for People

9	Country Club Management—(1,9)					Team Management—(9,9)		
8	Production is incidental to lack of conflict and					Production is from integration of task		
7	"good fellowship."					and human requirements.		
6					Dampened Pendulum—(5,5) (Middle of the Road) Push			
5	Impoverished Management—(1,1) Effective production				for production but don't go "all out." Give some, but			
4	is unobtainable because people are lazy, apathetic				not all. "Be fair but firm."			
3	and indifferent. Sound and mature relationships					Task Management—(9,1) Men are a commodity just		
2	are difficult to achieve because human					as machines. A manager's responsibility is to plan,		
1	nature's being what it is, conflict is inevitable.					direct and control the work of those subordinate to him.		
	1	2	3	4	5	6	7	8	9

Concern for Production

In several of the theories, the concerns for production and those for people will be seen in conflict. In three of the theories, the assumption is that attention to needs of individuals does not contribute automatically to production requirements.

Turning again to the chart (Figure 2-1), note that the lower right-hand corner represents the task management approach (9, 1). Here, primary concern is for output of the enterprise.[2] People are viewed solely in terms of their contribution to production.

This theory is based on the notion that a manager's central responsibilities are to achieve production objectives.[3] Those who are engaged in planning and controlling, in other words, are not those who carry out the actions directly. Under the task management

Behavioral Sciences

theory, a job is a job and someone has to do it. Like machines, people are seen as production tools, obligated to comply when told what they are to do.

At lower levels, concern for production may be thought of as actual output. Concern for production at the managerial levels is just as critical as units of output are among operators. The following kind of advice which was given to managers in the middle levels is characteristic of a task management orientation.[4]

> For an executive to challenge orders, directors and instructions, policy and procedures, rules and regulations, etc., smacks of insubordination or lack of cooperation. It shows his failure to understand the need for decisions at higher levels and for direction and control.

Thoughts, attitudes, and feelings are given little or no attention. When interpersonal conflict arises, the way to handle it (as well as all types of feelings) is to suppress it through disciplinary types of actions. Under the task management theory, if people don't comply after a certain amount of control has been applied, they should be replaced. Thus this theory has been referred to as an impersonal approach to managing.

The task management theory tends to produce unsound relationships among those who should be operating in an interdependent way within the organizational structure.[5] Since Chris Argyris has hammered home the negative consequences of this theory, little more needs to be said here.[6]

Production Incidental to Satisfaction

The country club management theory (see 1, 9, upper left-hand corner of the chart) is the reverse of the task management theory in a number of respects. In the former, production is incidental to satisfaction through social relations, good fellowship, and fraternity. The goal of this theory of management is to achieve harmony even though needs for output may suffer as a result. Assumption is that contented people will produce as well as contented cows, if given the chance.[7]

This theory is not well suited to organizations which strive for increased effectiveness. When, because of outside forces, the organization has to increase efficiency, it is frequently unable to respond to the demands. The country club theory does tend to flourish in organizations that approach monopolies and in bureaucratic structures.

Aware of the Need to Be Nice

Under the country club theory, too, executives are aware of the need to be nice. Moreover, the people in the system sense the phony quality in *good* human relations which are *not* related to conditions of work and production.

The be-nice approach leads to few overt production or personnel problems, because conflict is smothered and denied. The manager is seen as a likable fellow, a good Joe, or a big brother. He is prepared to make his subordinates happy, carefree, and satisfied at almost any economic price.

Also, the country club approach is unlikely to acheive any meaningful human relations gains, since conflict and frustration are smoothed over, not dealt with. This is the style which is sometimes called *soft* management, in contrast with *hard* management. Frequently the only alternative seen by hard managers is the country club theory.[8]

The impoverished management theory (1,1) is shown in the lower left corner of the chart. The manager at this position de-emphasizes concerns for production, with just enough being done to get by. He also disregards or diminishes the importance of human relationships. He takes the ostrich approach to feelings—instead of suppressing or denying them, he ignores them.

Within the executive level an impoverished management orientation can be found in circumstances in which a person has been passed by repeatedly. Rather than looking elsewhere, he adjusts to the work setting by giving minimal performance, seeking satisfaction elsewhere.

This managerial style is not so prevalent throughout an entire company. In a competitive economy, a company operated under this style is unlikely to stay in business long. On the other hand, many individuals who manage in the impoverished management style are able to survive in bureaucratic organizations,[9] and in country club managerial structures where the rule is that "no one gets fired."

Three approaches to management consider the task (9,1), the country club (1,9) or the impoverished (1,1) types simultaneously. All stem from the notion that needs of mature individuals and organizational requirements of production oppose each other.

The wide-arc pendulum theory, for example, describes a relationship between the task (9,1) and the country club (1,9) positions as: When management tightens up to get increased output, as often happens under recession pressures, it does so consistently with the task (9,1) attitude. Later, relationships become so disturbed that production suffers. Then management feels forced to ease off and to start being concerned with the thoughts, feelings, and attitudes of people.[10]

Output Pressures Are Lessened

Because of the negative results of the task approach, a swing in the direction of the country club position takes place. Output pressures are lessened to make people feel that management's intentions are good. Thus, the pendulum tends to swing from the one position to the other.

When a degree of confidence in management has been restored among the people, the tightening up occurs again to regain the losses in production suffered during the previous pendulum swing. Cracking down to get efficiency and easing up to restore confidence is the pendulum swing from hard to soft.

Another kind of pendulum swing within an organization is one associated with task management (9,1) pressures for production. In this cycle, as described by Robert L. Katz[11] top management's plans and commands lead to lower management's reacting in an impoverished manner (1,1), with indifference, apathy, and minimal output.

Tries to Tighten Up on Lower Levels

The failure of middle- and lower-levels of management to respond with concrete contributions to production is followed by top management's redoubling of its efforts to tighten up. Once successful influence from top to middle has been achieved, the middle tries to tighten up on lower levels. Thus the cycle goes from the task, to the impoverished, and back to the task management theory.

In the center of The Managerial Grid is shown the dampened pendulum (5,5) or the middle-of-the-road theory. This theory says, push enough to get acceptable production, but yield to the degree necessary to develop morale. When the happy medium is achieved, don't expect too much production. Recognize that one must be flexible and must give. Under the dampened pendulum theory, it is believed that by clever string-pulling, management can prevent either of the two concerns from blocking the complete attainment of the other.[12]

In a number of respects it would appear that managements which have abandoned the task theory tended over time to slide toward the dampened pendulum position. However, the shift is not a healthy one because the dampened pendulum position retains the theory of work direction contained in the task management position. The shift adds to the theory of human relations of country club management. The dampened pendulum approach does not solve the problem. Rather, a live-and-let-live situation is created under which the real problem is muted.[13]

Paternalistic management pushes for output in a task management way, but in time it recognizes feelings of alienation that separate the workers from the management. Therefore, an additional step is taken to satisfy lower-management levels and workers. Concern for people is expressed through *taking care of them* in a country club management fashion. Organizational members are *given* many fine things—good pay, excellent benefit programs, recreational facilities, retirement programs, and even low-cost housing.

However, these are not given to acknowledge contribution to output. They are given to *buy* subservience.[14] A paternalistic executive retains tight control in work matters, but is benevolent in a personal way. In other words, he treats his junior executives as part of his managerial family. Paternalism is a more or less stable mix of two anchor positions. Although it has failed repeatedly to solve the basic problems of getting production through people, it is still current in managerial thinking.[15-16]

Behavioral Sciences 93

Accepts Conflict as Inevitable

All the managerial styles just described accept conflict between concerns for production and concerns for people as inevitable. Each attempts to deal with the assumed basic incompatibility in a different way. The task management and impoverished management and pendulum theories are one or another in that the emphasis is either on production or on people.

The middle-of-the-road, or dampened pendulum, and the paternalistic modes try to achieve some sort of balance. Under the impoverished management approach, the manager does the minimum. Only under the team approach, description of which follows, is an integration achieved where the goal is production through people.

In the team management position (9,9), the building block in the organization is the team. It is not the individuals on a one-to-one basis, as though isolated from one another.[17] To a country club style manager, the style of a task manager and that of a team manager seem similar, since both express a high concern for production. However, the way in which production is achieved is vastly different.

A team manager describes his philosophy on work planning in this way:

> When a change is required, I meet with the work group, present the picture, discuss and get reactions and ideas, *build* in their ideas, commitment, and ownership. Together we set up procedures and ground rules and assign individual responsibilities. The work group sets goals and flexible schedules.

Planning, as but one example, is a product of team effort rather than of individual skill. In a well-integrated team operation, all voices carry weight in final positions.

Gaining effective integration among members of multi-level teams is seen as another important task of leadership. Here, joint effort is centered on production, and members share responsibilities for planning, directing, and controlling.

A third key objective of team management is that of linking teams into effective communication and problem-solving systems which have a peer relationship with one another. Union and management or correlated divisions in different departments would be examples.

Knowledge of human relations needed under the team management theory is far greater than most managers possess today. When problems of feelings and emotions arise in the working relationships, the manager who uses the team approach recognizes the problems and confronts them directly. He deals with and works through conflict as it appears.

Morale Achieved Is Task Related

The manager who uses the task management approach finds it difficult to distinguish team management from country club management because both reflect high concern for people. Yet the concern expressed by the country club manager is more in terms of satisfaction based on non-work aspects of the situation, such as good social relations. The team manager seeks to integrate people around production. The morale achieved under team management conditions is task-related.

Also, the manager who uses the team approach must have greater appreciation of and skill in unleashing individual motivation than managers using any other approach. Behavioral science theory and research findings have identified many of the conditions of team management.[18] In addition, the theory and ways of teaching are available for the manager who wants to shift from any of the other approaches to the team situation.

Production Improvement

Many organizational efforts of the past fifteen years have intuitively aimed toward improving the achievement of production through people in a team management approach. Without explicit managerial theory available to guide them, many efforts have been built on trying to shore up bad human relations brought about by the task management approach.[19]

One result has been what is called phony human relations. All too frequently, managers have adopted a managerial strategy in the country club area of The Managerial Grid as a soft approach.[20] They have been led down this path by the assumption that there must be a conflict between the organizational requirements and needs of individuals. Other managers have confused the emphasis on using the team as the building block with that of destroying individual self-expression and freedom.[21]

By now, in a number of instances, managerial effort has successfully employed behavioral science concepts and methods as the basis for moving toward a team managerial orientation. Some have used team training with specific development of The Managerial Grid for setting meaningful organization objectives to which all team members are committed.[22-23]

Sheds Light on Controversies

Understanding of The Managerial Grid is basic for appreciating present organizational structures and the relationships between organizational and human needs. In addition, the Grid sheds light on many of the day-by-day controversies present in managerial thinking.

Heaviest emphasis here is placed on the team management (9,9) position. Generally, the best long-term production is achieved and sustained when concerns for production and needs of people are integrated in the team direction, regardless of economic conditions.

When managements have tried to move away from the task management anchor position, without behavioral science training, the movement has tended to drift, unwittingly, in the country club direction. The movement frequently stabilizes somewhere around the dampened pendulum (5,5) position, where some push for production is balanced against some concern for people. The major difference between the task management and the dampened pendulum theories is in the increased recognition placed on people. The task management theory of work control is retained, however.[24]

The skills involved in moving toward the team management position are difficult to attain. The theory for integrating effective production through sound human relationships is complex in concept and application. Just as the physical sciences have become the indispensable underpinning of product development and technological innovation, a new era can be seen on the managerial horizon where the behavioral sciences are beginning to undergird the design of production structures.

Territorial Imperative*

Anthropologist Robert Ardrey, in his book the "Territorial Imperative,"[1] develops the thesis that in all animate beings there exists a genetically-determined form of behavior which compels them to identify with and defend a specific geographic area. He refines this generalization by stating that " . . . man is a territorial animal . . . ," postulating that human beings naturally assume a protective responsibility for the integrity and security of a definite piece of real estate. This concept, which the author states " . . . is today accepted beyond question in the biological sciences . . ." has become a fundamental theme in the Los Angeles Police Department's policy of progressive evolvement. It is clearly manifest in two of the Department's major programs: the Basic Car Plan, which has been in existence for about two years; and the "Line Decentralization Program," which is in its initial phase.

The fundamental premise of the Basic Car Plan is that a nucleus of nine patrol officers, led by a Senior Lead Officer (Policeman III + 1)[2] and complemented by detective, traffic enforcement officers, and accident investigators provides the basic police service to the citizens within a specific district. 24 hours a day, 7 days a week, 365 days a year. While this effort to cause our officers to identify with, get to know, and "protect and serve"[3] the citizens within their basic car district has an administrative rather than an hereditary genesis, the concept of police officers "belonging" to an area and feeling responsible for providing police service to the citizens who live and work in the area is consistent with Ardrey's "Territorial Imperative." Similarly, with the same degree of validity, the decentralization of the responsibility for providing the three "line" services (traffic, patrol, and detective) by phasing out the functionally-structured bureaus with city-wide responsibility and creating four "Operations Areas" to provide the basic police services is a direct application of the "Territorial Imperative" concept on a grander scale.

---

*Source: John A. McAllister, "Territorial Imperative," *California Law Enforcement Journal,* Vol. 7, No. 1, July, 1972, pp. 13-20. Reprinted by permission of the author.

Why Reorganize?

It is necessary to look first at the historical development of the Los Angeles Police Department to set the foundation for, and place in perspective, the necessity for its reorganization to meet the changing needs and service requirements of the citizens of Los Angeles. Until 1923 a history of the organization of the Department was a history of the expansion of a town Marshal's office to perform the multifarious duties required of a Metropolitan Police Department.

Between 1923 and 1929, the changes in the Department's organization merely amounted to efforts to consolidate service functions and provide greater administrative control, uniformity, and proper delegation of authority. In 1929, after almost fifty years of random development, the Department underwent a major reorganization by centralizing responsibility for the performance of each specialized function. At that time, scattered organizational entities performing similar functions were grouped together in bureaus, and the responsibility for providing traffic, patrol, and detective services city-wide was placed on the Bureau Commanders. Consequently, within each geographic division the performance of these functions was the responsibility of three division commanding officers under the direction of their respective bureau commanding officers. (An exception to this arrangement existed in the so called "outlying" divisions—those remote from the downtown area—where the patrol commanding officers, under the functional supervision of the Traffic Bureau Commanding Officer, had the traffic responsibility.)

During the 43 years since that last major reorganization, unparalleled changes have occurred in the city of Los Angeles: enormous population growth; the expansion of the population to the farthest reaches of the City's 465 square miles; the development of numerous discrete communities throughout the City, each with its unique characteristics and problems; and dynamic social changes. The Department endeavored to cope with the problems created by these phenomena by increasing the number of sworn personnel and by creating additional specialized organizational entities to perform administrative and technical roles complementing the basic line functions.

As the Department and the City grew in size and complexity,

the primary objectives of achieving maximum efficiency in the utilization of manpower at all levels in the three line services became increasingly difficult. The lowest hierarchical level at which total "line" responsibility rested and at which the coordination of the three "line" bureaus occurred was at the Assistant Chief rank, by the Director of Operations, whose command is comprised of 45 police divisions and over 76% of the Department's sworn strength of 6,999. Additionally, because of Departmental and City growth, the top command structure of the Department had become geographically and administratively remote from the basic field police personnel and from the people served by the Department. The nature, variety, and number of complex problems facing the Department and the different role the Department must play in the dynamics of today's society, as opposed to the 1920's, demanded that the organizational structure in the three "line" services be modified.

Which Way?

The proposed reorganization began on October 3, 1971, with the creation of the Valley Operations Bureau. See Chart 2-2. Continuing the program involves the phasing out of the Traffic, Patrol and Detective Bureaus, supplanting them with four operational areas and a Headquarters Operations Bureau. Each of the bureaus will be commanded by a Deputy Chief whose administrative office will be physically located within each area. The bureau commanders will continue to be directly subordinate to the Director of Operations. The Headquarters Operations Bureau will provide certain specialized detective services on a city-wide basis and will provide the vital inspection and control capabilities for the Director of Operations. In addition, the Helicopter Section and the 200 uniformed Metropolitan Division Task Force will be assigned to this Bureau.

Valley Operations Bureau

The 1,219 sworn and 141 civilian personnel of the Valley Operations Bureau combine their efforts to provide basic police service to the 1,250,000 citizens who reside in the 20 communities,

**Chart 2–2
Office of Operations**

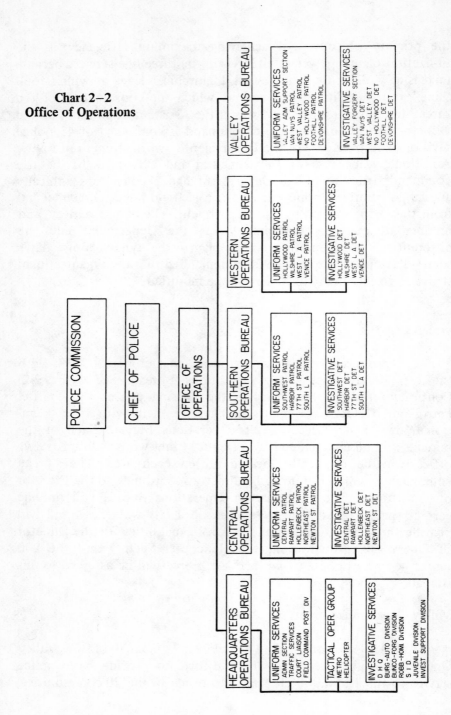

Behavioral Sciences

which occupy the 212 square miles of the city of Los Angeles generally known as the San Fernando Valley.

The command staff of the Administrative Office of the Valley Operations Bureau consists of a Deputy Chief and two Commanders. One Commander, the Commanding Officer of Investigative Services, has line command over five Captains who are Commanding Officers of the five Detective Divisions, and over the Valley Operations Bureau Forgery Section, which provides investigation of Worthless Documents for the entire Bureau. The other Commander, who is the Commanding Officer of Uniformed Services, has line command over five Captains who are Commanding Officers of the five Patrol Divisions, and over the Administrative Support Section in the Valley Operations Bureau office. Thus, the span of control of the Commanding Officer of the Valley Operations Bureau is two, and the span of control of each of the Commanders is six. (Each of the Commanders has a lieutenant reporting to him. One is the Officer in Charge of the Forgery Section and one is the Officer in Charge of the Administrative Support Section.)

Does It Work?

From the date of implementation, October 3, 1971, when the Valley Operations Bureau was created, it was obvious that many of the traditional problems which existed in the functionally-structured organization were eliminated in the new operationally-structured Bureau.

One of the key factors to the success of this first step in decentralization must be attributed to the ease of communication and the ease of direction and control existing in this close-coupled organization. Under the functional alignment, the three bureau commanders and their staffs are physically and administratively separated. Members of the Patrol Bureau staff occupied their suite of offices; the Traffic Bureau staff occupied similar separate facilities; and the same was true of the Detective Bureau staff. In the administrative office of Valley Operations Bureau the two Commanders responsible for all field services share the same office, utilize the same administrative support personnel, and view problems with a common operational perspective. Administrative inflexibility, provincialism, and poor communications between staff rank personnel in Valley Operations Bureau do not exist.

From the perspective of the Commanding Officer of Valley Operations Bureau a unique advantage of the area-operational concept over the city-wide functional structure, for a city the size and complexity of Los Angeles, is the relative ease with which emerging incipient problems can be identified and an appropriate police response be generated. His focus is confined to his prescribed area, and while his problems are in all three of the major functional fields of police work, including vice enforcement, juvenile programs, and community relations, his authority is commensurate with his responsibilities.

The accessibility of top ranking officers in the Department under the operations concept to the citizens in the area is increased immeasurably by decentralization. Rapport with the local press, civic and fraternal organization, school administrators, and with the whole array of individuals and organizations concerned with the problems in their community is enhanced. The question which was previously in their minds of who had the responsibility for solving a major police problem extended over more than one geographic community is clearly solved under a decentralized structure.

Who Gains?

Benefits from this reorganization will accrue to both the Department and to the citizens who live and work in the City. Internally, operational efficiency will be enhanced by the shorter lines of communication between the top command personnel of the Department and the division personnel because of the physical relocation of the Bureau Commanding Officers and their administrative staffs into the four areas. For example, the Commanding Officer of the Valley Operations Bureau, his staff of two Commanders, his Adjutant, clerical staff, and Administrative Support Section are housed in the geographic center of the San Fernando Valley in the Valley Headquarters Building some 18 miles from Parker Center, the main police headquarters in downtown Los Angeles. Thus, frequent personal contacts, individual and in regular staff meetings between the bureau Commanding Officer and his staff with the ten division commanding officers (five patrol, five detective), is facilitated. Of equal, if not of greater importance, is the increased decision-making authority which Bureau Commanding Officers will have in their area. For, unlike the Commanding Officers of the functional bureaus who have city-wide authority in only one of the three major line services, the area Commanding Officer has authorty to act in all three. Thus, coordination problems at that level within the area are eliminated. Through this reorganization three vital changes occur in the authority to direct, control, and coordinate the three line functions. It is reduced one hierarchical level from Assistant Chief, where it is now, to Deputy Chief. It is physically relocated in the area where it is exercised, and it is focused on a smaller than a city-wide area.

Skeptics?

Yes, there were and are skeptics about the feasibility and desirability of creating four major semi-autonomous Operations Bureaus. This skepticism surfaced early in the planning stages of the program, which officially began in August, 1970. However, the skeptics, both within and outside the Department, proved to be invaluable despite the fact that some considered decentralization pure iconoclasm. Many of their legitimate apprehensions provided the basis for changes which were incorporated into the final plan.

To reassure some of our skeptics, remember that the decentralization of the Los Angeles Police Department is not *total* decentralization with the creation of four completely autonomous police departments. The two other major organizational components of the Department, the Office of Administrative Services and the Office of Special Services are unaffected by this decentralization. The Planning and Fiscal Bureau, the Personnel and Training Bureau, and the Technical Services Bureau, for example, will continue to perform their department-wide functions in the Office of Administrative Services. Similarly, major narcotics and vice enforcement, the intelligence-gathering capability of the Department, and internal discipline will continue to be administered on a centralized basis in the Office of Special Services under the direction of an Assistant Chief, who reports directly to the Chief of Police. Additionally, the skeptics' anxieties should be relieved by the recognition that within the Office of Operations, in the Headquarters Operations Bureau, an exceptionally strong inspection and control capability will be established to ensure that Department policies and procedures are adhered to and that appropriate standards of performance are maintained throughout the City.

Another aspect of decentralization which gave rise to legitimate concern was the possibility of the loss of expertise in the functional specialties. Under the functional alignment expertise exists automatically at every hierarchical level from policeman to Deputy Chief, and the functional bureaus were the training grounds for personnel who were unsophisticated in these specialties. For example, in the Traffic Bureau, accident investigation officers have been traditionally supervised and trained by Accident Investigation Sergeants, who, in turn, were supervised and trained by Accident Investigation Lieutenants, and so on through the ranks to the Commanding Officers of the Traffic Bureau.

When the three functional bureaus are completely phased out on January 1, 1973, department-wide training programs will continue, but new responsibilities will be imposed on supervisors and Commanding Officers to develop their own expertise through planned careers. There will also be a greater emphasis placed on the divisional training programs. This renewed emphasis on specialized supervision has occurred in the Valley Operations Bureau. In each of the Valley Operations Bureau Patrol Divisions, a new "Traffic Sergeant" position has been created. It is his sole responsibility to develop training programs for accident investigators, traffic enforcement officers and patrol sergeants and officers, and to consistently monitor for the Division Commanding Officer the entire traffic program for the division.

How Far?

If the "Territorial Imperative" concept is valid for the Basic Car Plan and for an Operations Bureau, why would it not be valid at the divisional level? The answer, of course, is that it is entirely applicable both in theory and in feasibility. The Basic Car teams have their area and the Valley Operations Bureau has its "territory." The only difference between the two is size; thus, on January 1, 1973, the "Territorial Imperative" concept will be implemented on the divisional level by creating a Headquarters Command, commanded by a Captain III.[4] He will be superordinate to the existing Patrol

(Captain II) and Detective (Captain I) Commanding Officers. Thus, the coordination of the Patrol, Traffic, and Detective functions will be lowered two hierarchical levels—from Deputy Chief to Captain III. (The rank of Commander is in between the two.)

The benefits accruing to the Los Angeles Police Department and to the community from the continued decentralization of "line" authority by creating headquarters commands are identical to those derived from the creation of operational bureaus: One ranking police official will have sole responsibility for the performance of all of the field services in his division; thus, accountability and accessibility of police authority to the public is enhanced; and the *problems* of coordination between the three functions are eliminated at this level by placing complete decision-making authority in one individual.

Decentralization represents the natural evolution of a large metropolitan police department which strives to serve the needs of a diverse and sprawling community. The Los Angeles Police Department is seeking through decentralization of its operational forces to provide effective police management through clear identification of authority. The concept of the Territorial Imperative hopes to do just that.

Topics for Discussion

1. Discuss the advantages of including first-line supervisors in staff meetings.
2. Discuss the differences between system 1 and system 4 management.
3. Discuss the traditional and participative concepts of supervision.
4. Discuss the carrot-and-stick theory of motivation.
5. Discuss the managerial grid.
6. Discuss the implications of applying the "Territorial Imperative" concept to a police organization.

FOOTNOTES

Introduction

[1] Robert Presthus, *Behavioral Approaches to Public Administration* (University of Alabama Press), 1965, p. 21.
[2] *Ibid.*, p. 21.
[3] Rensis Likert, *New Patterns of Management* (New York: McGraw-Hill, 1961).
[4] Herbert Simon, *Administrative Behavior* (New York: MacMillan, 1961).
[5] Robert Presthus, *The Organizational Society* (New York: Knopf, 1962).
[6] Amitai Etzioni, *A Comparative Analysis of Complex Organizations* (New York: Free Press of Glencoe, 1961).
[7] Chris Argyris, *Interpersonal Competence and Organizational Effectiveness* (Homewood, Illinois: Dorsey Press, 1962).
[8] Chris Argyris, *Integrating the Individual and the Organization* (New York: Wiley, 1964).
[9] William G. Scott, "Organization Theory: An Overview and an Appraisal" in *Organizations: Structure and Behavior,* edited by Joseph A. Letterer (New York: Wiley, 1963).
[10] Resis, Likert, *op. cit.*
[11] Marshall E. Dimock, *A Philosophy of Administration* (New York: Harper, 1958), pp. 167-168.
[12] Harry L. Case, "Gordon R. Clapp: The Role of Faith, Purposes and People in Administration," *Public Administration Review,* June, 1964, p. 87.
[13] James V. Clark, "Distortions of Behavioral Science," *California Management Review,* Winter, 1963, p. 57.
[14] William G. Scott, *op. cit.*, p. 19.
[15] *Ibid.*, pp. 19, 20.
[16] Warren G. Bennis, Organizational Developments and the Fate of Bureaucracy, address given at American Psychological Association, September, 1964.
[17] Yehezkel Dror, "Muddling Through—Science or Inertia," *Public Administration Review,* September, 1964.
[18] Harold J. Leavitt, "Unhuman Organization," *Harvard Business Review,* July-August, 1962.
[19] Leon Radzinowicz, *The Role of Criminology and A proposal For An Institue of Criminology,* A Report presented to and approved by the special committee on the administration of criminal justice, The Association fo the Bar of the City of New York, 1964.
[20] Marray L. Schwartz, "A Hard Lesson for the Law," *Saturday Review,* November 13, 1965.
[21] *Los Angeles Police Department System Design Study,* System Development Corporation, Santa Monica, California, 1965.
[22] News item in the *Los Angeles Times,* May 16, 1966.

Police Management in Transition

[1] Dr. Kurt Lewin, "The Force Field Analysis Theory," *Field Theory in Social Science,* D. CArtwright (Ed.), New York: Harper, 1951.
[2] "Seven Step Plan for Problem Solving," University of Southern California, School of Public Administration, Center for Training and Development, Los Angeles, Calif.

A Behavioral Science Application to Police Management.

[1] National Training Laboratories Institute for Applied Behavioral Science, *Laboratories in Human Relations,* MTL-NEA, Washington, D.C., 1969, p. 36.
[2] *Ibid.,* pp. 36-37. McGregor's X and Y Theories were originally published in his book *The Human Side of Enterprise,* New York, McGraw-Hill, 1960.
[3] Gibb, Jack R. "Dynamics of Leadership," *Current Issues in Higher Education,* 1967, National Education Association.
[4] *Ibid.*
[5] Condensed from: "Union Carbide's Patient Schemers" by Gilbert Burck in *Fortune Magazine,* Dec. 1965, Time Inc.
[6] Sheldon, Davis A. "An Organic Problem-Solving Method of Organizational Change," *Journal of Applied Behavioral Science,* Vol. 3, No. 1 (Jan-Feb-Mar, 1967) pp. 3-21, Washington, D.C., NTL-NEA.
[7] An adaptation from: "Participative Management: Time For A Second Look, *Behavioral Science and the Manger's Role* pp. 198-214, National Training Laboratories-NEA, Washington, D.C., 1969. The question set originally appeared in *The Human Organization: Its Management and Value* by Rensis Likert, McGraw-Hill, 1967.

Managerial Grid

[1] The line of thinking that leads to the generalized version of The Managerial Grid is consistent with work by Rensis Likert, "Developing Patterns of Management," AMA General Management Series (New York, American Management Assn. No. 178, 1955), pp. 32-51; Edwin A. Fleishman, Edwin F. Harris, and Harold D. Burtt, *Leadership and Supervision in Industry* (Columbus, Ohio, Bureau of Educational Research, Ohio State Univ., 1955); Chris Argyris, *Personality and Organization* (new York, Harper & Brothers, 1957); and Douglas McGregor, *Human Side of Enterprise* (New York, McGraw Hill, 1960).

[2] W.R. Spriegel, *Principles of Business and Operation*, third edition (Englewood Cliffs, N.J., Prentice Hall, 1960), pp. 540-541.
[3] *Frederick Winslow Taylor, The Principles of Scientific Management* (New York, Harper & Brothers, 1947), pp. 36-38.
[4] Milton Brown, *Effective Work Management* (New York, MacMillan Co., 1960), pp. 14-15.
[5] Likert, *op. cit.*, p. 32.
[6] Argyris, *op. cit.*, pp. 123-163.
[7] N.R.F. Maier, *Psychology in Industry*, second edition, (Boston, Houghton Mifflin Co., 1955), pp. 138-139.
[8] Chris Argyris, "Organizational Effectiveness Under Stress," *Harvard Business Review*, (May-June, 1960), p. 138.
[9] Maurice Stein, Arthur J. Vidich, and David M. White, "The Meaning of Work in a Bureaucratic Society," in *Identity and Anxiety*, by B. Rosenberg and J. Bensman (Glencoe, Ill., The Free Press, 1960), pp. 182-185.
[10] Robert R. Blake and Jane S. Mouton, *Group Dynamics—Key to Decision Making* (Houston, Texas, Gulf Publishing Co., 1961), p. 7.
[11] Robert L. Katz, "Towards a More Effective Enterprise," *Harvard Business Review* (September-October 1960), pp. 80-102.
[12] W.H. Knowles, *Personnel Management* (New York, American Book Co., 1955), pp. 188-189.
[13] R.C. Davis, *The Fundamentals of Top Management* (New York, Harpers & Brothers, 1951).
[14] A.J. Marrow, *Making Management Human* (New York, McGraw Hill, 1957), pp. 75-76.
[15] Robert N. MacMurray, "The Case for Benevolent Autocracy," *Harvard Business Review* (January-February, 1958), pp. 82-90.
[16] John Perry, *Human Relations in Small Industry* (New York, McGraw Hill, 1954), pp. 139-131.
[17] McGregor, op. cit; Katz, op. cit.; and Rensis Likert, "Measuring Organizational Performance," *Harvard Business Review*, (March-April, 1958).
[18] Thomas G. Spates, *Human Values Where People Work* (New York, Harper and Brothers, 1960), pp. 176-187, and A. Zaleznik, C.R. Christensen, F.J. Roethlisberger, *The Motivation, Prodictivity and Satisfaction of Workers* (Boston Harvard Univ. Press, 1958), pp. 387-432.
[19] Robert R. Blake, "Applied Group Dynamics Training Laboratores," *ASTD Journal* (1960) 14-21-27, and Blake and Mouton, *op. cit*, pp. 1-26.
[20] Robert Tannenbaum Irving R. Weschler, and Fred Massarik, *Leadership and Organization* (New York, McGraw Hill 1961), pp. 6-22.
[21] Malcolm P. McNair, "Thinking Ahead What Price Human Relations," *Harvard Business Review* (March-April, 1957), pp. 15-22.
[22] *An Action Research Program for Organization Improvement* (Ann Arbor, Mich. The Foundation for Research on Human Behavior, 1960) p. 71.

[23] A.C. Bidwell, J.J. Farrell, and Robert R. Blake, "Team Job Training—a New Strategy for Industry," *ASTD Journal* (1961) 15, 3-23.
[24] Frederick Lesieur, *The Scanlon Plan* (New York, John Wiley & Sons, Inc., 1958).

Territorial Imperative

[1] Robert Ardrey, "The Territorial Imperative," 1966, Kinsport Press, Inc.
[2] On January 1, 1971, the Los Angeles Police Department revised and expanded its rank structure. The new rank structure and pay grade system known as the "Jacobs Plan" provides for higher pay grade positions and bonus paid positions within civil service classes. Thus, within the class of Policeman there are three pay grades: Policeman I, Policeman II, Policeman III, and a bonus paid position Policeman III + I. The bonus pay of a Policeman III + I is one step (5½%) over a Policeman III. The higher pay grade and bonus positions are indicative of the more demanding and responsible assignments within a civil service class. Attainment to these positions is dependent upon experience, expertise, outstanding performance and demonstrated leadership qualities.
[3] "To Protect and To Serve" is the official motto of the Los Angeles Police Department.
[4] Full implementation on January 1, 1973, is subject to the approval of the Los Angeles City Council during the 1972-73 Fiscal Year budgetary deliberations.

SELECTED READINGS

Blagg, Clifford D., "Management Theory in Law Enforcement," *Police,* Vol. 14, No. 2, November-December, 1969, pp. 41-43. Stresses that police management must cope with the rapid changes occurring in society. This can be accomplished by decentralization, specialization, creation of an internal review board and the establishment of a lay advisory board for external communications.

Blake, Robert R. and Jane S. Mouton, *The Managerial Grid,* Houston: Gulf Publishing Co., 1964. Provides an operational blueprint of change and development for both the individual and the organization. Includes descriptions of managerial theories, styles of managerial action and a diagnostic system.

Foss, Laurence, "Managerial Strategy for the Future Theory Z Management," *California Management Review,* Vol. 15, No. 3, Spring, 1973, pp. 68-81. Presents a theoretical frame of reference for a new managerial grid which include: concern for people, concern for production, and a concern for the organization. Stresses the relationship between organizational life and individual motivation.

Gourley, G. Douglas, *Effective Municipal Police Organization.* Beverly Hills: Glencoe Press, 1970. This text reviews the classical school of thought as it is applied to municipal police organizations. Includes detailed proposals for local agencies based upon department size.

Munro, Jim L., *Administrative Behavior and Police Organization,* Cincinnati: The W.H. Anderson Co., 1974. Views police administration from an open systems theory approach. Includes chapters on leadership, motivation, the police personality, training, planning, and concludes with a model for police organization.

Murphy, Patrick V. and David S. Brown, *The Changing Nature of Police Organization,* Washington: Leadership Resources, Inc., 1973. The authors point out that new tasks, new problems, new goals and objectives, new values held by the police and by other community leaders—all these require new police organizational response and innovation.

Whisehand, Paul M. and R. Fred Ferguson, *The Managing of Police Organizations,* Englewood Cliffs, N.J.: Prentice-Hall, 1973. A multidimensional approach to the theory and practice of managing a police organization draws heavily from behavioral research and existing innovative management styles.

 **The study of this chapter will enable you to:**

1. List the basic rules in systems analysis.
2. Describe two types of surveys.
3. Define the term "systems analysis."
4. Identify the steps which should be considered in drawing a systems analysis.
5. List the three sub-systems of the police system.
6. Describe the iterative process.
7. Write a short essay describing the Police Complaint Processing System model.
8. List the three basic classes of cost/effectiveness.
9. List four major parts of a complaint system.

3

Systems Analysis

Introduction*

A LARGE AMOUNT of information has been expounded by authorities on the subject of management. Much of this information has not been disseminated by the average Chief of Police because it is not within his particular field of training. Daily, however, it is becoming increasingly important for the Chief to select the best way to build strong organization and also have good management. He must utilize and adapt new available management information to the field of law enforcement.

A possible solution in this area is a new discipline called *systems*. This theory of management is being used more and more and is proving to be an efficient and effective solution to management problems. As in any other form of management, no statement in this treatise should be defined as being a hard and fast rule; rather it should be used to investigate further use.

Several basic objections arise to the use of other forms of management in police work. The primary objection is that police work, by actual definition, is and must be flexible enough to cope

---

*Source: Eugene G. Columbus, "Management by Systems," *The Police Chief*, Vol. 37, No. 7, July 1970, pp. 14-16. Reprinted from the Police Chief magazine, with permission of the International Association of Chiefs of Police.

with any situation that arises in any given period of time. In business and government, where other methods are used, management itself requires, in order to prove effective, that rules be fixed and followed by all members of the organization. Police work, too, must follow certain rules, but it is facetious to say that a rule can be set for every situation that the police manager will face. However, he must be able to plan, based on prior knowledge of certain events, and understand how the systems of which he is a part will react or should react.

Management by systems has a basic rule. *Each system must be understood and considered a separate entity within the overall structure of the organization.* Basically, the systems as a group make up the organization. Knowledge of this will afford the manager the ability to follow the span of control, the chain of command or communication and enable him to assign like tasks to related work groups in the time and area that they are needed. Without this knowledge, he must search for these things, costing the department time which it cannot afford.

It is important that the use of data processing be completely understood and the application of the hardware into the systems be regulated if it is to be a useful tool and not a burden. It also must be understood that the application of management by systems does not depend on the output of a high priced device that computes. The computer and the related components are a tool.

A similarity may be drawn in this regard to an adding machine. It would be foolish to base and operate a management technique on an adding machine—and so it is with a computer. The adding machine may be considered a tool, a part within a cycle. Without the tool, it may be more difficult to establish a base of operation or a required output. So it is with a computer. Another analogy may be drawn by stating that it would be foolish to base the operation of a squad of men on one sergeant. What would happen if the sergeant should become ill? Would the men all go home? No. This is the same application as a computer. The system then is not based on the device, rather, the device is a tool within the system which aids in the many complex computations that are made to assist the manager in making a decision.

This decision-making process is assumed to be necessary on all levels of supervisory command. No device will answer all the needs of management. Some areas of information such as personal traits and characteristics may be programmed into a computer to establish who is the best suited person to complete a task. This would, however, be a very involved program and would not be completely accurate. The human mind, in the long run, must come into play because a human being, namely the manager, will make the final decision in any given problem. Systems can aid in this decision by contributing all of the necessary facts available in the proper sequence to permit the decision to be made with some mathematical or scientific probability. Masses of data can be assimilated by machinery and rapid, accurate answers obtained which are based on probabilities more accurate than the human mind. No personal feelings are involved and the machinery can operate much faster. Speed is thus gained. This application to police work is apparent. Besides routine administrative duties, the police manager at all levels is required to answer questions requiring immediate attention, many of which are of multiple variety and some involving human life. The speed necessary is evident. These decisions will be made on a background of experience, facts known, data and human emotions. The ability to eliminate the areas of facts and data by relying on a machine to produce these allows the police manager to balance his final decision with his experience and human judgment. This lessens the actual burden on the manager and allows for accuracy never before obtained. In some cases this type of decision making is referred to as quantitative analysis.

What then are the steps to be taken to organize and operate a management approach of this type? Several basic rules are needed.

The rules may be divided into the following categories: surveys, analysis of results, synthesis of system, the documentation of systems, implementation of systems and the continuing necessary planning and research. These are described below. The assumption is made that these functions will be performed by an experienced technician. No attempt should be made to accomplish these tasks without proper training.

Survey

A survey must be made of all phases of the administrative duties of the department. This survey must include, in detail, processing techniques of all documents and forms presently in use. Also included should be the organizational structure with the titles, authority, and responsibilities of the groups that comprise the entire organization. Job descriptions for all employees, indicating their responsibility as they see it, should be obtained.

There are several ways and methods by which a survey may be conducted. Walking through the operations is one and collecting data that has been produced earlier is another recommended method. Other ways include personnel interviews and research of existing policies and procedures During this time, the surveyor must be careful not to recommend any changes in the existing operations. He must act as a fact finding agent only.

Analysis

Collation of the collected data is the next logical step. The data must be categorized and studied to determine the basic operations of all phases of the organization. In order to expedite the overall job, part of the analysis should be done as the survey is being conducted. Caution must be exercised not to make any conclusions with this limited amount of data, however. Flow charts should be made during this period to explain the present operations. This will aid in the measurement of effectiveness of the new operations when they are operational.

Synthesis

After completion of the survey and analysis of all phases of the organization, the synthesis begins. The synthesis must be complete in detail to avoid overlooking any phase. It will require considerable time and effort in planning and coordination. The organization's interaction must be taken into consideration and the existing management should be consulted at each step to gain cooperation and to promote understanding. Failure to gain this cooperation can mean fatality at the time of implementation for even the best plan. All of the new concepts that can be employed should be fully discussed and understood. Programs for retraining should be established where necessary. In the course of this synthesis, the entire organization is "reassembled," so to speak, using the proper techniques, hardware and personnel. New forms and procedures must be created and secured.

Documentation

After the synthesis has been completed, all new concepts along with any of the old methods that were retained must be documented. Documentation may take several forms. The line personnel will require documentation necessary for completing a task, using proper techniques and forms. The operating personnel will require enough documentation to show clearly the entire operation and the processing necessary of any forms or other documentation involved in the system. Flow charts showing the processing steps of any system will also be included. These must be clear and concise. All documentation should be assembled in proper sequence with allowances made to correct or change any area that may later develop.

Implementation

When the preceding steps are completed, implementation is the next step. This should be done on a carefully planned timetable to assure slow phasing into the overall operations. A close review is necessary at this stage to assure compliance with the rules and the new techniques developed. It is recommended that a PERT program (Program Evaluation Review Technique) be developed and followed. Some changes will be required as the systems go into operation. These changes should be made with all members of the department understanding how and why. This will foster cooperation and, in many cases, excellent suggestions will be received. Emphasis must be placed on the care needed in the initial phases of implementation.

Planning and Research

Continuing research must be done to improve the systems. The good manager will assign one person to be responsible for any necessary planning and research. This endeavor will encompass a review of all systems as a collateral by-product of the research. Failure to continually review will cause an impasse and the organization will lapse into a state of suspended animation and become ineffective. New methods and procedures must be reviewed in order to implement them where useful and effective.

Additional Benefits

Many by-products will be realized from this form of management. Forms design and control may be accomplished easily. This will create more usable forms for the man on the street and thus it will reduce his time spent with forms. Forms control can be a major factor in reducing operational costs in time and material. Further, more complete reports for estimating tasks and better statistical information will be available.

Use of Data Processing

Data processing is a tool of the manager. If properly used, this tool can afford him much help. For this to be done, the rules shown earlier must be followed. After completion of the survey and analysis, the synthesis begins and at this point consideration of the use of data processing must be made. Key questions will arise and must be answered. For example, how much will it cost? What will it do that a human cannot do? What equipment will be needed and how much training will it require to operate? Is it worth while?

It would be difficult to consider these questions in this paper because we are speaking of generalities and not any specific department or organization. It is enough to state that the benefits of data processing equipment have been found to be economical and justified in a large majority of installation. The question we are required to answer here is, What value is this to the public administrator or manager?

Assuming that key equipment will be available to the manager, we may proceed to explain in what areas he may use the equipment. Today, the average police department is saturated with paper work. Much of this has been brought about by the demands of the public for better and more reliable services. If measures can be taken to assist in the processing, filing and retrieval of data, they should be used. Some of these measures actually allow for better distribution of manpower in a given area at a set time. To do this, facts of events must be properly processed. Case record reporting, daily activity reports of the officers, budgets and personnel accounting are but a few of the many areas that may be utilized within the scope of this method.

Other uses are many and varied. A citizens group comes to the police and asks what the accident rate is near a given intersection. A large school or housing development is being planned for this area and they feel that it would increase their local problems because of the additional traffic load. They need the accident information to aid in argument against this development. Most departments would state that they cannot supply this data. Why not? Isn't this a public service? Shouldn't the police know what the accident rate is at any given place within their jurisdiction? They should, not only to supply information but to know their own problems! The use of machine searches for this type of data is very simple. It is not simple without the data in machine form.

Another use would be in the securing of adequate manpower. How many chiefs can go to their governing body and state *exactly* why they need more men? Most will say their problems have increased; they do not have enough men to do the job requested. The members of the body listen politely and think of the number of times they have seen a policeman sitting in a coffee shop drinking coffee. What they don't know is that the officer they have seen may very well have been working for twelve hours without relief to complete his task. Data processing can and does give facts on manpower expended in hours and on cases and other related factors that cannot be argued with. Without such information the administrator is not prepared to debate a budget problem. Data processing can give him this material.

I believe that police as a group may expect management by systems to come more to the front when it is realized that this area will give answers, unavailable at this time by other methods, to many questions.

Systems Analysis Applied to Law Enforcement*

The problem of resource allocation for a police system is similar to that of many other public systems, namely: (1) A lack of agreement regarding the objectives of the system, and their relative

---

*Source: Ernst K. Nillsson, "Systems Analysis Applied to Law Enforcement," *Allocations of Resources in the Chicago Police Department,* Washington: Law Enforcement Assistance Administration, March 1972, pp. 1-12.

importance; (2) A lack of knowledge of alternative means for accomplishing goals, either within or outside the system; (3) A lack of agreement defining the criteria of performance; and (4) A lack of knowledge of transfer functions which would enable the prediction of output from any given set of inputs.

The police system has to be studied as a distinct system within the social structure of society. Optimizing easily quantifiable relationships is likely to obscure the important qualitative aspects.

"The legitimate point (can be made) that police systems can be understood only as institutions in interaction with the rest of the social structure."[1]

Identification of Objectives

The Police System objectives are related to law enforcement, order maintenance and public service. Though everyone might agree as to the desirability of the first objective, there is disagreement on what to enforce and how.[2]

"No policeman enforces all the laws of a community. If he did, we would all be in jail before the end of the first day. The laws which are selected for enforcement are those which the power structure of the community wants enforced."[3]

The second objective, order maintenance, designates the police system as a buffer for the social system. This is bound to involve conflict situations in which there is no consensus as to what constitutes order and the propriety of the methods of enforcement employed. The function of public service is much less controversial, but constitutes a large drain on police resources. Often these services could be more efficiently performed by other public or private organizations.

Even if an objective such as crime prevention has been agreed upon, it is important to know the alternative methods which can accomplish the objective. Often the most important aspect of improving a system is the generation of good alternatives. In addition, each null alternative has to be investigated. Instead of devoting additional resources to a police system, they might produce better results if allocated to the courts or correctional agencies, or if used for social work or community building. Thus, it is necessary to consider alternatives outside the police system proper.

Criteria of performance represent the means by which a system is to be evaluated. They should provide a way of measuring how well objectives are being accomplished. For example, is an average response time to a call for service a good criterion; is the number of traffic citations issued by each officer a good indicator of traffic management?

Lastly, there is a lack of quantitative descriptions of the police system. This holds true for descriptions of the system and its environment as well as transfer functions for different activities (a transfer function relates inputs to outputs for a given activity). An input-output guide should permit an indication of, for example, the number of policemen needed to control a mob of 200 people or how many police cars must be in service to achieve a certain response time to high priority calls and how response time relates to the probability of arrest.

Answers are sought for the questions posed earlier, and this work has three objectives:

1. To define the Police System (its objectives, its interfaces with other systems, and measures of effectiveness).
2. To develop a new structure for allocating costs (an accounting system). This structure should facilitate the development of production models and the evaluation of benefits.
3. To develop production models for the Response Force in order to evaluate alternatives.

Meeting the first objective is partly solved by the presentation of a conceptual model of the Police System.

The discussion proceeds from the meta-system level down to models of specific activities. First the Police System, its objectives and criteria are defined. Secondly, to make the resource allocation problem manageable, a structure is developed for cost-benefit analysis. This structure is called a Resource Analysis Budget and necessitates a whole new accounting system. The present allocation of resources are calculated for this new accounting structure. Lastly, production models are used to determine efficient combinations of resources.

Using Systems Analysis

The systems approach is a rational framework for complex problem solving emphasizing hierarchies of systems and their interrelationships. Most often the problem is ill-structured and the objectives are unknown or undefined.

"The systems approach is one in which we fit an individual action or relationship into the bigger system of which it is part, and one in which there is a tendency to represent the system in a formal model."[4]

The systems approach is the methodology used to develop a conceptual model of the police system. The model specifies the objectives and the outputs of the police system and consequently permits determination of output categories (programs) for the Resource Analysis Budget. The systems approach offers a tool for structuring the analysis, and consequently some protection against erroneous suboptimizations.

The Police System as well as the Criminal Justice System is a largely uncharted area. Suboptimizations are ever present hazards; in fact, the optimization of Police System performance is itself a suboptimization.

"A system may be defined as a set of objects, either fixed or mobile, and all relationships that may exist between the objects. All systems are composed of sub-systems and are members of a higher system."[5]

For example, the Police System is in part a member of the Criminal Justice System which is part of the Social System within which our society exists. The Police System, in turn, is a set of sub-systems.

For resource allocation analysis, these sub-systems are a set of mission oriented (output oriented) sub-systems. These sub-systems are usually called programs. The cost structure of the system, with respect to the given programs, is called The Program Budget. The analyst tries to select a set of sub-systems which:

1. Are consonant with the plan of the decision maker;
2. Have operational objectives and measures of performance;
3. Are as independent as possible;
4. Facilitate cost-effectiveness analysis.

Figure 3–1
Systems Analysis Paradigm

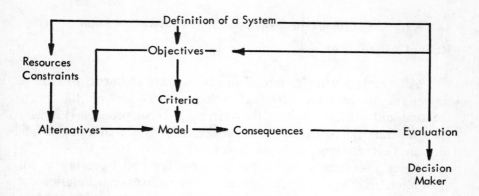

An environment may be defined as a set of objects outside the system. It is the aggregate of external conditions which affect the system.

The systems approach can be succinctly exhibited in a paradigm. The following steps should be considered in drawing a systems analysis. (See figure 3-1)[6]

1. Define the desired goals.
2. Develop alternative means for realizing the goals.
3. Develop resource requirements for each alternative.
4. Design a model for determining outputs of each alternative.
5. Establish measurements of effectiveness for evaluating alternatives.

After a system and its environment have been specified, the analyst should consider the objectives of the system and the resources and general constraints which are present. Resources are the total available material which can be allocated. Constraints are limitations imposed on the system.

The objectives express what the system is trying to achieve and to what end resources should be applied. An objective should be defined in such a way that an operational, quantitative measure of performance is possible. It is of little use to have an objective which cannot be quantified.

Equally important are measures of performance. They permit evaluation of how well the objective is being achieved.

Role of Alternatives

Alternatives offer different means of using resources to achieve objectives. Developing alternatives represents one of the more creative and crucial steps in the systems analysis process. It is here that the analyst seeks to define new alternatives that can provide increased effectiveness over previously considered alternatives.

Once alternatives have been specified, the cost of resources for each alternative has to be determined. This involves considerations of risk, time and different types of costs. To arrive at the benefits of an alternative, a model is necessary. The model determines the output to be derived from a given amount of resources.

Lastly, the cost and benefit of each alternative has to be evaluated to select the optimal alternative. The criterion function relates costs and benefits to system objectives and provides the basis for selection.

"It is my experience that the hardest problems for the systems analyst are not those of analytic techniques . . . . What distinguishes the useful and productive analyst is his ability to formulate (or design) the problem: to choose the appropriate objectives; to define the relevant, important environments or situations in which to test the alternatives. to judge the reliability of his cost and other data, and finally, and not least, his ingenuity in inventing new systems or alternatives to evaluate."[7]

This point cannot be emphasized enough.[8] The great danger in systems analysis lies in not spending enough effort in defining what the system under study should be, and instead seeking to optimize the effectiveness of a given system. The big payoffs are likely to come from a construction of new world views of problems, rather than optimizing current structures.

This point is illustrated in Figure 3-1 by the arrows drawn from the evaluation phase to the objectives and the alternatives.

An Art in Infancy

The current state of the art, with respect to police resource allocation optimization, is in its infancy. Most research into the criminal Justice System has dealt exclusively with the social dimensions. Analytical contributions have appeared only during the last five years.

A systems analysis approach was used by the President's Commission on Crime and Law Enforcement to define the scope of the Criminal Justice System problem, possible research approaches, and technology that could be applied.

"Because of the enormous range of research and development possibilities, it is essential to begin not with the technology but with the problem. Technological efforts can then be concentrated in the areas most likely to be productive. Systems analysis is a valuable method for matching the technology to the need."[9]

Blumstein and Larson recently published an article which looks at the flow of people through the Criminal Justice System.[10] It is not a systems analysis as they do not discuss objectives or measures of effectiveness, but rather a descriptive model of the flows. This step is important, however, as it provides a quantitative description of a portion of the real world.

Describing a Police System

From a general point of view, a police system is a service organization. Its clientele are people who have broken the law as well as people in need of help. It is a twenty-four hour, city-wide, dual purpose service force.

The police system is not part of the market mechanism. Its output is not a good sold in the market in competition with other enterprises; it is a public service good. The community devotes a certain amount of resources to the system and expects an output, which never is too well defined. Even if the inputs and the outputs of the system were given, the internal process of a police system is difficult to optimize. Very little is known about the transformation of inputs into outputs (the transfer functions). Consequently, tradeoffs between different methods of controlling crime (for example, using more or fewer detectives or using one-man or two-man patrol units) are not known. This is a serious drawback in trying to allocate resources and develop a departmental budget.

The metropolitan police force is usually a paramilitary system. It is characterized by strong internal controls and centralized decision making. Its organizational goals, as pointed out in the Field Study of San Diego,[11] are primarily oriented towards the crime-fighting function.

The Police System does provide two separate services: Crime control and public service. The former is the main focus of activity as will be shown in the Program Budget. This crime control function is part of the efforts of the Criminal Justice System; the public service function is part of the City Government.

The Police System is a set of sub-systems which are part of higher order systems. (See Figure 3-2) The Police System is a member of the Criminal Justice System (CJS). Its function is preventing criminal events and, failing this, to identify and apprehend the offender. There are other members of the law enforcement agencies in addition to metropolitan police departments; these include federal, state, county and special police, such as Burns, Brinks, etc.

The Police System is also part of the City Government. Its public service mission is a function of the twenty-four hour, city-wide availability of the police force. Part of this function could be carried out by people with no police training. This function

Figure 3–2
Systems Analysis of the Criminal Justice System

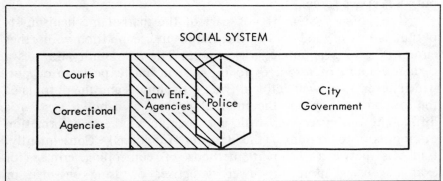

includes actions such as animal rescue, locating missing persons, and ambulance service, all of which could be performed by other city agencies or private groups.

The Police Department has another objective, Community Support. The generation process of individuals who may choose a criminal career is deeply rooted in social, psychological and economic variables over which society has some control. Crime is the responsibility of society, and its control cannot be delegated solely to a Police Department. The Police Department responsibility is to deter and apprehend offenders. The Criminal Justice System can effect deterrence, but this is effective only to the extent that society (or the social group to which the potential offender belongs) disapproves of criminal acts.

Community support implies the willingness of the community to fight crime, both by giving support, help, and resources to the Police Department, and by creating means to affect the crime generation process. Instead of actively seeking community support, police departments have often, in their desire to be professional, tended to become systems isolated from the community. This has had some detrimental effect on police effectiveness.

The investigation of the crime control problem will proceed by first analyzing the Criminal Justice System and then, in more detail, the Police System. This will permit the specification of objectives for the Police System.

Criminal Justice System

To help specify the Police System, which is the focal point of the analysis, it is necessary to consider the higher order system. The Criminal Justice System (CJS) has been charged by society to regulate and control certain classes of behavior. These classes of behavior are determined by the legislative branch of government and interpreted by the courts.

The sub-systems of the CJS: The Police, the Courts, and the Correctional Agencies. The police identify misconduct and apprehend the offenders. The courts determine the facts of the case and rule on its disposition. Correctional Agencies administer prisons and supervise the parole system.

**Figure 3–3
Conceptual Model of the Forcing Function**

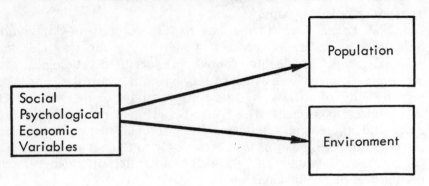

How does the CJS affect the generative process of criminal events? The structure of the crime control function is exhibited by a conceptual model. It displays the pertinent sub-systems, decision points and mechanisms for change. It permits an analysis of how the CJS can affect the potential criminal's decision making and how the impact of crime can result in community response.

The model postulates that the forcing function of the crime generation process is a function of social-psychological-economic variable. (See Fig. 3-3)

These variables affect the individual's utility function and consequently affect his propensity towards a criminal career. They also affect the distribution of opportunity by altering the generating mechanism. A discussion of the specific mechanisms is outside the scope of this paper.

Welfare programs provide family assistance which gives children a better start, thus reducing the likelihood of their pursuing a criminal career. Job training programs and increased employment opportunities will provide an alternative to crime for an income. For example, people might demand stricter legislation (i.e., cars must have theft proof locks) or elect voluntarily to lock their cars. In either case, the underlying mechanism generating opportunities has been altered.

Two factors are necessary to create a criminal event. There has to be an individual or group of individuals and a specific set of opportunities. A specific opportunity is defined as a factor of:

1. Type of opportunity (theft, robbery, etc. This leaves open the question of the appropriate classification);
2. Gain (usually in dollars);
3. Availability (this dimension measures the probable degree of difficulty of execution associated with the specific opportunity. This permits differentiation between a car that is locked and unlocked, located in the street or in an underground garage);
4. Location (in space);
5. Time (interval of time when opportunity exits).

For a given type of opportunity, distributions can be generated with respect to location and time. The set of all opportunities is called Environment.

The population considered in the model is the total population of the community. It is a set of individuals characterized for our purposes by the following attributes:

1. The individual's perception of the environment. The model chooses to maintain an actual environment and vary the individual's knowledge of the actual opportunities. The value of this attribute would fall between 0 and 1. That is to say he has incomplete knowledge.
2. The individual's knowledge of deterrence. Deterrence is the expected value of negative benefits that the Criminal Justice System contributes to a given type of opportunity. It is a function of the probability of arrest for a given type of opportunity based on past performance by the police system, the chance of being sentenced, and the length of the consequent jailterm and amount of fine. Again the value would fall between 0 and 1. (These benefits would be pure number to which a utility transformation would be applied).
3. The individual's utility function. The coefficients of this function are determined by past social-psychological-economic effects. The utility function concept will permit an explanation of how past states of the individual will influence his present decision-making. If an offender committed a successful crime (i.e. large monetary reward, not apprehended) one day, he is not likely to attempt

another crime the next day. His attitude towards the risk or estimation of his own abilities may have changed as a result of his success. The utility concept also permits analysis of "crimes of passion." The individual puts a low estimation on negative benefits or the positive benefits are very large. That is, the utility function encompasses, among other things, past experience, needs and behavior towards risk.

The decision making process resulting in a criminal event can be viewed as a two-step decision process. This allows distinguishing between inputs, which are a function of the past performance of the CJS, and inputs at the moment of execution.

First, the individual is permitted to contemplate the opportunities known to him and make an *apriori* decision to actually commit a specific crime. The relevant input from the CJS deterrence, as defined above, of which the individual has varying degrees of knowledge. Knowing the individual's utility function, the opportunity having the greatest utility can be determined and a "go, no-go" decision made.

The second decision point is present immediately prior to the execution of the planned criminal event. The potential offender evaluates the actual circumstances of the opportunity and makes a go, no-go decision.

The first stage was an *apriori* decision based on the probable circumstances surrounding the event. The second state becomes the actual sample reflecting: (1) the juncture of the probable circumstances, and (2) action taken either by private groups, (persons) altering the generation of opportunity distributions and/or their factors, or police actions affecting deterrence or opportunity distributions. For example, a person might decide to break his habit of not locking his car or the police department may employ a new tactic against CTA bus robberies.

For many events, commonly called "crimes of opportunity," the time interval between the decision points is very small. However the interval could be measured in days. Figure 3-4 summarizes this discussion in pictorial fashion. It has been said that there is a formula for crime: "Desire plus opportunity equal crime."[1,2]

**Figure 3-4
Conceptual Model of the Decision Phase**

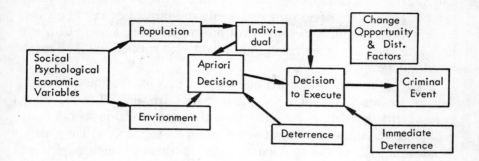

Criminal Justice System Response

What is the CJS reaction to the criminal event and how can it affect the crime generation process?

The Police sub-system responds to the criminal event seeking to identify and apprehend the offender. Police strategy and tactics can influence the decision to execute.

The generation process of crime is affected by deterrence. Deterrence was defined as the expected value of negative benefits which are a function of the risk of arrest, chance of sentencing, length of jailterm, and fines for different classes of criminal events.

The Courts and Correctional Agencies may either emphasize deterrence or rehabilitation. Rehabilitation is the effect the CJS has on the individual as he is processed through the CJS, resulting in a change in his utility function. The Police contribute through special handling of juvenile offenders, the courts by the sentence they provide, and the correctional agencies by programs which seek to integrate the individual into society.

There is a tradeoff between deterrence and rehabilitation. By rehabilitating the offender the CJS lowers the deterrence effect. The negative payoffs cannot be as large with a satisfactory rehabilitation program.

Community Response

There are usually two parties to a criminal event: the offender and the victim. (The exception is "crimes without victims" such as gambling.) The set of victims represents the impact of crime on the community. This becomes input for private and civic action. Citizens may arm themselves, private group might hire special police to react to criminal events.

The community (individuals, civic groups, businesses) may decide to react through the democratic process. That is, have government legislate new programs to alter social-psychological-economic variables or commit more resources to the CJS. They may, in addition, affect the opportunity distributions through laws (cars shall be locked, banks must have detection cameras) or by their own behavior.

Figure 3-5 summarizes the Criminal Justice System discussion in an expanded, integrated schematic that approximates the interactions of sub-systems within the Criminal Justice System.

Model of a Police System

This section focuses in more detail on the police contribution to the crime control function. Police System impact on the crime process occurs at four points: (1) Forcing function, (2) *Apriori* decision, (3) Decision to execute, and (4) Criminal event.

It will be convenient to analyze the major activities of the police system in terms of three sub-systems:

- Response Force
- Preventive Force
- Follow-up Force

Police response to a criminal event can be differentiated with respect to the detection process. Detection is defined as the identification of a criminal event. The criminal event detection by a person or by the police. In the model all non-police detection will be considered as person originating. When a person detects a crime, he initiates a call for service to the police department. If the police, through offensive tactical patrol, detect a crime-in-progress, the

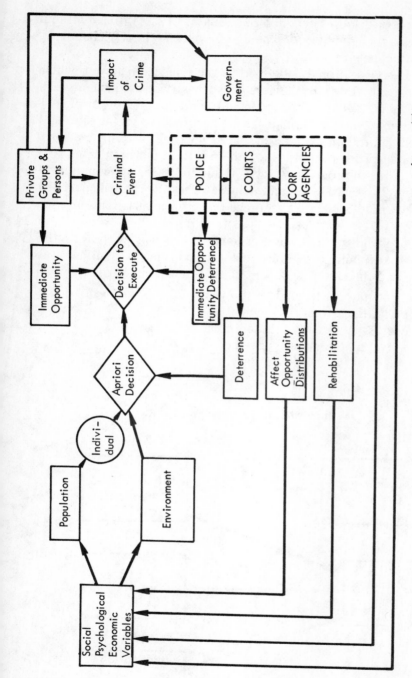

Figure 3–5
Expanded View
of the
Criminal
Justice System

person feedback loop need not be actuated. For "crimes without victims" (gambling, etc.), the detection process is carried out by specialized police unit.

The Response Force is defined as the police sub-system which responds to calls for service. These calls for service are generated by criminal events, public service demands and reports of suspicious activities. Public service demands consists of calls such as sick and injured transport, animal rescue and locating missing persons. Reports on suspicious activities are an important factor in being able to detect crime-in-progress. It also is an indicator of community cooperation in fighting crime. Chicago has a campaign called "Operation Crime Stop" to encourage this citizen participation. (See Figure 3-6)

The probability that the Response Force will apprehend the offender is a function of the time elapsed since the crime was committed and the tactic used. The elapsed time consists of:

(1) Time until citizen detects event and initiates call to the police department.
(2) Processing time by the Communications Center.
(3) Travel time for the assigned cars.

Figure 3—6
Inputs to the Response Force

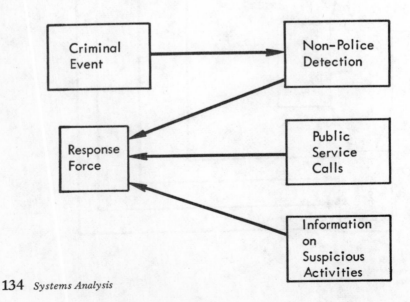

It has been shown that the apprehension probability is a decreasing function with respect to elapsed time.

It is possible to initiate campaigns which stimulate citizens to be sensors for the police department and impress upon them the necessity of transmitting the information in a timely manner. This activity might very well have a larger potential payoff than optimization of police detection or response.

Analysis of the effectiveness of the Reactive Force is of great importance. Police departments are being offered hardware such as car locators and computerized communications centers but have presently no means to evaluate the benefits. How much will the proposed hardware decrease response time and how will this affect the probability of apprehension? Finally, how much is an increase of the probability of apprehension worth?

The Preventive Force is the offensive force in the combat against crime. It interacts with the crime process in two ways. It seeks to detect misconduct and apprehend the offender. It also influences the decision to execute a criminal event by affecting the perceived presence of police. For example: have policemen in uniform and marked cars or by otherwise giving the potential offender an impression of police omnipresence.

This can come about through actual presence as a result of successful positioning of forces in time and space or through propaganda.

The Preventive Force also may affect the decision to execute by restricting actual opportunity, either by removing it completely or changing the factor of availability. This would be done through premise check, checking parked cars for valuables, removing drunks from the street, etc.

The Follow-up Force is the third sub-system. Its function is to apprehend criminals through the investigative process. It also includes the actions on a case following the booking of an offender.

Fig. 3-7 illustrates the interactions of these three forces in police functions.

Figure 3—7
Further Development of Police Systems Inputs and Outputs

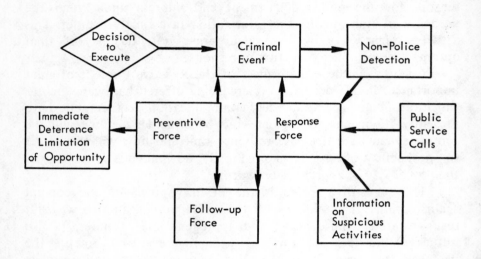

Police Systems Outputs

The outputs of the Reactive Force are arrest and public service. The probability of arrest was expressed as a function of elapsed time and tactics used. The Preventive Force outputs are arrests and impact on the decision to execute. The probability of apprehension is a function of elapsed time, probability of detection (i.e. being at the scene of the event, and recognizing that an event did in fact occur) and tactics used. Follow-up can be characterized by the probability of arrest through investigation. It is dependent on elapsed time and methods used. All of the above functions are also dependent on the type of crime. The tradeoff between the Response and Preventive Forces, given a criminal event, is that the latter may detect an event with a low probability but may have a higher probability of apprehension (due to shorter elapsed time).

Deterrence is an input to the *appriori* decision point. The Police System variable is the probability of arrest for the system (i.e. the combined efforts of all three sub-systems).

136 *Systems Analysis*

The Police System does affect the forcing function by changing the mechanisms generating opportunities. It can also affect an individual's utility functions through rehabilitation measures, mainly with respect to juveniles. This group of offenders is given special attention in order to influence them away from a criminal career. For example, special youth officers handle the cases and often a station adjustment is made.

The conceptual model can account for Community Relations programs. The Police System can influence the crime generation process by devoting resources to communication with private groups and individuals. These measures would influence community support and, hopefully, encourage the community to assist the police in the apprehension process and, even more importantly, affect the generative process of crime. These communication links can be called Human Relations with respect to individuals, and Community Relations with regard to groups.

For a more thorough discussion of these phases of police activity, we recommend "Dilemmas of Police Administration" by James Q. Wilson in the September-October (1968) issue of *Public Administration Review*.[13]

An effective Community Relations program seeks to explain the crime generation process to the community, what the police role is, what it can be expected to do, and what the community can do.

There is also a link to other branches of Government, for the sake of completeness, to emphasize that police departments have to make other city, state and federal officals cognizant of Police problems, results and limitations.

In summary, the outputs of the Police System are:
1. Apprehension of offenders.
2. Impact of immediate environment on the criminal event.
3. Impact on *apriori* decision.
4. Rehabilitation measures.
5. Changing opportunity distribution.
6. Public service.
7. Community support.

The array of these relationships between the Police System, the larger Criminal Justice System and other governmental systems is illustrated in Fig. 3-8.

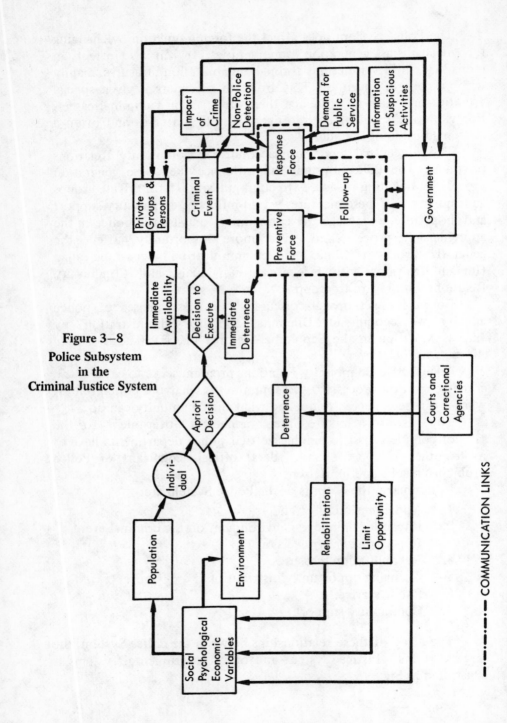

Figure 3–8
Police Subsystem
in the
Criminal Justice System

138 *Systems Analysis*

Police System Objectives

Three missions and specific outputs have been identified for the Police System. It remains to specify the objectives of the system.

The first mission is Protection of Life and Property and Maintenance of Peace and Order. It becomes convenient to subdivide the broad notion of crime control into two classes of events as criminal events differ in degree of seriousness and the nature of police response. Crime will be defined as index crimes and hit-and-run accidents.

A second category of misconduct can be called Quasi-Criminal, whose objective contains activities devoted to the enforcement of city ordinances to a large degree. These are crimes of lesser seriousness than index crimes and the maximum sentence is a year in jail and/or a fine. The main offenses are disorderly conduct and drunkeness.

Maintenance of Peace and Order can be subdivided into an objective called Public Peace and one called Traffic Regulation. The Public Service and Community Support objectives conclude the list.

Mission	*Objective*
Protection of Life and Property	1. Crime Control
Maintenance of Peace and Order	2. Quasi-criminal Control
Public Service	3. Public Peace
Community Support	4. Traffic Regulation
	5. Public Service
	6. Community Support

These objectives can be compared with lists of objectives found in the literature.

The International City Managers Association listed five police objectives:[14]

> Prevention of Criminality
> Repressions of Crime
> Apprehension of Offenders
> Recovery of Property
> Regulation of Non-criminal Conduct

Systems Analysis **139**

Another list includes:[15]

>Prevention of Crime
>Investigation of Crimes
>Apprehension of Violators
>Presentation of Criminals for Adjudication
>Services to the Public
>Enforcement of Non-criminal Ordinances
>Regulation of Activity within the Public Way

Peter Szanton defined the following objectives:[16]

>1. Control and Reduction of Crime
>2. Movement and Control of Traffic
>3. Maintenance of Public Order
>4. Provision of Public Service

The first two lists are not output-oriented in an independent manner and consequently would be difficult to use in a resource allocation analysis. Szanton's list is excellent but neglects the goodwill aspect. It has been said that a bulldozer is an effective crime fighter. This proposition would be a feasible alternative if there were no objective to represent the social system. For example, repressive police measures might prevent crime, but if individual's rights are destroyed in the process there should be a way of indicating this.

Systems Analysis: Aid to Decision Making*

Alain C. Enthoven, Assistant Secretary of Defense, systems analysis, once told the special subcommittee on the Utilization of Scientific Manpower that "systems analysis is nothing more than quantified or enlightened common sense." A simple statement at first glance, but if scrutinized, this definition turns out to be complete.

Enthoven was constantly criticized by congressmen, military

---

*Source: Richard L. Shell and David F. Stelzer, "Systems Analysis: Aid to Decision Making, "*Business Horizons,* December, 1971, pp. 67-72. Reprinted with permission of the publisher.

stategists, and the general public for trying to run the Pentagon with a computer. The most vicious attack came from the military. General Thomas D. White, former Air Force Chief of Staff, described the systems analysts as "often overconfident, sometimes arrogant, young professors, mathematicians, and other theorists who might not have enough worldliness or motivation to stand up to the kind of enemy we face." Vice-Admiral Hyman G. Rickover remarked that the systems analysis studies "read more like rules of a game of classroom logic that like a prognosis of real events in a real world."[1]

How could common sense cause such an uproar? Possibly, documenting and quantifying our human decision-making process only manifests how often we do not make decisions rationally. This is only conjecture, but let us look at systems analysis and see exactly what it is and why it is needed.

What is Systems Analysis?

The systems analysis approach describes many means by which problems are analyzed to find the most effective and efficient solution within certain constraints. Although there are many variations, the analysis is composed of nine basic steps:

>   Define the problem
>   Define the objectives
>   Define the alternatives
>   Make assumptions concerning the system
>   Define the constraints
>   Define the criteria
>   Collect the data
>   Build the model
>   Evaluate the alternatives.

The Basic Acts

Defining the problem may appear to be a needless step since this is the basic reason the analysis is being undertaken, but it is probably the most important step in the procedure. In the problem definition, there must be an accurate description of the present situation showing some sort of disparity that must be eliminated.

The process of determining exactly "what is wrong" can be the most consuming part of the entire analysis and the most critical, for the most perfect solution to the wrong problem does nothing to solve the real problem.

Objectives must be defined to provide a structural framework and over-all goals for the systems analysis. Clearly stated objectives are also useful for establishing limits and guidelines for the remaining basic steps.

The *definition of alternatives* should be exhaustive even though some alternatives are obviously inferior. The reasoning behind this is that some new constraints might arise, making the superior alternatives impossible to implement. These alternatives represent the competing "systems" for accomplishing the objectives. They present opposing strategies, policies, or specific actions, including the required fiscal and physical resources. "They need not be obvious substitutes for each other because the objectives should be general enough to be accomplished by many different strategies."[2]

Assumptions must be made about the larger system within which the alternatives will work. This system should include anything that affects the problem situation or the alternatives. Of course, facts are much more desirable than assumptions, but it is not always possible to know or predict precisely how things will be in the future. The statistical sensitivity of these assumptions will have to be tested when the model is built. This can be accomplished by modifying different assumptions and observing the effect on the desired output.

It is difficult to *identify all constraints*. However, more information about problem restrictions will improve the presentation of analysis and will prevent inappropriate evaluations. The first and most obvious constraint in most cases is money; after this, the list becomes hazy. Constraints do not have to be physical or even measurable, but they do have to be recognized.

One constraint that must be considered early in the analysis is top management's philosophy towards scientific management. The perfect solution of an important problem could be arrived at through systems analysis and yet never be implemented because of top management's distrust (or fear) of scientific management. Other constraints are psychological, sociological, technical, traditional, administrative, political (both office and national), and, of course, physical (men and equipment).

The *definition of criteria* is important to the analyst, for these are the rules or standards by which he ranks the alternatives in order of desirability. They must be relevant to the problem area, include consideration of all major effects relative to the objectives, and, ideally, be adaptable to meaningful quantification. It is important to remember that, in some cases, the mere mention that analysis is being undertaken or action is under consideration may be enough impetus to significantly alter the problem situation.

The *collection of data* is somewhat mundane and even boring, but is obviously as important as any other part of the analysis. It is mandatory that all pertinent data of each alternative be collected in a usable format, and through a method that will not bias the solution.

Building a model is not always necessary in every analysis, but in complex problems for which a vast amount of data exists for each alternative, it is desirable to have one. A model is generally needed because experimenting with the real system is either impossible, economically infeasible, or quite dangerous, as in some defense projects. A model can also serve as an aid to thought and communication, a tool for production, and an aid for control purposes and for training and instruction.

The *evaluation of the alternatives* is the "putting-everything-together" step. This can be done through many analytical tools, using the predetermined criterion as a measuring stick. Two of the most publicized evaluation methods are cost-benefit analysis and cost-effectiveness analysis. In cost-benefit analysis, the cost of implementing each alternative is compared with the dollar value of the benefits accrued from implementation. Cost-effectiveness analysis compares the cost of implementation of each alternative with its real benefit (not in dollar terms).

Most systems analyses have these steps incorporated within their approach. Of course, some authors will have different terminology or explicit documentation, but the above nine steps are fairly basic to any analysis.

The Tenth Step

One additional step must be included. It should be understood that it is not necessarily the choice of the best alternative. As E.S. Quade puts it:

Unfortunately, things are seldom tidy: too often the objectives are multiple, conflicting, and obscure; alternatives are not adequate to attain the objectives; the measures of effectiveness do not really measure the extent to which the objectives are attained; the predictions from the model are full of uncertainty; and other criterion that look almost as plausible as the ones chosen may lead to a different order of preference.[3]

Thus, an iterative process begins with reexamining objectives, finding new alternatives to a slightly different problem definition, and testing the statistical sensitivity of new assumption. This is desirable for laboratory research or philosophical thought, but in practical cases the decision maker has to come up with a solution for implementation. The constraints he has defined, specifically time and money, normally will not allow him to follow the well-known theoretical problem-solving model to the perfect solution with no uncertainty as shown in Figure 3-9, the spiral model of problem solving.

Figure 3—9
The Spiral Model of Problem Solving

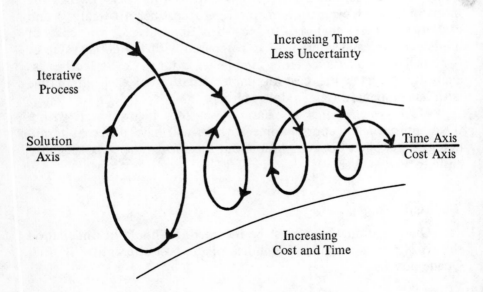

After the analyst has evaluated the alternatives, the decision maker must identify the best solution by considering facts, assumptions, and uncertainties for his problem. In certain situations, the systems analyst and the decision maker are the same person, but usually the analysis is performed by someone on the staff of the decision maker.

If the alternative identified is considered acceptable by the decision maker, the plan is implemented; if not acceptable, the decision maker must evaluate his constraints in order to determine if he has the time and/or money to continue in the iterative model as shown in Figure 3-10. In many cases, he may be forced to implement a less than optimal plan. As implied earlier, the constraints placed on the analyst are frequently so restrictive that often his solution is nothing better than an "intelligent guess."

Something should be said about what systems analysis is not. It is not a panacea for every decision maker. It does not tell the decision maker which alternative to choose. It is a method of investigating, not solving, problems. All of the various components of the analysis are defined by man and often based on many untested assumptions. A correct decision cannot be assured even if the analysis is carried out to perfection.

Systems analysis is a tool of the decision maker, and it is not a bad tool in and of itself. Some of the general criticism of the systems analysis approach alluded to earlier seems unfounded, or as James R. Schlesinger said, "It would be a mistake to turn over a new proverbial leaf—and generally find fault with the tools rather than the craftsmen."[4]

Why is Systems Analysis Needed?

The constraints imposed upon the decision maker are often subtle and troublesome. Although there were many constraints imposed upon the decision makers of the early 1900's, they are not comparable with the ones that have been added during recent years.

As labor unions grew in strength, management found that it could no longer assume that its employees were just another resource to be utilized in the production process. Any decision involving the labor force has to be scrutinized to ensure that a contract has not been broken. There have been many examples in which an entire

**Figure 3–10
The Iterative Cycle**

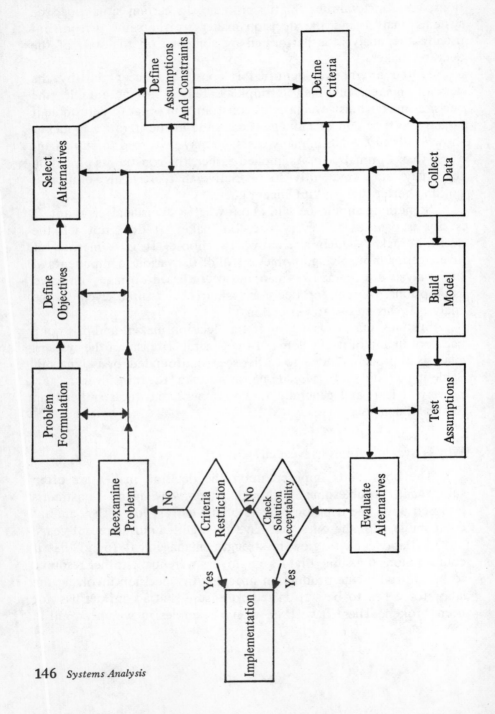

factory has been closed down by a strike called because one man was fired. Legislation has recently been passed (the Occupational Safety and Health Act of 1970) that requires an employer to assure that his employees have safe and healthy work environments, and legislation has been introduced that would require the employer to pay the major part of a general health insurance premium for all of his employees. The spiralling cost of labor in itself is enough to warrant the careful analysis of all alternatives.

"Consumerism" has come to light in the past few years, obliging industry to recognize its responsibility to its customers. Regulatory agencies, formerly puppets of the industries they regulated, have found new strength and courage in the public outcry for better and safer products. Many states have recently revised product liability laws to favor the consumer. Ronald A. Anderson, professor of law and government at Drexel University, predicts that there is "reason to believe that eventually liability will be imposed upon the producer or merchant seller of a product without regard to whether there is any defect."[5]

Since the end of World War II and the introduction of television, the vastness of the federal government has become more and more evident to the general public. The defense department can no longer give out lavish contracts to researchers to develop weapon systems that possibly will never be used. An agency or bureau can no longer hide major mistakes from the public. It is known in governmental circles that the public will view an area of government more critically if it discovers a major mistake within it. This is a great constraint placed on many governmental heads today.

The increasing concern about ecology has prompted many manufacturing industries to add pollution control equipment. However, the public wants ecology improvements without price increases. The systems analysis approach will aid in the development of lower cost antipollution manufacturing operations.

These are some of the recent constraints placed upon the decision maker, and more will probably develop. Decisions cannot be made without knowing their full impact on the public as well as on the various markets.

Another area that has changed for the decision maker and complicated his decision process is the number of alternatives available to him to attain a certain objective. Recently, technology has moved so rapidly that no man could possible remain completely up-to-date even in his own field. Of course, if the best alternative is not known, the decision maker can only make an inferior decision.

The other extreme (considering too many alternatives) is demonstrated by the fact that a new breed of professional has entered the scene—the consultant salesman. In many cases, these people could be called "alternative creators." Without a systematic method to evaluate all of the various alternatives presented to him, the decision maker may choose an alternative which fulfills the consultant's objectives, but not his. Things are complicated even further when a computer is a possible component of some or all of the alternatives.

It is difficult to imagine a complex systems analysis today which does not involve the services of a computer either as part of an alternative or in the analysis of the alternatives. To a decision maker not knowledgeable in the computer field, the various design alternatives incorporating computer technology can boggle the mind. But this should not cause him to reject the computer, because it can both improve some alternatives and expedite the analysis.

The basic factor to remember is that the computer is just another tool by which the analyst or decision maker obtains his desired objectives. Also, a computer salesman who begins to describe the various characteristics of his hard and software systems is talking about means and not ends. The computer specialist who begins the conversation by asking what output is needed or what objectives are sought is the man who is going to help.

Time is a constraint in most decisions. In a market place where fad products may have a life cycle of six months, or in a world where total destruction could be complete within a day, a decision maker can ill afford not to take advantage of any tool which will speed up his decision without adding uncertainty.

Systems analysis is a technique for structuring common sense in problem evaluation. Problems confronting decision makers have become more complex by the addition of new constraints such as pollution control, more stringent product liability laws, and a more observant public. In summary, the use of systems analysis will improve the decision maker's problem-solving capability. The following points should be remembered:

Systems analysis is (or should be) the documentation of a method for analyzing problems.

Systems analysis can be applied to many business areas outside the scientific-technological field.

Systems analysis is not a panacea; the decision maker must ultimately select and implement the best alternative.

Systems: Application to an Agency*

The usefulness of systems analysis in small, as well as major metropolitan, community police departments is demonstrated by a study of police response to calls for service in a medium-sized Connecticut community (population 45,000).

As a formal approach to problem solving, the technique is usually associated with scientific research and the management of complex government, military or major industrial operations. A study, reviewed here, was conducted to help demonstrate the applicability of the scientific "systems" approach to the study of municipal police problems, particularly those of a fairly small community. In the test-case police department, this led to more economic and effective deployment of men and equipment.

The purpose of systems analysis is to understand a complex system sufficiently to predict the consequences of its continued operation, both as it is presently operating and as it might be restructured or operated differently. The method of systems analysis is to describe the components and interactions within the system sufficiently to construct an abstract working model, or simulation, of the system and then to experiment with the model. *Figure 3-11* illustrates the development and use of a system simulation model.

In this study a simulation model (an abstract representation) was developed to identify the system components (functions and

---

*Source: Richard M. Davis, "Police Management Techniques for the Medium-sized Community," *The Police Chief,* Vol. 37, No. 7, July, 1970, pp. 44-50. Reproduced from the Police Chief magazine, with permission of the International Association of Chiefs of Police.

**Figure 3–11
Development and Use of System Simulation Model**

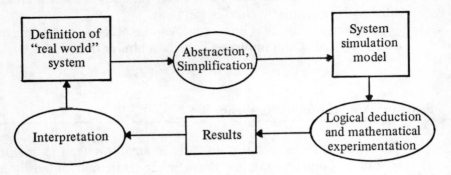

activities) and the interactions (relationships) of the "real world" complaint-processing system in a case-study police department.

The design and use of the Police Complaint Processing System model follows the basic structure of systems analysis, which usually involves these steps:

- Study, define, and describe the system (municipal police complaint processing), including its objectives, functions and other components that influence its performance.
- Analyze qualitatively and quantitatively, where possible the characteristics of the system, with particular attention to the characteristics of the resources and the demands (inputs) and services (outputs) of the system.
- Develop a diagram of the system and its components.
- Develop an operational model to represent the system.
- Test the validity of the model using "real world" data to determine its accuracy.
- Operate the model experimentally to determine the performance and behavior of the system under different conditions, to evaluate alternative resource allocations.

Once criteria of performance and costs are specified, the model can be used to answer three basic classes of cost/effectiveness questions. These are:

- Class I—Which system design will produce a stated level of performance at least cost (i.e., fixed performance-variable cost)?
- Class II—What level of performance can be obtained from a fixed expenditure (i.e., variable performance-fixed costs)?
- Class III—What is the optimal combination of performance levels and expenditures (i.e., variable performance-variable cost)?

A Case Study

The usefulness of the Complaint Processing System model was fully realized only when it was tested and applied to "real world" data. The police department of a medium-sized Connecticut community (population 45,000) was selected for a case study. The department's communication network and its internal and external complaint handling system were analyzed in detail.

The case-study police department has four patrol beats, unequal in size but designed to equalize the demand for police services among them. Consequently, the business area of the community constitutes the smallest beat. Each beat has one motorized unit twenty-four hours a day. The unit assigned to each beat is its *primary* field service facility and is principally responsible for: (1) responding to calls for service; (2) providing inspectional duty; and (3) providing repressive patrol as time permits. Two supplementary "random patrol" units are available and can be deployed during high demand periods. These are *secondary* field service facilities, in that they: (1) provide "cover" or fill-in for the primary unit, if this unit is not available to handle a call for service; (2) concentrate repressive patrol on high demand streets; and (3) provide traffic services. Thus there are four patrol units in the field at all times, with two more available for assignment as necessary.

Calls for police services within these beat areas are serviced by the primary and secondary patrol units. Calls are classified according to fifty-seven categories of crimes and services, in six operationally similar groups:

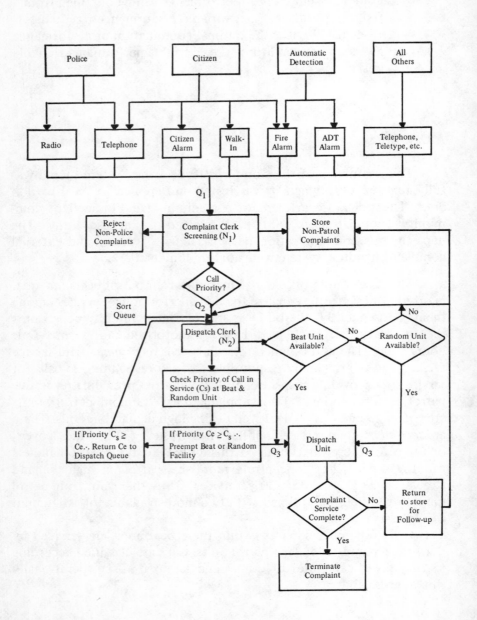

Figure 3–12
The Complaint Processing System

152 *Systems Analysis*

	Priority
Part I crimes	1
Part II crimes	2
Traffic accidents and violations	2
Bank and funeral escorts	3
Equipment and traffic violation warnings	3
Other	4

Figure 3-12 shows the paths of incoming complaints, from the time they are received by the department until they have been acted upon and terminated in the system. The path of a complaint, and its processing time, depends on (1) the order in which calls enter the system, (2) the locations at which the calls originate, and (3) the priority of each call.

Model Uses

The model was developed as a tool for departmental manpower planning and management. As such, the model identifies relationships among the various parts of the complaint processing system and the performance of the system. For example, typical measures of system performance are:

- the amounts of time required by the system to respond to calls for service,
- the amounts of time required for calls of different types and priorities to be processed through specific parts of the system, such as the dispatch function,
- the capacity of the system to prevent work pile-ups, and
- the availability of manpower and other resources to perform required tasks.

Costs are expressed as manpower, vehicles and other resources required, or as penalties incurred. Penalties, where they can be identified, are measured in equivalent dollar values lost when the system's performance does not meet specified levels. For example, consistently slow response would probably reduce police effectiveness in deterring armed robberies; thus, the penalty for slow response would be an increase in this type of criminal activity.

The Model

The Complaint Processing System model comprises two component programs (see *Figure 3-13*) written in the Fortran programming language.[1] The Complaint Analysis Program (CAP) analyzes police complaint records to provide statistical information about the number, duration, and place of origin of calls entering the system. This information is entered into the second program, the Complaint Processing Simulator (CPS), which traces the call through the various activities of the department's complaint system to its disposition by the uniformed patrol division. These two programs together form a model that can be manipulated to depict alternative future systems as well as the systen now in use.

**Figure 3-13
The Complaint Processing System Model**

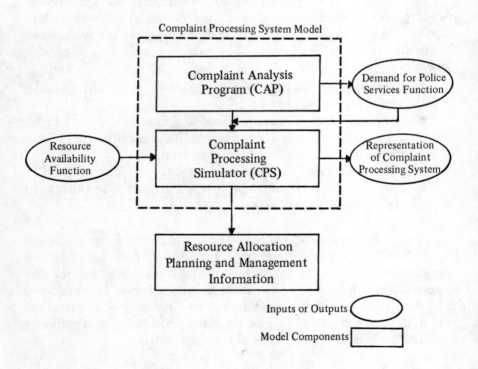

154 *Systems Analysis*

A computer program (General Purpose Simulation System/360) traces the complaints through the information system (see *Figure 3-14*). Each function in the computer program represents a real processing event by changing the rate and or the direction that the call moves to a subsequent function. The computer program was checked using real data to make sure it represented the complaint system accurately.

Experiments

The final step in the research program was to design and conduct experiments to evaluate the performance of the police complaint processing system under conditions of varying demand and available resources. It was assumed that the performance of the complaint system depends mainly on the resources (especially manpower) available. The three major parts of the complaint system (the complaint clerk, the patrol unit dispatcher, and the individual field patrol units) were evaluated according to such criteria as:

- use,
- total number of calls processed,
- average processing time,
- average queue length (waiting line length),
- maximum queue length.

The overall complaint processing system was evaluated by these criteria:

- number of calls processed through the entire system
- average time to transmit the system
- frequency distribution of total processing times.

Each experiment represented a 24-hour simulation period, comprised of three 8-hour shifts. Two principles of police theory governed the alternatives of resource allocation. First, the patrol car is an emergency unit and should be available to respond to calls in a minimum of time. Second, according to professional opinion,[2] patrol units should spend half their time on inspectional and repressive patrol and half responding to calls for service.

Nine experiments were designed: The first three compared alternative deployments under a constant (present) demand. The

**Figure 3–14
Computer Simulation of Information System**

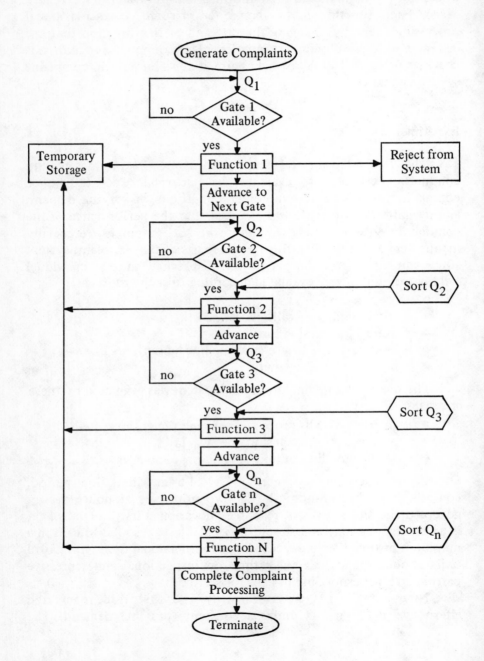

next three compared the effects of increasing priority 1 demands on a constant (present) level of resources. The final three experiments compared the effects of increasing the total, overall demand on the same (present) level of resources. These nine experiments were evaluated against a "base" run measuring system performance against present demand and resources. These experiments are summarized as follows:

Experiment number	Demand variable	Number of patrol units, 3 successive shifts per day	Total Number patrol units per day
Base run	Present demand	4, 5, 6*	15
1	Present demand	4, 4, 4	12
2	Present demand	4, 5, 5†	14
3	Present demand	4, 6, 6	16
4	Priority 1 x 2††	4, 5, 6	15
5	Priority 1 x 5!	4, 5, 6	15
6	Priority 1 x 10	4, 5, 6	15
7	Present demand x 2	4, 5, 6	15
8	Present demand x 5	4, 5, 6	15
9	Present demand x 10	4, 5, 6	15

* Present deployment.
† Suggested deployment.
†† Denotes twice the present number of priority 1 calls.
! Denotes five times the present number of priority 1 calls.

Model Results

The experiments were performed according to the principles and criteria above, and the results were analyzed.

Current Deployment. The suggested deployment of four units on shift 1, five units on shift 2, and five units on shift 3 would, on the average, handle incoming complaints as efficiently and effectively as the current 4, 5, 6 deployment plan. The suggested plan would save the cost of one patrol man and one patrol unit per day ($40/day or $14,600/year). The addition of one man with his complement of equipment is a significant expenditure for a department in small and medium-size communities. In the test community it has been demonstrated that the current demand for police service does not warrant the continuation of the deployment 4, 5, 6 plan as originally thought. Rather, it would be better to reassign one man from the routine 4-12 midnight shift to a split shift covering the peak demand period in the early evening with the remainder of his time devoted to nonroutine functions such as tactical operations or even community/police relations. Union regulations and other institutional factors must be confronted in redesigning work schedules, however the police administrator has supporting evidence to back up his case for modifying shift assignments.

Future Deployment. However, the demand for police services is continually shifting. One distinct advantage of the management model developed is the facility with which it permits us to investigate the impact of future demands on administrative policy. For example:

Priority I Calls. If the number of priority 1 calls were doubled, the present deployment plan would be marginal; if priority 1 calls were increased five or ten times, present deployment would be wholly inadequate.

Total Calls. The current deployment strategy would be marginal at best if the total number of calls were doubled and would be inadequate if they increased five times. The current complaint system would become saturated if the total demand increased ten times. Under saturation conditions, performance would deteriorate rapidly: only priority 1 and 2 complaints would be processed, and priority 2 complaints would take an abnormally long time. The number of nonpriority complaints handled would be reduced severely; those processed would take seven hours, on the average, to complete.

Conclusion

These are merely selected results of the experimental analyses. They serve to demonstrate that the management information derived from these experiments can provide the police administrator with basic knowledge useful in the design, evaluation, and the effective allocation of resources, by function, within the complaint system of a municipal policy department. The results further show that the systems approach can benefit the small and medium-size department as well as the larger municipal agencies, providing useful information for more effective day-to-day operations and long-range planning of the department's activities.

Topics for Discussion

1. Discuss the technique of documentation of systems.
2. Discuss the use of data processing in systems analysis.
3. Discuss the systems analysis paradigm.
4. Discuss the two-step decision process which results in a criminal event.
5. Discuss the differences between a response force and a preventive force.
6. Discuss the nine basic steps of systems analysis.

FOOTNOTES
Systems Analysis Applied to Law Enforcement

[1] Arthur Niederhoffer, *Behind the Shield: The Police in Urban Society* (New York: Anchor Books, 1967), p. 13.
[2] Jerome H. Skolnick, *Justice Without Trial: Law Enforcement in Democratic Society* (New York: John Wilery and Sons, 1966).
[3] Dan Dodson, speech delivered at Michigan State University in May, 1955; reported in *Proceedings of the Institute on Police Community Relations,* May 15-20, 1955 (East Lansing: The School of Police Administration and Public Safety, MSU, 1956), p. 75 and quoted in Niederhoffer, (Ref. 1), p. 12.
[4] Charles Zwick, *Systems Analysis and Urban Planning* (Santa Monica: Rand Corp., 1963).
[5] Kenneth Heathington and Gustave Rath, "The Systems Approach In Traffic Engineering," *Traffic Engineering,* June, 1967.
[6] For further reading, see: G.H. Fisher, "The analytical Basis for Systems Analysis," Rand Corp., May 1966, p. 3363; A. Hall, *A Methodology of Systems Engineering* (Princeton, N.J.: D. Van Nostrand Co., Inc, 1962); Van Cort Hare, *Systems Analysis: A Diagnostic Approach* (New York: Harcourt Brace and World, 1967); Charles Hitch and Roland N. McKean, *Economics of*

Defense in the Nuclear Age (Cambridge: Harvard Univ. Press, 1963); E.S. Quade *Analysis for Military Decisions* (Chicago: Rand McNally & Co., 1964); E.S. Quade, "Some Problems Associated with Systems Analysis," Rand Corp., June, 1966, p. 3391.
[7] C.J. Hitch, *Decision Making for Defense* (Berkeley: University of Calif. Press, 1965), p. 54.
[8] Lindsey Churchill, "An Evaluation of the Task Force Report on Science and Technology," Russell Sage Foundation mimeograph, 1968.
[9] The President's Commission on Law Enforcement and Administration of Justice, *Task Force Report: Science and Technology* (Washington, D.C.: Govt. Printing Office, 1967) p. 3.
[10] A. Blumstein and R. Larson, "Models of a Total Criminal Justice," *Operations Research,* Vol. 17, No. 2 (March-April, 1969).
[11] The President's Commission on Law Enforcement, the Police and the Community, *Field Surveys LV,* Vol. 1 (Berkeley: Univ. of Calif., October, 1966).
[12] Allen P. Bristow, *Effective Police Manpower Utilization* (Springfield: Thomas Press, 1969).
[13] James Q. Wilson, "Dilemmas of Police Administration," *Public Administration Review,* September-October, 1968.
[14] *Municipal Police Administration* (Chicago: International City Managers Assn., 1961).
[15] F. Leahy, *Planning-Programming-Budgeting for Police Departments* (Hartford: Travelers Research Center, Inc, 1968) from a budgeting workshop sponsored by Florida Institute for Law Inforcement in 1966.
[16] Peter Szanton, "Program Budgeting for Criminal Justice Systems," *Task Force Report: Science and Technology,* see Ref. 9.

Systems Analysis: Aid to Decision Making

[1] Herbert Chesire, "The Whiz Kids at the War Council," *Business Week* (Jan. 30, 1971), p. 6.
[2] Barry G. King, "Cost Effectiveness Analysis: Implications for Accounting," *Journal of Accounting* (March, 1970), p. 41.
[3] E.S. Quade, "Systems Analysis Techniques for Planning-Programming-Budgeting," The Rand Corporation, 1966, p. 8.
[4] James R. Schlesinger, "Uses and Abuses of Analysis," a memorandum prepared for the Subcommittee on National Security and International Operations, 1968.
[5] Ronald A. Anderson, "Product Liability—What is the End of the Road?" *Case and Comment,* LXXVI (January-February, 1971), pp. 50-55.

System: Application to an Agency

[1] Author: Fortran (*Formula Translation*) is an automatic coding system that is a compromise between the language of the computer and that of the scientist that allows the scientist to utilize the computer for problem solving.

[2] Interviews with G. Goldstein and others, of various Connecticut municipal police departments.

SELECTED READINGS

Buckley, Walter, *Sociology and Modern Systems Theory,* Englewood Cliffs, New Jersey, Prentice-Hall, Inc., 1967, pp. 42-81. Critiques the current "social system" theory and constructs a new model identified as a sociocultural system. Discusses systems parts, systems tension, and feedback.

Cleland, David I. and William R. King, *Management: A Systems Approach,* New York: McGraw-Hill Book Co., 1972. Focuses on the application of systems concepts to the management process, rather than to individual management problems. Provides both a synthesis and a generalization.

Freeman, Sydney, "A Systems Approach to Law Enforcement Training," *The Police Chief,* Vol. 35, No. 8, August, 1968, pp. 61-69. Describes in detail the experiences and problems in applying the systems approach to curriculum development for law enforcement training.

Olson, Bruce T., "Conflicts in Values in Criminal Justice: A Proposed System Analysis," *Police,* Vol. 14, No. 2, November-December, 1969, pp. 44-48. Describes the model of a criminal justice system using some of the concepts of general systems theory. Includes a systems taxonomy and a discussion of value conflicts.

Riesau, Victor D., "An Integrated Justice System," *The Police Chief,* Vol. 36, No. 2, February, 1969, pp. 48-54. Describes the development of an information system built around the needs of a total criminal justice system, rather than the separate sub-systems.

● **The study of this chapter will enable you to:**

1. Identify the salient characteristics of existing police structures.
2. List the major causes of poor morale among police employees.
3. Write a short essay describing a democratic police model.
4. Define the term "planning."
5. Identify four general types of plans used in law enforcement.
6. List the five steps utilized in the scientific approach to planning.
7. Identify two administrative tools of fiscal control.
8. List the four major components in PPBS.
9. Identify the purpose of a program budget report.

4

Managerial Functions

Introduction*

Theories are developed to facilitate understanding; however, social science theories, unlike theories of the physical sciences, are complicated by the fact that their subjects think and act. Human decisions and actions have a multiplicity of causations including past experiences, influences of culture, and expectations about the future. Consequently, social theories are particularistic. Their usefulness is restricted by specific values and perceptions that determine the characteristics of behavioral patterns and rationality.

Social science theory about the organization and management of bureaucracies is no exception; it is also particularistic and must be situationally conditioned. Therefore, since rationality is culturally or normatively determined, conclusions about "proper" or "improper," "right" or "wrong," and "good" or "bad" organizational arrangements cannot be absolute. The values and expectations of the social system within which the organization exists will define the norms and boundaries for organizational arrangements and managerial practices.

---

*Source: John E. Angell, "Toward an Alternative to the Classic Police Organizational Arrangements: A Democratic Model," *Criminology*, Vol. 9, No. 2 and 3, August-November, 1971, pp. 185-194. Reprinted with the permission of the publisher.

The structural model most frequently advocated and utilized in American police endeavors was first implemented in the Anglo-Saxon world by the Metropolitan Police Act, which established the London "bobbies" in 1829. This model follows closely the tenets of the classic organizational theory, which is an ideal-type based on pre-twentieth-century organizations. In spite of the tremendous changes in society, its culture and values, no significant changes have occurred in the approach to police organization and management since 1829. However, just as it took the traumatic realities of Panzer warfare in World War II to cause the United States Army to abandon the horse cavalry and horse-drawn artillery, the inability of the police to deal effectively with the social problems of the past decade has alerted perceptive people to the need for changes in police administration.

Characteristics of the Existing Police Structures

The structures of modern American Police organizations are rationalized, hierarchical arrangements that reflect the influence of classic organizational theory as promulgated by Max Weber (Bendix, 1962; Gerth and Mills, 1958: 196-244; Henderson and Parsons, 1947: 329-340). The most salient characteristics of these departments, as with all organizations that are based upon classic theory, are:

(1) formal structures are defined by a centralized hierarchy of authority;
(2) labor is divided into functional specialties;
(3) activities are conducted according to standardized operating procedures;
(4) career routes are well established and have a common entry point; promotions are based on impersonal evaluations by superiors;
(5) management proceeds through a monocratic system of routinized superior-subordinate relationships;
(6) status among employees is directly related to their positions (jobs) and ranks.

These characteristics result in a firmly established, impersonal system in which most of the employees and clients are powerless to initiate changes or arrest the system's motions. While this organizational arrangement has generally been afforded high esteem among prescriptive authors in the police field, questions are increasingly being raised concerning its adequacy (Kimble, 1969; Myren, 1960).

Classic Bureaucracies in General

Criticisms of classic bureaucracies in general and police organizations in particular are plentiful (Bennis, 1966). The most common criticisms fall into four categories (Argyris, 1957: 1-24; Bennis and Slater, 1964).

(1) *Classic theory and concepts are culture bound.* Weber's normative conclusions about organizations were founded on his observation and studies of early military organizations, the Catholic church, and the Prussian army. Therefore, his theoretical concepts quite naturally reflect the authoritarian biases of such systems.

(2) *Classic theory and concepts mandate that attitudes toward employees and clients be inconsistent with the humanistic democratic values of the United States.* Managers in organizations adhering to classic philosophy are expected to view employees and clients of the organization as "cogs" that can be relatively easily replaced. The individual value of each person, a fundamental assumption of American democracy, is foreign to classic organizational concepts.

(3) *Classic structured organizations demand and support employees who demonstrate immature personality traits.* Employees of classic organizations are analogous to children in a family—they are expected to obey orders and carry out assignments. This situation is best illustrated by traditionalists among military officers who are fond of telling their enlisted subordinates, "You're not paid to think, you're to do as you're ordered." Employees who do not question, but blindly obey every regulation and order are rewarded, whereas the mature person who raises legitimate questions about the organization and its

activities is often ostracized and punished. Such behavior discourages attitudes of independence that are characteristic of a mature adult personality.

(4) *Classic organizations are unable to cope with environmental changes; therefore, they eventually become obsolete and dysfunctional.* The hierarchical organizational structure and related classic theory power arrangements stifle communications and restrict information about both the internal and external environments of the organization; therefore, they find it difficult to detect and respond to changes. In addition, the emphasis upon routinization of organizational activities creates inflexibility in employee and organizational behavior and reduces the organization's ability to adapt to change.

These criticisms are as relevant to police departments as other bureaucracies. In addition, police departments have specific problems that are peculiar to their operations and can be traced to classic organizational concepts.

Problems Related to Police Bureaucracies

Although many police organizational problems can be related to the basic bureaucratic theory as operationalized by police managers, three of the most significant are appropriate as illustrations. These problems are: (1) the state of police and community relations where well-developed police bureaucracies exist, (2) the state of morale among police employees, and (3) the lack of communication and control in law enforcement agencies (President's Commission on Law Enforcement and Administration of Justice, 1967).

Police and community relations. Since increased efficiency is a basic goal of classic organization theory, consolidating small organizations and centralizing control over them is always at least rationally justified. Their concern for efficiency and economy has caused police administrators to develop a myopia to side effects that accompany increased centralization of police departments. For example, consider the side effects from attempting to develop the one "best" procedure for enforcing an abandoned vehicle law in a large jurisdiction with an economically heterogeneous population. Assume that those who have the greatest economic advantage and the most political influence feel a need to eliminate inoperable vehicles from the city. Since they are politically powerful they have no difficulty impressing upon the equally middle-class police management the importance of enforcing this law. According to classic theory a uniform policy is developed and officers are instructed to enforce the law in a nondiscriminatory fashion (that is, they cannot make exceptions to the enforcement policy), and they carry out the policy in a highly impersonal manner.

Although not blatantly apparent, this kind of enforcement is highly discriminatory. First, the lower-income citizens are generally the only people who have inoperable vehicles where the police can detect them; second, lower-income people cannot afford to maintain their cars in as good a state of repair as can higher-income people; and, third, lower-income people need the parts from their inoperable autos to repair the ones they are currently driving. In addition, an abandoned vehicle law has no social utility for people with lower incomes if they are not disturbed by the presence of inoperable cars. The value of having such a vehicle may be greater to them than a tidy backyard.

The centralized authority and responsibility of the police to develop a policy facilitate their manipulation by the powerful persons and groups, while it precludes other less powerful groups from legitimately influencing these policies. In other words, classic organization concepts do not facilitate adequate policy flexibility. If the police adhere to classic principles, they cannot develop their policies to meet legitimate needs and values of individual subcultures or groups—even when these variations might improve justice, as in the preceding example. This inflexibility is detrimental to police-community relations.

Classic theory also supports police reformers who insist that police departments be isolated from politics. As police departments become more refined and move nearer this goal, they move further away from another basic goal of democracy—guaranteeing every citizen access to and influence with governmental agencies. Under a highly developed police bureaucracy, nearly all citizens view their police department as essentially beyond their understanding and control. Where the police department is a highly developed traditional bureaucracy, its structure and its philosophical underpinnings will eventually cause the organization to become socially irrelevant and ineffective. This situation in turn will have a profoundly damaging effect upon police and community relations.

Once a negative police-community relationship begins to develop, communications problems increase and may worsen geometrically. The classic organization impedes improvements by further restricting communications and by failing to facilitate personalized attention to subgroup problems. This situation makes it impossible to develop a significant level of role consensus between minority groups and the police.

Police employee morale. The classic organization model appears to support a perpetual state of low morale among employees of bureaucracies. Max Weber is quoted as having condemned this aspect of bureaucracy (Bendix, 1962: 464):

> It is horrible to think that the world could one day be filled with nothing but these little cogs, little men clinging to little jobs and striving toward bigger ones—a state of affairs which is to be seen once more, as in Egyptian records, playing an ever-increasing part in the spirit of our present administrative system, and especailly of its offspring the students. This passion for bureaucracy is enough to drive one to despair. It is as if in politics we were deliberately to become men who need "order" and nothing but order, who become nervous and cowardly if for one moment this order wavers, and helpless if they are torn away from their total incorporate in it. That the world should know men but these, it is in such an evolution that we are already caught up, and the great question is therefore not how we can promote and hasten it, but what can we oppose to this machinery in order to keep a portion of mankind free from this parcelling out of the soul, from this supreme mastery of bureaucratic way of life.

If employees in a democratic environment are to be satisfied, they must be more valuable than cogs in a machine. Their jobs must be challenging and rewarding enough so that they have a sense of pride and self-importance in performing them.

The division of labor and organization structure should be such that an employee can be content doing his job well. A good patrolman or specialist should not be dependent on a promotion to a supervisory position for increases in pay or status. It is irrational to train and coach an officer until he obtains a high degree of competence in performing his specialty, and then promote him into a supervisory position where his skills will be useless to him (Peter and Hull, 1969). Yet in a police department employees are hired for one level but they are expected to strive for promotions to completely different kinds of jobs in supervisory positions. Consideration is seldom given to the fact that a good patrolman may be a poor manager. In fact, under classic concepts it is improper to reward a patrolman with a salary equal to that of a top administrator even though both may be equally important to the successful operation of the organization.

Another major cause of poor morale among police employees is the conflict between generalists and specialists (Wilson, 1963). In police organizations the most important people in the organizations, the generalists or patrolmen, tend to become nursemaids to the specialized officers such as investigators, and juvenile and traffic officers. This situation creates tension between police generalists and police specialists, and results in a lack of cooperation toward the accomplishment of common goals. The reason for this conflict is apparent. The patrolman's duties mandate that he be highly skilled and knowledgeable in handling a wide range of human behavior. However, he is accorded law status and pay, whereas the specialist receives much more of both. Patrolmen often believe the public thinks that detectives solve large numbers of serious crimes (in fact they usually have a far lower arrest rate than patrolmen and a large percentage of their arrests are based upon information provided by patrol officers); and that juvenile specialists are the only police officers who are concerned about helping children. On the other hand, they believe that the public credits uniformed patrolmen only with victimizing citizens and starting riots. These perceptions may so affect the motivation of the generalists that they shunt responsibility onto the specialist at every opportunity.

In addition, low morale among police employees is also caused by feelings about their inability to affect their own working conditions. As the educational level of police employees rises, they insist that they have a right to be involved in decision-making processes of the police organization. Educated officers believe that they have the ability to make decisions about their jobs. Police activism has increased and a number of jurisdictions have recognized the legitimacy of police employees' groups and unions. Such recognition is contrary to the tenets of monocratic, classic theory, which holds that the ultimate decision-making authority rests with the chief of police and flows from the top of the hierarchy to the positions at the bottom; however, the trend toward employee involvement in decision-making processes is not likely to cease. The continued utilization of classic autocratic managerial techniques by traditional managers only increases employee hostility and dissatisfaction.

Communication and control. According to the postulates of classic organization theory the chief administrator is responsible for controlling the personnel of the organization. It is surprising that although students have noticed that certain types of control are almost always absent, they have failed to deduce the basic weaknesses in the theoretical foundation of the existing organization structures (Angell, 1967). Apparently, a chief of police of a large department cannot remain loyal to classic principles and at the same time exercise informed control. The explanation for this situation can be traced back to a long recognized problem of bureaucracies—communication.

The hierarchy of authority through which communications travel distorts and filters communications both deliberately and unintentionally (Tullock, 1965). The chief administrator seldom gets a true picture of what is occurring in his department. When he issues a directive to correct a situation (which he probably has already perceived inaccurately), his communication will most likely be distorted as it travels through the hierarchy; therefore, it will not have the impact he originally intended. Even with improved communication, the assumption that formal authority to command can force compliance from subordinates appears to be questionable. As Chester Barnard (1968) pointed out many years ago, authority rests with the subordinates rather than with the supervisor. In other words, if the subordinates are not disposed to accept them, orders will receive little or no compliance.

Various functional supervision units established to guarantee employee compliance with departmental expectations have not been notably successful (Angell, 1967). Therefore, it is questionable whether in our culture, traditional bureaucratic principles can accomplish the objective of adequate control.

Weber's concepts of organization may not be adequate for organizing police departments in the United States. The police problems appear to indicate that gaps in expectations and role perceptions exist between actors who are significantly related to police organizations. Further application of the tenets of classic theory at this time in this society would serve only to heighten tension and increase conflict. Therefore, the basic hope for correcting the dysfunctional trends of American police organizations lies in the development of a new model that will be compatible with the values and needs of American society.

Organizing: A Democratic Model*

The following model is an attempt to develop a flexible, participatory, science-based structure that will accommodate change. It is designed for effectiveness in serving the needs of citizens rather than autocratic rationality of operation. It is democratic in that it requires and facilitates the involvement of citizens and employees in its processes. It is designed to improve decision-making and role consensus among citizens and employees by increasing the exchanges of information and influence among the people who are related to the organization.

Overview of the Structure

The basic model is an organization with three primary sections: (1) General Services Section, (2) Coordination and Information Section, and (3) Specialized Services Section. This arrangement would not be structured in a hierarchical fashion with formal ranks and formal supervisors. In order to improve communication and

---

*Source: John E. Angell, "Toward an Alternative to the Classic Police Organizational Arrangements: A Democratic Model." *Criminology,* Vol. 9, No. 2 and 3, August-November, 1971, pp. 194-206. Reprinted with the permission of the publisher.

increase the flexibility of the organization all supervisory positions, as they have been traditionally defined, have been abolished. Similarly, military titles and ranks are not used.

The controls in this system are varied, in contrast with the single chain of command control required by classic concepts. Although the control responsibilities will be well defined, no single section or individual will be totally responsible for controlling the entire organization. The control system is defined as a system of checks and balances in which one section of the organization has authority in a second instance, and the third section in the third. The General Services Section would consist of teams of generalists decentralized to work in a small geographic area. On the other hand, the Coordination and Information Section would be centralized and might even include many jurisdictions (e.g., a regional or state level). Within the Coordination and Information Section would be those activities related to the coordination of activities and housekeeping of the organization (e.g., those activities presently called administrative and staff functions). The Specialized Services Section would contain those specialized activities currently classified as line units (e.g., investigative, juvenile, and traffic functions).

General Services Section. The General Services Section of the organization would consist entirely of police generalists, who would have equal rank and would have no formally assigned supervisor. The leadership is expected to develop situationally as the circumstances dictate. In other words, team members can determine who will lead them, and the person who occupies the leadership role may change as the situation changes.[1]

Officers will be assigned in teams of five positions by the Coordination Section to work within an assigned geographic area.[2] Except for broad guidelines that would prevent extreme or deviant behavior by the generalists, no procedural guidelines will be imposed on these teams by administrators in the organization. The freedom from rules is intended to permit: (1) the local teams to adopt goals and policies consistent with the needs and desires of the people in their community, and (2) the teams to develop their own methods for handling the problems within their geographic area.[3] However, the goals, policies, and procedures established by the teams could be registered with the Coordination and Information Section.

A team will be expected to work closely with the community of its area. Each team should be responsible for maintaining a community office that will be its local headquarters. Informal meetings involving team and community members should be held periodically to discuss the policies, procedures, problems, and conduct of the community and police. Attempts should be made in these meetings to get consensus on various police responsibilities and procedures.

The teams are also expected to involve specialists from the other sections of the organization in these community meetings. These specialists will be able to provide additional information that can be used to identify and solve community problems as well as serve as communication links with their fellow specialists.

Team members should also attend intraorganizational communications and training programs and meetings. These activities will be designed to eliminate organizational conflicts and misconceptions and to improve the abilities of team members. In addition, study and supervision groups that include team members may be established, ad hoc or permanently.

The officers on each team will have considerable flexibility in deciding when to utilize the services of the Coordination and Information Section and the Specialized Services Section of the organization. In other words, the generalists within each area will be expected to decide how far to go in each particular situation. For example, they will have the right to decide when they will call an investigator to assist in the investigation of a crime. However, no organizationwide regulations will prohibit them from completing an investigation alone. If, on the other hand, a team member decides that he needs the assistance it will be available. This freedom from organizationwide policies does not prohibit individual teams from establishing guidelines for their own members.

The evaluation of these teams could be multiphased such as:

(1) team evaluation whereby each member of the team would evaluate all other members of the team;
(2) evaluation by Coordination and Information Section, which could determine whether the teams are adequately accomplishing the organizational objectives: this section probably should not evaluate the procedures that the team uses in doing the job, but rather it would assess how far

the team has gone in adequately meeting its obligations and accomplishing its goals;
(3) evaluation by the community served by a team;
(4) functional evaluation by specialists in their areas of expertise.

Coordination and Information Section. The Coordination and Information Section of the organization will be made up of functional supervision units (including an ethics unit to ensure that the teams meet the established standards), a planning unit that will provide assistance for the teams, and certain other support units such as detention, records, communications, and training. In other words this section will contain units that are concerned not with policy development, but essentially with providing coordination, support, and minimal supervision for the other sections of the organization.

One of the most important responsibilities of this section is the definition of communities and the assignment of team members to them. This activity involves determining areas where the citizens have relatively homogeneous value systems and assigning teams to these areas in such a way that their workloads are equal.

In the assignment of team members, this section should attempt to provide each team with generalist members who have complementary skills and attitudes. Every team should be monitored and evaluated constantly, so that improperly assigned officers can be reassigned to other teams and areas where they can perform better. When evaluation reveals that the teams as a whole are not meeting their objectives or not complying with standards, the Coordination Section should be responsible for breaking up the team and reassigning the members. A new team must then be selected and assigned to the area.

The Chief Coordinator of this section should play a role analogous to that of a hospital administrator. It will be his job to:
(1) represent the organization on occasions when a spokesman for the entire organization is needed;
(2) oversee the continual updating of the organizational philosophies and long-range objectives;
(3) coordinate the activities of the various segments of the organization and settle conflicts and duplications of efforts;

(4) provide employees and teams with maximum, yet equal, support within organization resources.

This official should be selected for the job because of his expertise and abilities in coordinating and managing human organizations. He should have a contract for a definite number of years and be considered for reappointment at the end of his term. Members of the organization as well as the public should be involved in his selection and evaluation. Obviously, his success will depend on his ability to maintain the system.

Specialized Services Section. The third branch of the organization, the Specialized Services Section, will employ and house specialists. These specialists will be available to support and assist generalists. The generalists will have the freedom to establish their own team policy regarding the utilization of specialists. Their policy may leave such decisions up to the individual generalist.

When called by a member of the General Services Section, a specialist will serve at the pleasure of the generalist or team who called him. He is responsible for performing his particular specialty without interference from the generalist in much the same way as an X-ray technician or other medical specialist performs services for a doctor. Therefore, while a specialist will work at the discretion of and as an assistant to the generalist, he is nevertheless responsible for doing a professional job. Once he has completed his work his only responsibility is the submission of a report to the generalist who requested his service. Decisions regarding further action are the responsibility of the generalist.

Specialists may be evaluated by their fellow specialists, functional supervision units in the Coordination and Information Section, the generalists or teams for whom they work, or any combination of the preceding. If they are not performing adequately, the authority for retraining lies with the Coordination and Information Section; however, these people should have received most of their professional education and training prior to their employment. This procedure will permit the hiring of skilled specialists who are already qualified professionals.

In addition to the traditional specializations of police work, new specialties could be developed to improve the quality of policing. For example, if a decision to disarm the General Service officers were made, a specialized unit might be established to assist the generalists in situations involving explosives or firearms. Such a unit could be kept on reserve in much the same way as firemen are kept ready. Members of this unit could be highly trained and adequately equipped to respond immediately anywhere in the jurisdiction. The potential for such specialization is limited only by the imagination.

Although the specialists will have a considerable amount of freedom to perform as they see fit, it should be reemphasized that they will be working for the generalists rather than for an independent supervisor as they do now. On the other hand, these specialists may be expected to evaluate the quality of the activities performed by generalists in their area of expertise (e.g., investigators might evaluate the quality of investigations performed by the various teams).

Expected Advantages of this Model

This organizational arrangement offers the potential for solving many of the problems that elude solution under the classic police organizational arrangement. First, by eliminating formally assigned supervision and by providing officers with more control over their own jobs, it should increase the morale and effectiveness of employees. In addition, it should result in improvement in the ability of the groups to establish and achieve goals.

The decentralization of these groups will give citizens more influence in policy decisions. This influence should provide the generalists with more direction than police officers currently have for the establishment of socially relevant enforcement policies as well as the exercise of discretion. It will also provide more flexibility in policy, so that the various segments of the population will receive more appropriate police service. These factors should improve the relationships between the public and police officers, and thereby improve the effectiveness of the police.

This organizational structure increases the professional standing and prestige of the generalist without damaging the status of the specialist. It recognizes that the generalist is important and has the intelligence needed to make decisions about his job. On the other hand, the importance of the specialist is recognized and rewarded.

Since this plan permits hiring from outside the organization, highly skilled persons could be selected for specialized jobs, thus saving the organization training expenses.

In essence this organizational structure destroys the formal classic hierarchy. It establishes a system of checks and balances somewhat like that of the federal government. Some aspects of this arrangement are analogous to a hospital situation, and other features are very similar to those used in universities. It provides for increased communications and more adequate information about perceptions and expectations of the various actors in the system. It provides employees with the authority and responsibility necessary for attaining professional status. It facilitates citizen involvement and organizational responsiveness, the hallmarks of democratic institutions. And it should encourage the development of more mature employees and citizens.

Related Changes

In order for this model to work substantial changes in aspects of management, other than the organization per se will be necessary. One of the most important of these changes will be the modification of police training programs. The traditional training method utilized in police academies—the lecture—is inadequate to prepare employees for the decision-making responsibilities they will encounter in this type of organization. For this system to function properly personnel must have a broad education that emphasizes improving their decision-making abilities. The generalists will, because of structure, have to be well versed in human relations and dealing with people, politics, management, and a broad range of other activities, with which they are not too familiar at the present time. Training activities for General Services officers may utilize techniques such as the crisis intervention program that has been developed by Mort Bard in New York City (Bard, 1969) and the community-based training program that has been implemented in Dayton, Ohio (Igleburger and Wasserman, 1970). The training of the specialists will be facilitated because people can be hired from outside into these positions; therefore, they will be trained to perform at a high level of efficiency before they enter the organization. The Coordination and Information Section can be staffed by management specialists, planners and other specialists who have the appropriate training and education. Many such positions might be staffed by students, researchers, and educators to increase the relevance of higher education and draw on the ideals of its personnel.

A second area where changes may be necessary is salary. Under this organizational structure there should be no significant differences among the salaries of the various people. A patrolman or a generalist is equally important in his particular area as an administrator is in his area, and the investigator is as important as an administrator or a police generalist. Therefore, the starting salaries might be in the neighborhood of $9,000 or $10,000 with merit increases to about $20,000 a year. Since the work of people in all three areas is equally important, they should also receive merit and seniority salary increases up to the same maximum. Such salaries would eliminate the need for patrolmen to be promoted to a supervisory position for salary increases, and it would improve efficiency by allowing employees to concentrate exclusively on their jobs.

A third important activity that might be necessary is the establishment of an ethics committee. This committee might consist of employees and citizens who would hear complaints regarding the behavior and decisions of various members in sensitive, well-defined areas. It would have the power to take certain kinds of action to correct misbehavior and improper decisions by professionals within the organization. Part of the value of this unit would be in the information it would provide for decision about training and organizational improvements. Although this unit might be part of the Coordination and Information Section of the organization, it does not necessarily have to be located in this section. Other arrangements, such as an independent unit outside the organization, may be equally desirable.

However, the most important change necessary is in attitudes. Police executives will need to place more confidence in their officers, and the officers will have to accept increased responsibility and deal with it realistically. Executives will be tempted to impose their values about how things should be done on the teams and they will have to resist this temptation and evaluate teams on their results. Team officers will be constantly frustrated by the ambiguous, unstructured, insecure nature of their jobs and they will need to learn to accept and deal with the problems inherent in such an environment. Naturally, people currently holding middle-management positions will have the biggest adjustment to make in order to understand why their positions are unnecessary. Members of the organization and many citizens will have difficulty adjusting to this arrangement. Democratic ways are never tension-free or easy.

Conclusions

The problems related to the utilization of the military or classic model for municipal police organizations that have heavy service goals are numerous. I have attempted to present an alternative organizational model which I believe would be an improvement on the existing approach to police organization. This model is democratic in that it rejects the dictatorial, classic approach and facilitates the involvement of many people in its decision-making processes. On the other hand, it also provides for control of its employees and coordination of their activities. It provides methods for making policy differentials as well as for maintaining organizational efficiency. Significantly, it increases the probability of the police officer's job becoming a full-fledged profession.

Implementation of the recommendations should be well planned and tied to a program of evaluation. A department might be reorganized to continue the existing approach in all but two experimental areas where teams constructed along the lines recommended would be implemented. The communities where these experimental units are implemented could be subjected to pre- and postevaluations. Comparisons of attitudes and changes related to the operation would be essential. Such evaluation would require an extensive interdisciplinary research effort of a caliber never before undertaken in the police field.

Planning*

Every police department, whatever its size, needs and utilizes planning, whether the process is recognized as planning or not. In the large, metropolitan departments this task is often handled by a specialized group of highly trained, well paid individuals. In other departments the planning function is usually handled by the chief or his deputy. The difference is quantitative. There is equal need for planning in both large and small departments; only the volume of activity and the form it takes differ.

Why does every police department need planning?

No situation is static. Even in the smallest community manpower needs, operating costs, crime rates, the regulatory work load, and public service requirements change over time. Communities grow in population size, or in some cases, diminish. And even where no noticeable change takes place in the size of the community, changes in composition of the population occur. Changes in age and sex composition, changes in the way people earn a living, changes in the income status of residents, and physical changes in the community all are factors affecting law enforcement.

The chief whose department keeps pace with the community must engage in a constant process of evaluation and revision. When he finds it necessary to request additional manpower, he must be in a position to go before the local legislative body and justify that request. When he needs additional equipment and facilities or funds for training, he must be able to explain these needs in factual terms, meaningful to those who make decisions as to the allocation of public funds.

This requires planning. The financial support and cooperation from legislative and administrative bodies that is essential to effective performance of police services can best be obtained on the basis of factual information, relating police operations to changes in the community. The chief must be able to show evidence of the need for action, to evaluate alternative methods of meeting that need, and to present his recommendations forcefully by demonstrating that requested funds will be purposefully directed to the solution of real

---

*Source: John Ashby, James L. LeGrande, and Raymond T. Galvin, "The Nature of the Planning Process," *Law Enforcement Planning,* Washington: U.S. Government Printing Office, 1968, pp. 1-17.

problems. The rapid increase in the cost and complexity of law enforcement in recent years has made systematic planning a necessity in even the smallest communities.

A Definition of Planning

Planning, in simplified terms, involves the marshalling of facts relevant to the solution of particular problems, evaluation of alternative solutions, and detailing (in advance) the action to be taken toward implementing desired changes. The problems to be solved may range all the way from procedural problems in the field to administrative problems which have an important bearing on the future of the entire department. Determining what is *relevant* is a primary and crucial aspect of good planning. It must be followed by systematic analysis of the pertinent facts, existing conditions, the problems or inadequacies to be met, and potential improvements.

Four general types of planning enter into the provision of law enforcement services: management planning, operational planning, procedural planning and tactical planning. These can be highly specialized functions in a large organization, but must be integrated by the police administrator in a small community.

Management planning is department-wide in scope and is directed to the setting of policy and determination of departmental goals. Budgeting, personnel management, and departmental organization are the major aspects of management planning. Budgeting affects all operations; personnel management must deal with recruitment, promotional standards, and training; and the development of job descriptions, and operational units are organizational problems related to budget and personnel considerations. Overall planning at this level delimits the framework in which operational, procedural, and tactical planning take place.

Operational plans deal with such matters as beat or area boundaries, shift assignments, patrol schedules, and equipment maintenance. Systematic planning of operations is all too often neglected, operations being conducted haphazardly until some crisis calls attention to inadequacies or problems that could have been avoided with a little forethought.

Procedural planning is also an often-neglected aspect of law enforcement activity, though in the area of day-to-day procedures in the field, planning can usually make a very valuable contribution. Carefully considered procedures can do much to improve police efficiency and safeguard the lives of officers in the field. Most deaths and injuries to police personnel in the line of duty occur as the result of poor procedures in the field. Sound, written procedures with which every officer is thoroughly familiar are the backbone of efficient, effective field operations. These procedures require frequent and regular review to insure that the methods in effect are up-to-date with current conditions in the community and departmental goals and objectives.

Office procedures are equally important to the overall efficiency of the department. The records required in the day-to-day routine of a department for dispatching, booking, preservation of evidence, and control and maintenance of departmental property need not be burdensome in quantity and complexity, but they must be systematic. Clear, precise, functional records which are easily completed and fully understood by all personnel responsible for their handling are essential. The forms used should be reviewed periodically to evaluate how well they serve the required purposes and whether they can be designed more efficiently in terms of cost, completeness of information recorded, ease of completion, and time required by record procedures.

Tactical planning deals with operational and procedural plans for the handling of crises and emergencies. Police departments must be prepared to handle many events which may never occur, such as airplane crashes, bomb threats, riots, civil disorders, and catastrophic conditions like floods and tornado damage. The successful execution of emergency tactics often requires the assistance and cooperation of other agencies. Well laid procedural and operational plans, based upon strategy sessions with neighboring departments, sheriff's offices, state and federal agencies, and, sometimes, private security and welfare agencies, prepare police departments to act effectively in emergency situations; thus, tactical planning can avoid confusion and, often, disaster when such situations arise.

The planning task must be approached in terms of long-range, intermediate, and short-term plans. Long-range planning deals with the future of the department over an extended period of several years. Future personnel and equipment needs must be estimated on the basis of projected expansion of the community. Nearly all operational and management planning as well as tactical planning must be done in the light of long-range considerations. Intermediate planning deals wth the continuing and recurring problems of routine procedures and operations. Short-range plans are mainly tactical in nature, dealing with the handling of particular situations such as parades and sporting events or downtown congestion due to peak shopping periods.

Departmental Goals and Planning

Planning is a process for developing predetermined courses of action which offer the greatest potential for obtaining desired goals.

The traditional police goals are frequently stated as the prevention of crime and disorder; the preservation of the peace; the protection of society; and the protection of life and property while guarding personal liberty. The major methods or strategies for accomplishing these goals are often enumerated as the prevention of crime, the repression of crime, the apprehension of criminals, the regulation of non-criminal conduct, the providing of services, and the protection of personal liberty.[1] It is well to recognize, however, that each community has unique characteristics, attitudes and police problems. The local environment may be such that a broad statement of police objectives and strategies gives little guidance to the structuring of the individual police organization and the planning process therein.

It becomes necessary for each police administrator to specifically determine the primary goal of his organization, along with any sub-goals, and the priorities among them. The department should then be organized and structured according to these goals; planning based upon them, and departmental operations governed by them. There are no golden means concerning the activities of a police department or the various legitimate methods that police officers should utilize in performing their tasks. The organization and operation of a police department should be based upon the police administrator's assessment of the local situation and the local need. Although the police literature can give him guidance, the ultimate determination of what the police department is to accomplish and generally how it is to accomplish it will be dependent upon the administrator's decisions as to goals.

The administrator's goal determination should not be conducted in a vacuum. He must inventory the local environment to evaluate his course of action. He must look to the various spheres of influence that control and affect the police department. His inventory should include:

Documents of Legal Creation—What authority and limitations have been placed upon the police department by statute, ordinance and charter? Has the purpose of the police department been specifically stated or implied by the legislative body? What did the legislature view as the major police goals?

City Manager, Mayor, Councilmen—This group has a knowledge of pressing city problems. They constitute the basic power structure above the chief. What do they expect from the police department?

Departmental Personnel—What do various departmental personnel feel the police goals should be? What problems exist? How do they feel they can be solved?

Other Criminal Justice Officials—What do other officials believe the police goal should be? The prosecutor, judges, sheriff, state police, attorney general, city attorney, neighboring chiefs of police all are in a position to give ideas and evaluations.

Departmental Records—What trends and problems are reflected in departmental records?

Community Groups and Organizations—What are the beliefs and ideas of various community groups concerning the operation of the police department and what do they expect? The news media, service and social groups, businessmen, leading citizens, clergy, school officials, minority groups—including militants, and professional men are going to be influential in all city affairs. Their counsel may be valuable to the police department.

After these various consultations the police administrator should be in a position to evaluate what is expected of the police organization. Utilizing his education, training, experience, and professional expertise and judgment, he is in a position to establish goals, both general and specific, which give direction to the department, guide organizational methods, provide a base for planning, and allow for the effective concentration of police operations.

In reality, the determination and establishment of stated organizational goals furnishes a definition of purpose and a set of perspectives to which one can refer in structuring one's organizational conduct. Such goals can offer a foundation for the effective disciplining of different individuals by providing them with a uniform set of performance standards. Even more important for purposes of the planning process, the expressly stated organizational goals give the planner a firm base from which to operate. Without a realistic statement of goals and priorities, the planner must, of necessity, approach each individual problem in an *ad hoc* manner and with the hope that his resolution of the problem and the subsequent plan conform with the unstated departmental philosophy. Thus, for planning to function at its best and accomplish the most, it is necessary that it have articulated goals and priorities to use as points of reference.

The Process of Planning

Virtually every police problem lends itself to analysis by application of the scientific method. The scientific approach is recommended by all of the authoritative literature in the police planning field.[2] This approach involves (1) the discovery of the problem, (2) the isolation and clarification of the problem, (3) the collection and analysis of pertinent data and opinions, (4) the identification and evaluation of alternatives, and (5) the selection between alternatives. Subsequently, it may be necessary to "sell" the plan, arrange for its execution, and evaluate its effectiveness.[3]

Discovery of the Problem. Problems in the police organization may be discovered in many ways. By the nature of the police function many problems come to the attention of the administrator from external sources, e.g., the mayor, a citizens' group, a civil rights organization, a traffic safety council, the press, or individual citizen complaints. Internally, the administrator learns of problems from periodic reports through the chain of command, personal conferences with department personnel, staff meetings, events or incidents which reflect an improper police response, and similar sources.

Most major problems will eventually make themselves known through internal or external complaint. When problems are raised in this manner, they frequently require an immediate reaction. Procrastination until problems reach a crisis stage tends to account for the fact that police planners find themselves in a continual state of fire-fighting, and frequently treating only the symptoms of the problems rather than the true causes.

Perhaps, many of the crisis planning situations could be avoided if the administrator attempted to establish various types of "early-warning" systems to discover the existence of problems before they reach a stage where sophisticated planning and thought are made virtually impossible by the nature of the circumstances.

There is no doubt that the chief of police is ultimately responsible for departmental planning. However, to be totally effective, every member of the department should be encouraged and solicited to engage in constructive criticism of departmental operations. Every member of the department participates in and observes the operation of various plans, policies, procedures and practices within the department. Each member of the department is thus in a prime position to evaluate the various shortcomings of operations. The secret of early discovery of problems is to establish various channels of communications within the department which will move the information concerning these shortcomings to the level of management which can effectively cope with them. The chain of command obviously is the basic means for accomplishing this. However, consideration should be give to other methods, such as frequent staff conferences at *all* levels, the establishment of a departmental suggestion system, and the other appropriate means to get the necessary "feedback" to the individuals within the department who can take corrective action.

The administrator might also encourage comments or complaints from external sources. Citizen complaints, against which police departments have traditionally taken defensive positions, may well serve as an untapped source of problem discovery. If a complaint by a citizen were regarded as an aid to police administrators rather than simply as the commencement of an action against a police defendant, it might not only improve public reaction but serve as an effective means of problem discovery.[4]

If the problem is not discovered and the need for a plan not recognized, obviously no plan will be developed nor will a solution be forthcoming. It is very important therefore that extensive attention be given to the discovery of problem areas and that action be taken long before the crisis stage is reached. Otherwise, essential planning will not be done and the planner will find himself constantly devising stopgap plans in critical situations.

Isolation and Clarification of the Problem. The problem isolation and clarification stage may well be the most difficult part of the entire planning process. The planner must take adequate precautions to assure that he is actually studying the problem itself, rather than simply dealing with a symptom of the problem.

Symptoms may not be at all indicative of the true problem. For instance, assume that a prosecutor is not receiving supplementary investigative reports from the police department's detectives on criminal defendants until four days after the commission of the crime. Initial analysis of the problem might lead one to believe that either the present detective force was not competently performing its job or that it was understaffed. An immediate plan or remedy without extensive consideration might be to tighten supervision in the detective bureau and assign additional personnel. However, a study of the reporting system might disclose other problems which would not be reflected in a cursory review of the problem of late investigative reports. A study might show that the detectives do not receive the initial reports made by patrol officers until two days after the offense occurred and therefore cannot prepare their supplementary reports. This would lead to a study of the staffing patterns and personnel assigned to the report clerks' office or clerical pool. This in turn might lead to the disclosure that patrol officers wait until the end of their tours of duty to make all reports, thus placing a high volume of reports on the clerks at one time. The end result of this study might dictate the establishment of a field reporting system for patrol officers, a revision of the staffing pattern in the clerical pool or possibly the addition of more clerical personnel in order to correct the true problem. By initially adding more personnel to the detective bureau and tightening up supervision, a symptom would be treated but the true problem would not be corrected. The recognition of the true problem not only resulted in the elimination of the prosecutor's difficulties but also increased the efficiency of the police operation. It is important to note that the benefits obtained from solving the true problem were substantially greater than those that would have been obtained by solving the limited problem originally stated.

Collecting and Analyzing Pertinent Data. The collection and analysis of data play a key role in the planning process. A plan generally is no better than the information utilized in its formulation or the judgment exercised in its preparation.[5]

The process of data collection involves the acquiring of all information that may be of value in preparing the plan. There are a vast number of sources to which the police planner may turn. Depending upon the nature of his problem, the planner may use one or more of the following sources.

—*Review of the Literature.* Whether the problem is organizational, operational or technical a review of the literature in the field will serve in most cases as an effective first step in data collection. From this review the planner may develop a prospective or "frame of reference"[6] concerning the specific problem.

The literature obviously includes books relating specifically to police administration and operations, as well as the various police journals.[7] The review, however, should not be limited exclusively to these areas. The sophisticated planner recognizes that the word "police" does not have to be in the title of a book or article to make its contents applicable to his problem. For instance, comparatively little material has been published on the subject of *police* personnel management, but there is a wealth of such material in the general administrative, military, and business literature. A planner attempting to establish an efficiency rating system may find only a few effective guidelines in police texts, but he may be able to locate numerous useful sources in books related to other disciplines. Thus, the planner who has at his disposal a well developed police library still should not overlook the tremendous resources to be found in the public, university, or law libraries in his community.[8]

—*Records, Reports, and Documents.* An all-inclusive definition is applied to this category of information. With the exception of the formal literature, all existing data in written form comes under this designation.

The records system within the police department serves as a primary source of data—particularly relating to operational activities involving problems of crime, delinquency, vice, and traffic. The planning of police operations must of necessity be based upon critical estimates of need involving statistical and analytical interpretation of records data. With proper internal records it is possible to predict, with a high degree of accuracy, the operational needs of the department and develop plans based upon these relatively scientific predictions.[9]

Other internal sources of written information should also be utilized by the planner. Such material as personnel records, radio communication logs, jail booking registers, officers' activity reports, disciplinary action transcripts, opinions from various legal officials, letters, and numerous other types of documents may give the planner great insight into the specific problem. Much of this type of material may not be maintained by the department's records unit. The planner must develop techniques and methods for locating this information from the numerous independent files of various other departmental units.

External reports may provide considerable relevant information. Such material as the *Uniform Crime Reports, Statistical Abstract of the United States, Municipal Yearbook,* reports of the city planning commission, minutes of the city council meetings, consultant surveys, annual reports of other departments, special studies or projects completed in other law enforcement agencies, and many similar documents may minimize the time and effort expended by the planner. Materials produced by business and industry should also be reviewed for applicability.

—*Interviewing.* The opinions, ideas, and suggestions of those individuals directly involved in the problem area may be of indispensable value to the planner. The individuals who have experienced the problem firsthand may be able to give the planner direct insight that he could obtain in no other way. As an example, if there is a problem concerning the dispatching of radio units, the dispatchers should be interviewed at length, as well as other officers directly involved with the problem. They may have personal ideas and solutions which are both unique and practical, again saving the planner time and effort. Professional planners have frequently been labeled "brain-pickers" because of the thoroughness with which they apply the interview technique.

The planner should not overlook potential interview sources outside the police department. Quite frequently local business or industrial concerns have experienced the same type of problems as police departments. It is highly possible that an examination of how business solved a similar difficulty will give the planner considerable assistance in his search for an acceptable solution.

Professors from local colleges and universities may provide a tremendous resource for the planner, especially if the institution has a police administration program. Even if the professor does not personally have material relating to the problem, he probably can refer the interviewer to valuable references.

—*Questionnaires.* When a quantity of data is needed from numerous sources, either internal or external, the written questionnaire may prove to be an effective means of data collection. When the planner utilizes the questionnaire, he is requesting a large number of people to contribute their time to this project. Therefore, before resorting to this technique, the planner should make certain that the data is not otherwise available. When constructing the questionnaire the planner should be certain that it can be answered with reasonable convenience, that it is not vague and ambiguous, and that it is constructed to efficiently obtain all the data needed to accomplish its purposes.

—*Observation.* Where detailed information is not available concerning how specific tasks are performed, the length of time required to perform them, the use of equipment, or the processes involved, the planner may have to observe the specific operation. By viewing and recording the way an operation is done, the researcher may acquire a considerable volume of information not otherwise available.[10] For instance, if the planner were assigned to scientifically determine the number of men needed to staff the booking desk and jail, he might find it necessary to observe the actual operation for a reasonable period of time on various working shifts to determine the volume of work and the specific amount of time required to perform it.

Where another law enforcement agency or business firm has encountered a similar problem and satisfactorily solved it, direct observation of the other unit's operation may be highly beneficial.

—*Experimentation.* Experimentation is probably the most advanced method of data collection since it creates new information. Normally, experimentation is utilized to gather information when other means are inadequate.[11]

A police department considering the adoption of a certain type of vehicle for patrol, the use of cadets or policewomen for parking enforcement and control, the use of patrolmen for certain types of criminal investigations, or the adoption of some form of non-lethal weapons, may find that little data is available concerning the particular application it desires to make. Therefore, a "pilot study" may be conducted, utilizing the new procedure, method, or equipment on a *limited* basis to obtain data for comparison purposes.

—*Identification and Evaluation of Alternatives.* Perhaps occasionally a problem will have only one logical solution. More than likely, however, the planner will be required to choose the "best" solution from a number of alternatives, which have become evident during the data collection and analysis stage. It becomes the planner's responsibility to thoroughly evaluate all significant alternatives.

A number of guidelines may be used in evaluating the various alternatives:

1. Which alternative best accomplishes the specific objectives of the plan?
2. Which alternative is most compatible with existing policies, procedures, customs, traditions, practices and techniques of the department?
3. Which alternative is most practical when available resources (manpower and equipment) both present and future are realistically appraised?
4. Which alternative may be most effectively implemented considering the time available before implementation is imperative?
5. Which alternative will be most acceptable by the city's governmental power structures and the various segments of the community?
6. Considering all factors, which alternative has the best liklihood of successfully correcting the problem?

—*Selection between Alternatives: The Final Decision.* As a part of his staff responsibilities the planner has the task, after consultation with other departmental personnel, to completely work out the details of the specific plan and present a single, coordinated proposed plan of action. This requires him to select from the various alternatives the course of action which has the highest probability of success under the existing conditions. In this context the planner becomes a decision-maker. Of course, the actual final choice still belongs to the chief.

Although a number of guidelines have been cited for evaluating alternatives, it is well to point out that they are not all-inclusive. In making the final decision concerning the proposed plan of action the planner must consider the nature of the individual problem, the uniqueness of the local community, the internal composition of the department, and numerous other local variables.[1,2]

PERT and a Planning Problem*

Planning is a process that everyone in a law enforcement agency, patrolman to the chief, must engage in from time to time. A useful tool of planning, a spin-off from the military and industrial sector, in projects and certain classes of operations is the PERT chart. The following is an explanation of PERT and an example of its use in a problem that has been faced by most agencies sometime in their history.

PERT is short for Program Evaluation and Review Technique. The concept grew out of the works of Gantt and Taylor of the scientific school of management and refined by the moderns. It can easily be mastered by any administrator or field officer and applied to most planning problems faced by an agency. To gain a full understanding of all the potential applications of PERT, the planner might want to look into training programs provided by private firms such as Halcomb Associates of Sunnyvale, California, or higher education institutions involved in management training for police like San Jose State or Long Beach State.

---

*Source: Peter C. Unsinger, "PERT and a Planning Problem," Paper presented at the Northern California Police Planners Association, May, 1974.

The Concept

The basic idea behind PERT is simply this: a major element in all planning efforts is correctly estimating time. If the planner can accurately forecast this element then the likelihood of the plan going "according to plan" will be increased. We know that many activities needed to complete a project or operation can occur simultaneously while others cannot be undertaken until other activities or events have been completed. This is exactly what we do when we sit down with the staff and "map" out the general outlines of our plan. We simply set down in writing the step by step scenario, or chronological order if you will, the actions that will have to be taken in order to achieve the objective we set down for ourselves. Once this is done, the activities can be grouped together into those activities or events that can be worked upon and accomplished simultaneously and those activities that are dependent upon the completion of other prerequisite activities. The next step is asking those who have the experience what their estimates are for the time it will take them to do the task. You're already PERTing.

The inquiry as to how long it will take people to complete the task has probably serviced you pretty well in the past. What is recommended here is that you go a little further and begin PERTing the plan out. Instead of general estimates you ask for three time estimates: the optimistic time (to), the most likely time (tm) and the pessimistic time (tp). The optimistic time (to) is the time the executor of that portion of the plan estimates it will take to complete the task if all goes right and he can go at it with "open throttle"; the most likely time (tm) is the estimate of time given that normal activities that involve him are competing for his attention; and the pessimistic time (tp) is the time estimate of how long it will take to get it done if it occurs during a period of abnormally hectic activity.

There are some problems involved in getting these times. If the one asking for the estimates is the chief you can imagine how optimistic all three estimates will be in comparison with the solicitations executed by a patrolman ranked planner. The human element in estimating times should not be overlooked. It might even be that time cannot be accurately estimated since no one may have the experience to give reasonable estimates. Some things just can't be estimated like a break in a big case. With experience and a willingness to learn from prior mistakes, PERT becomes quite useful to the agency and helps the planner in finding the weak "links" in the planning and execution chain.

After all the activities have been laid out and the time estimates have been received, the planner "maps" out the activities into a network. Now, let's use an example that most departments have faced in the past.

The Problem

A new subdivision is planned and police services must be available the day the new homeowners move into the subdivision. Research data indicates that three new units should be available to meet the increased calls for service.

In determining the elements necessary to provide units, you identify the following elements: manpower, equipment and patrol vehicles. Under manpower you find that you'll have to recruit, select, equip, send to academy, place under field training officers (FTO) and allow them to "bloody" themselves on the streets before you consider the manpower ready. For equipment you'll have to order for the manpower and the vehicles. Patrol vehicles will require the writing up of specifications, getting/awarding bids, receiving the cars, equipping them in the corporation yard and then getting them and the manpower together for the "solo" or "bloody" phase. Once all these tasks are done the units (man and machine) will be ready to provide those services the day the subdivision opens for business.

Table 4-1 shows all the events necessary to meet the problem. Each event is given a number for the eventual PERT chart. Some prefer using letters while others will write the event out in full. Each according to his own! The event is described. The time estimates are provided for the optimistic time (to), most likely time (tm) and the pessimistic time (tp) and who provided the estimates. Note that they have been grouped according to those that can occur simultaneously (i.e., manpower, equipment, vehicles) and those that cannot begin until something else has occurred (i.e., can't train 'em until you've selected them). Once you've got all the activities set down you map them out like in figure 4-1.

Table 4–1
PERT Event Table

Event #	Description of the event	time to-tm-tp	Estimates provided by
10	Recruitment over/begin selection process	4-4-4	Personnel and Training
20	Selection completed	3-6-9	Personnel and Training
30	Begin academy	0-2-4	Personnel and Training
40	Complete academy	15-15-15	Personnel and Training
50	Complete field training	4-6-8	Field Training Office
60	Begin solo experience (event 150 complete solo phase)	2-6-10	Patrol Division Commander
70	Order equipment for vehicles & personnel	1-1-1	Purchasing Department
80	Receive equipment for vehicles	4-6-8	Manufacturers Representative
90	Receive equipment for personnel	3-4-5	Manufacturers Representative
100	Publish specifications and solicit bids on patrol vehicles	1-3-5	Technical services and Purchasing Department
110	Receive/Award bids	4-4-4	Purchasing Department
120	Receive vehicles	12-16-20	Purchasing Department
130	Corporation yard begins installation of equip.	0-0-0	Corporation Yard
140	Vehicles ready for service	1-3-5	Corporation Yard
150	Department ready to provide patrol services for subdivision	0-0-0	

Looking at the PERT chart there are some figures there. They provide us information. For instance, between event 20 (personnel selected) and 30 (begin academy) are the figures 0-2-4. The 0-2-4 is simply the to-tm-tp. The estimates from Personnel and Training are that an academy class begins every four weeks. If they're selected on Friday they can begin the following Monday so zero weeks is the optimistic time. It might even be four weeks before a class begins and that would be the pessimistic side of them showing. Most likely they estimate it to be two weeks. The figure below the to-tm-tp is what is

Figure 4–1
PERT Chart

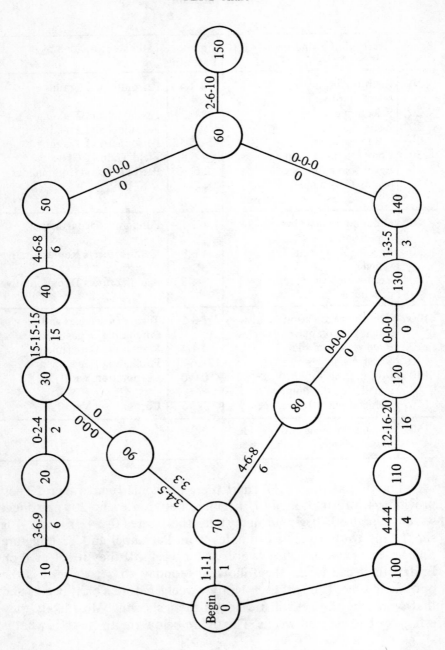

196 *Managerial Functions*

called the actual estimated time (te). This figure is derived from the use of a statistical formula that is based on empirical research and simply the time estimate that you are just as likely to err 50% of the time too high or 50% of the time too low. The formula is:

$$\frac{to + 4(tm) + tp}{6}$$

and in the case of the 0-2-4 it would be $0 + \frac{4(2) + 4}{6}$ or $\frac{12}{6}$ or 2.

The te, or expected time, is simply the weighted average of all three of the estimated times.

Now that we have all these times, what can we do with them? We could figure out when we should start recruiting the manpower and ordering all that hardware. Again, putting the numbers in table form, we know when things should happen or when they should be completed. Looking at path 10-20-30-40-50-60-150 from the begin-

Table 4—2
PERT Path Table

Path 0-10-20-30-40-50-60-150				Path 0-70-80-130-140-60-150			
to	tm	tp	te	to	tm	tp	te
4	4	4	4	1	1	1	1
3	6	9	6	4	6	8	6
0	2	4	2	0	0	0	0
15	15	15	15	1	3	5	3
4	6	8	6	0	0	0	0
0	0	0	0	2	6	10	6
2	6	10	6	8	16	24	16
28	39	50	39				

Path 0-100-110-120-130-140-60-150				Path 0-70-90-30-40-50-60-150			
to	tm	tp	te	to	tm	tp	te
1	3	5	3	1	1	1	1
4	4	4	4	3	4	5	3.3
12	16	20	16	0	0	0	0
0	0	0	0	15	15	15	15
1	3	5	3	4	6	8	6
0	0	0	0	0	0	0	0
2	6	10	6	2	6	10	6
20	32	44	32	25	32	39	31.3

ning we should begin recruiting about 39 weeks before we move into the subdivision. For path 100-110-120-130-140-60-150 we should begin 32 weeks before the great day. Notice that everything is in weeks. It could have been in days or minutes—again the choice is yours. Whatever time unit you use, be consistent and don't mix them up!

What if you wanted to hurry things up? The path that takes the longest is the critical one, or CPM in jargonese. So you put the "heat" on those involved in path 0-10-20-30-40-50-60-150, not on the corporation yard. You might also want to know how long it takes. Simply adding up the te and from beginning to end is 39 weeks so 39 weeks after you started you should be finished.

Don't throw away your plan and its PERT material. After its all over it should be critiqued so that the experience acquired isn't lost. PERT networks can also be developed after to reflect the times that actually were consumed. The documentation is the agency's experience or memory.

Fiscal Management*

A police executive is not simply a crime fighter or a policeman of special and superior rank. He is also a business manager who should accept full responsibility for fiscal management of his agency. Moreover, he is a fiscal planner, responsible for developing the future expenditure requirements for personnel, equipment, facilities, and programs necessary to accomplish his agency's goals and objectives. While he has an obligation to follow all prescribed procedures, he should, where necessary, urge government officaials to improve the jurisdiction's fiscal policies and practices, and he should seek to raise the level of his own fiscal management to achieve the goals and objectives he has established. His role should not be passive; it should be active and progressive.

When the city administrator and the city council begin their task of allocating funds to all departments and units of government

---

*Source: National Advisory Commission on Criminal Justice Standards and Goals, *Police,* Washington: U.S. Government Printing Office, January, 1973, pp. 132-133, 136-141.

for the following fiscal year, the police chief executive will find himself in a highly competitive relationship with other municipal department heads. The council has difficult decisions to make and is concerned with effecting an appropriate allocation of limited resources according to public needs as expressed and demonstrated by various department heads. Such decisions can never be precisely accurate nor entirely acceptable to public constituencies, and the council is subject to many influences.

Carefully developed budgets with adequate justification of all major items, especially on their first appearance in a budget document, are an important responsibility of the police chief executive. In view of typical limitations of revenue which most local governments face today, the police chief executive must be persuasive but objective with administrators and councilmen alike. He must demonstrate sound judgment in his budget planning in order to gain acceptance of his recommendations.

The budget document, as it leaves the office of the police chief executive, is the position statement on money needed to initiate, maintain, or expand programs, functions, and activities of the police agency. Although the executive should not prepare the budget unassisted (except in very small police agencies), the responsibility for a sound fiscal document is his.

While the police chief executive may assign fiscal management tasks to subordinates, accountability for all aspects of his agency's fiscal policies, processes, and control is his. Within the framework of his jurisdiction's governmental structure, his accountability is judged and his practices reviewed by the jurisdiction's chief executive and, ultimately, by its legislative body.

Procedures in fiscal management must support the policies established by the governing body; if they do not, they should be modified or discarded. Fiscal management procedures should be carefully developed to guide agency employees in budget preparations. Every fiscal affairs officer should write fiscal procedures and, after approval by the police chief executive, distribute them to management personnel. Agencies with no fiscal officer should have an agency employee prepare fiscal procedures. In either case, the fiscal affairs officer of the jurisdiction should be consulted, and procedures should be compatible with those established by the jurisdiction.

Budget Planning and Preparation

Annually, before budget preparation begins, the chief executive of the city usually forwards a budget message to all agency heads. It is written to lay ground rules for budget preparation, establish time frames for the process (often in accordance with State law), explain the general financial condition of the city, detail how certain costs (such as personnel) are to be itemized, and set forth requirements for justification. Properly prepared, it is of material assistance to the police chief executive in preparing his own budget message to responsible police agency personnel.

Although budget preparation should be a matter of serious concern 12 months of the year to police agency personnel, formal budget development usually is scheduled over a period of 6 to 8 months prior to the mandated budget adoption date. Certain budget review and presentation periods are established informally, some may be prescribed by the city council or the city executive, while others are required by State statutes or regulatory bodies or agencies. Table 4-3 illustrates a schedule of action periods and dates for a large city.

Ultimate responsibility for departmental budgets lies with the municipal legislative body: the city council. All departmental budgets are formally presented by the mayor or manager, usually following extensive preliminary discussions. The budget form is prepared so that even at this last point of decision certain modifications can be made. A very simple budget document is depicted in Table 4-4, showing how such modifications may be indicated.

At the final hearing before the council, the mayor or manager may be supported by the police chief executive and, in large departments, by certain of his staff, as well as by the city fiscal officer.

Fiscal Control

Annual budgets should be developed in cooperation with all major organizations within the agency. Commanding officers of the agency's bureaus—patrol, traffic, and detectives—must weigh their needs and present budget estimates that include reasonable and economically sound requests.

Table 4-3
Preparation Calendar on Fiscal Year Basis

What Should Be Done	By Whom	On These Dates
Issue budget instructions and applicable forms	City Administrator	November 1
Prepare and issue budget message, with instructions and applicable forms, to unit commanders	Chief of Police	November 15
Develop unit budgets with appropriate justification and forward recommended budgets to planning and research unit	Unit Commanders	February 1
Review of unit budget	Planning and Research Staff with Unit Commanders	March 1
Consoldiation of unit budgets for presentation to chief of police	Planning and Research Unit	March 15
Review of consolidated recommended budget	Chief of Police, Planning and Research Staff, and Unit Commanders	March 30
Department approval of budget	Chief of Police	April 15
Recommended budget forwarded to city administrator	Chief of Police	April 20
Administrative review of recommended budget	City Administrator and Chief of Police	April 30
Revised budget approval	City Administrator	May 5
Budget document forwarded to city council	City Administrator	May 10
Review of budget	Budget Officer of City Council	May 20
Presentation to council	City Administrator and Chief of Police	June 1
Reported back to city administrator	City Council	June 5
Review and resubmission to city council	City Adminstrator Chief of Police	June 10
Final action on police budget	City Council	June 20

Schedule is for a large department requiring 8 months to develop and process the budget and gain administrative and council approval for it. Other major departments may be on different schedules as a matter of convenience for the administrator and council.

Managerial Functions

Table 4-4
Budget Document Showing Approval Process

	A 1971-72 Budget	B 1972-73 Request	Department C Manager's Recomm.	D Council Decision
Personal Services:				
Office of the Chief	23,000	25,000	23,000	25,000
Lieutenants	30,000	40,000	30,000	40,000
Sergeants	72,000	72,000	72,000	72,000
Patrolmen	400,000	430,000	445,000	430,000
Metermen	16,000	16,000	10,400	16,000
Secretarial and Clerical	24,000	24,000	20,000	20,000
Overtime	22,000	26,000	28,000	26,000
Temporary Help	2,400	3,600	3,000	3,000
Purchased Personal Services	2,000	3,000	2,000	2,500
Maintenance	12,000	12,000	10,000	10,000
Total Personal Services	603,400	651,600	643,400	644,500
Current Expenses	72,000	86,000	76,000	80,000
Equipment	4,000	6,000	6,000	6,000
Total Direct Expenses	679,400	743,600	725,400	730,500
Staff Benefits:				
Group Ins. (Detail to be				
Retirement provided by				
Other finance officer				
after budget				
adoption				
Total Staff Benefits	59,600	63,500	64,600	63,700
Total Budget	739,000	807,100	790,000	794,200

In police agencies large enough for functional organization (i.e., requiring development of bureaus, divisions, units, or offices), budget development should begin at the lowest supervisory level and should be a consolidation of proposed unit budgets. Sergeants, lieutenants, captains, and persons of higher rank in large agencies should be involved and must assume responsibilities in budget development. Thus, in the staff services division of a large agency, the unit commanders of personnel and training, planning and research, and others must develop and justify budgets for their operations.

In small agencies, watch or shift commanders should be involved in the budgetary process. Decisions are then made at each level of command as the budget is processed. In very small agencies the chief alone may prepare the final budget. In larger ones with a planning unit, and perhaps even a fiscal officer, the unit or the officer should process recommendations and prepare the final document.

In large agencies the police chief executive's final decisions on controversial programs or items will be influenced by full staff discussion of a tentative final budget.

Police managers often complain that they do not receive the necessary financial support to achieve their objectives. However, those same managers seldom prepare budget requests with written justifications. Or if justifications are written, they are often poorly prepared. Detailed justifications for a budget item should provide the reasons each item is needed.

A police manager who recognizes a problem of needed financial support should prepare a written justification setting forth the need for the added expense and justifying it. If possible, cost-effectiveness should be demonstrated. To identify problems, every police supervisor should inform his superior of the need for additional personnel, equipment, or supplies if they are necessary to successfully perform his task. Every major division or bureau head should consolidate proposed unit budgets to form the division or bureau budget. It is essential that every supervisor and every manager participate in determining budget needs, and it may be helpful if participation extends to the lowest level in the hierarchy. Managers should prepare justifications, but details such as projective costs of fringe benefits should be the responsibility of the fiscal affairs officer.

Division or bureau budget requests should be scrutinized by the police chief executive and his top staff to assign priorities to items of the agency budget. The fiscal affairs officer should assist in the staff review by providing staff assistance to the police chief executive.

Police agencies occasionally initiate new programs and then fail to evaluate them, especially on a basis of cost-effectiveness. A vigorous analysis of programs should be conducted periodically to evaluate program achievements and program costs.

As an exercise in determining priorities, police agencies should construct a budget at 80 percent of the current operating budget. To merely project a 20 percent across-the-board cut may not be the best way to cut costs and would do nothing to establish priorities. Evaluation of programs for this exercise may suggest abandoning, reducing, or modifying some activities within the hypothetical reduced budget. When the budget is reworked on a 100 percent basis, the new order of priorities may substantially influence it and provide essential support for top priority functions. It may result in a more efficient and effective police agency.

Total Agency Involvement

Without adequate control over allocated funds, police agencies could run out of funds needed to carry out their programs. For this reason, in addition to fiscal controls established by municipal controllers, police agencies should develop and adopt well-designed fiscal controls to inform management on the status of the various salary, expense, and equipment accounts, and take remedial action if necessary, to bring these accounts into balance.

The Los Angeles Police Department uses two administrative tools for fiscal control, the Annual Work Program and the Expenditure Plan. The former reflects the planned production for the year, the total number of personnel required to perform the work by deployment, and relevant projected work units and statistics. The Expenditure Plan plots out the money allotted to each salary, expense, and equipment account for each of the thirteen 4-week periods of the fiscal year.

At the completion of each allotment period, actual expenditures, etc., are compared to the two model plans. This comparison reveals how closely the agency is following projected fiscal planning.

Unforeseen situations invariably arise where the need for additional police services cannot be anticipated by prior fiscal planning efforts; therefore, it is essential that interaccount transfers be available to the police chief executive and major element commanders. Examples of such situations range from additional funds needed to compensate for overtime expended during natural disasters or civil disturbances, to funds needed to purchase material not authorized in the operating budget.

To provide for such contingencies, various mechanisms of adjustment should be available within the budget system. These options should include transferring funds from a later funding period to the present period, transferring funds from an account that has a savings to one that requires additional funding, and requesting that additional funds be granted for police needs.

Establishing a highly flexible budgetary transfer process, however, increases the need for prompt and critical review of periodically prepared summaries of expenditures, balances, and interaccount transfers. These summaries should consist of reports on allotments and encumbrances, and should be reviewed promptly by the police chief executive and by the head of each major organizational element within the agency.

By exercising the proper administration of fiscal controls, the police agency fulfills its civic responsibility to provide prudent fiscal management of the taxpayers' money without neglecting the necessary level of police services expected by the community.

Program Planning Budgeting System*

The complexity of modern police operations exceeds the ability of a police administrator to make intuitively all the right decisions. A police administrator today must use his intuition in conjunction with a problem-solving methodology. The goal of having an effective and economical police organization is worthy of the best management concepts available. For this reason, the Dayton Pilot Cities Team decided to assist local criminal justice administrators in implementing the *Planning, Programming, Budgeting System* (PPBS). This management concept is based upon two important principles of good management—management by objectives and management by exception.

The system provides a police administrator with a systematic problem-solving methodology which allows him to delegate to a team of specialists the work of identifying problems and generating decision-making information. In today's society, a police administrator no longer has the time necessary to generate the information for total decision-making; therefore, he must convert to procedure what he has done before by intuition. A specialist has a particular skill, and can be given the time necessary to perform the work called for by the procedure. Under the guidance of the police administrator, a team of specialists can pool their knowledge, judgment and skills to help solve complex management problems. Such a team will make the powerful tools of mathematical analysis, data processing,

---

*Source: Gary Pence, "Program Planning Budgeting System," *The Police Chief*, Vol. 38, No. 7, July, 1971, pp. 52-57. Reproduced from the Police Chief magazine, with permission of the International Association of Chiefs of Police.

and independent estimates available to the police administrator as a means for reducing uncertainty of assumptions.

The primary advantage of this problem-solving method is that it identifies the problem and generates the information needed to select the best solutions and to justify the logic of that solution to a higher authority. Information that is based solely on intuition is seldom subject to logical scrutiny and formal justification. Unless a city commissioner, county commissioner, mayor, township trustee, city manager or county administrator can be convinced by rationally acceptable information that a chosen plan offers the most advantageous combination of effectiveness and economy, the plan will not be approved. The penalty for not using a systematic problem-solving methodology can mean delay after delay of approvals until the plan is cancelled.

A plan is a predetermined course of action to solve a problem. Problems exist within a police department when either the achievement of program objectives is less than desired, or program costs are more than can be afforded. A problem is solved by finding the optimal solution—i.e., the course of action with the greatest combined effectiveness and economy. This should be a police administrator's criterion for determining which changes to implement.

PPBS Requirements

The development of PPBS requires the police administrator to first identify and define the major goals and objectives of his organization. This is the beginning step in the development of a formal procedure which allows an administrator to couple his intuition with analytical planning.

The PPBS consists of the following components:

1. A *Program Structure* which identifies measurable achievement of program objectives.
2. A *Budget Structure* which identifies the cost of achievements of program objectives.
3. A *Program Budget Reporting System* (management infor-

mation system) to identify relevant problems and costs of achieving program objectives.
4. A *Planning System* to formulate, evaluate, and select optimal solutions for the relevant problems identified.

Methodology

The development of the program structure requires an administrator first to identify and define the major goals and objectives of his organization. This is the beginning step in the development of a formal procedure which allows an administrator to develop a design for productive change.

An administrator, in order to accomplish this task, must deal with two primary factors:

1. The services the public needs, and;
2. The resources (tax dollars) available to pay for that service. These two factors must be expressed in measurable terms. This is accomplished through definition. The services the public wants are divided into, and defined in terms of, *service areas*. The characteristics of a service area are that it is *multi-organizational* (no single city department is responsible for a service area), *multi-funded* (no single fund pays for a service area), and *multi-dimensional* (no single dimension measures a service area).

The Dayton Police Department was fortunate in developing its program structure in that the Office of Management Services had provided the city departments with an excellent framework within which to work (Chart 4-1). The following is an example of the services the Dayton citizens want, broken into service areas:

1. Security of Persons and Property.
2. Community Environment.
3. Human Resources.
4. Transportation.
5. General Government.

These service areas are then divided into categories and programs which are related to measurable objectives. It is, therefore, possible

**Chart 4-1
Dayton Police
Program Structure**

Service Area—Security of Persons and Property
  Category—Maintenance of Law
    Program—Crime Control
      Program Objectives:
        1. Decrease unreported crimes.
        2. Decrease notification time.
        3. Decrease apprehension time.
        4. Increase clearance of reported offenses.
        5. Increase successful prosecution of persons arrested.
        6. Increase recovery of stolen property.
    Program—Traffic Control
      Program Objectives:
        1. Decrease reported traffic congestion.
        2. Decrease accidents caused by traffic violations.
        3. Increase successful prosecution of persons arrested for traffic violations.
  Category—Maintenance of Order
    Program—Conflict Management
      Program Objective:
        1. Decrease the number of crimes committed as a result of tension incidents (disorder) within the community.
    Program—Family Crisis Intervention
      Program Objective:
        1. Decrease crimes against individuals by individuals within the same family unit.
  Category—Community Service
    Program—Emergency Police Service
      Program Objective:
        1. Decrease response time for calls.
    Program—Non-Emergency Police Service
      Program Objective:
        1. Decrease response time for accepted service calls.

to detail a service area in terms of its measurable objectives. The program structure provides the police administrator with a job definition. The following is the Dayton Police Department's program structure:

A police department's objectives provide direction to the activities of the various segments of a police department and serve as a means by which multiple interests are combined into a joint effort. Each part of a police organization can contribute toward department-wide objectives if it understands its relationship to those objectives and can determine, through measurement, its progress. Objectives must be realistic and stated in terms of the specific end result desired. A police administrator who has clearly defined the objectives of his department has laid the foundation necessary to perform the various functions of management required of him. The realization of objectives provides the basis for decision-making and a direction for the actions of a police agency.

There are no "golden rules" that a police administrator should follow in developing his department's program structure. However, objectives should in some measure reflect the community's needs, and decisions should not be made in a vacuum. A police administrator, in developing his program structure, should check various reference points which could affect his structure. Legal documents, such as a city charter, may have given police express responsibility for certain functions. It is also helpful to obtain the judgments of other members of the community and to review the department's past acitvities. The following is a list of possible reference points that a police administrator should check: (1) government officials, (2) pressure groups, (3) department personnel and records, (4) other criminal justice administrators, and (5) community leaders.

Priorities

The program structure is a definition of the police job. It provides the basis for productive change. Having developed a program structure, it is now possible to discuss the priorities of programs. It was asserted earlier that the administrator must deal with two primary factors, the services the public wants and the money or resources available to provide those services. Since most criminal justice administrators are confronted with limited resources, priorities must be established in the allocation of resources.

The relative value of programs may be determined in several ways. Priorities can be established through public needs surveys, which are expensive, or line management judgments, which could be biased. It is also possible to consult community organizations, elected officials and community leaders in order to obtain their judgments. In Dayton, a public needs survey will be used in conjunction with the judgement of the police administrators to establish priorities.

Budget Guidance

Regardless of the methodology used in establishing priorities, the procedure for the allocation of resources remains the same. The Dayton Police Department has now established the basis for a budget guidance procedure through its program structure, based on the principle, "Dollars follow need." This is represented as equal effectiveness par last dollar spent. This would be represented mathematically in the following manner:

$$\frac{\text{Effective Whole}}{\text{Budget Whole}} = \frac{\text{Effective Part}}{\text{Budget Part}}$$

Example:
  Ideal effectiveness of service area = 100%
    (Security of Persons and Property)

  Ideal effectiveness of the particular program = 30%
    (Conflict Management, for example, may be regarded as a program that is 30% of the task involved in the entire service area.)

  Anticipated budget of service area = $1,000

$$\frac{100}{1000} = \frac{30}{Bp}$$

  Bp = $300

  That is, $300 should be assigned to the particular program under consideration.

In other words, a police administrator should employ his resources where they provide him with the greatest overall gain in achieving stated objectives.

The budget guidance procedure provides the administrator with information necessary to evaluate the allocation of resources based upon his own judgment. This information may provide the foundation for the reorganization of police resources through a plan that is more consistent with stated objectives.

Budget Structure

Once the program is developed, it is necessary to develop the budget structure. The purpose of the budget structure is to identify and relate the cost of achievement with the program objectives. The budget structure is dealt with in terms of budget activities. An "Activity" is a manageable work package which requires accountability of resources (dollars) expended. Each activity results in the creation of a "cost center" (ledger) which requires an administrator to account for his expenditures and credits. There are only two reasons for an activity: (1) a division within a department does work for one or more programs, and (2) program work is performed by two or more departments. If an administrator confuses budget activities with the collection of management information, he can create an accounting problem. In other words, it would be logical for an administrator to want to know the amount of time his men spend on counseling, but impractical to budget for this as a separate activity under each program area. Therefore, the budget structure, which relates costs to achievements of program objectives, would consist of the following activities under each program:

1. Office of the Director.
2. Field Service Division.
3. Evaluation and Coordination Division.
4. Staff Services Division.

The Program Budget Structure allows an administrator to review his budget in terms of consumption factors which should equal physical output. The police administrator in Dayton is no longer budgeting for paper clips, but is planning and budgeting for programs that will accomplish stated objectives. A line item budget provides an administrator with information pertaining to how many paper clips he bought. It does not tell him what they were needed for nor how great the need was. The PPBS relates resources to programs which have clearly defined objectives and established priorities. This provides the police administrator with the capabiltiy to project the cost of his programs based on established relationships and is one of the major advantages that the PPBS has over the traditional line item budget.

Police Program Budget Report

The Program Budget Report will be the basis of the Pilot City Team's police development program. It will identify problems which have been defined, such as when a program's cost is more than can be afforded or achievements are less than desired.

In order to develop the Program Budget Report, it is necessary to define each objective and specify what data is needed to measure the Department's achievement in relation to that objective. The Pilot City Team is presently working in conjunction with the Police Department in developing the information system, which will provide that data.

Purpose of Program Budget Report

The purpose of the Program Budget Report is to provide the police administrator with information necessary to identify problems. Today's police administrator is confronted with a veritable wealth of information. However, it is not all relevant to management problems. The police administrator cannot afford the luxury of reviewing all the information available to him from the use of modern data processing techniques. Anyone who has ever attempted to interpret the reams of quantitative data which are presently available to most municipal police administrators will recognize the sense of futility that develops. A police administrator can only achieve optimum control through identifying the critical or limiting points within his operation and directing his attention or resources to these points of adjustment, which are identified through variation from the norm. The ability of a police chief to interpret the meaning of measurements of past performance is necessary if the program budget report is to be properly utilized. A police chief must be able to determine what variation is significant. If he cannot, he will make corrections when they are not needed or fail to make corrections when they are needed.

The Program Budget Report provides the police administrator with the important relationships without the minutiae that normally accompany them. This allows a police administrator to employ the concept of management by exception.

Criteria for Evaluation

It is impossible to measure the worth of a policeman's smile, unless it is reflected through the achievement of specific and measurable objectives. A measurement must be relative to something; therefore, the only thing that can be measured is an increase or decrease. An administrator's effectiveness is reflected in whether or not he is improving in terms of an increasing or decreasing measurement. This can be explained in terms of a police objective such as decreased apprehension time. A police administrator need only know what his apprehension time was for the preceding time period in order to be able to ascertain whether is he decreasing or increasing relative to that time period. (The Dayton Police Program Budget Report will be produced in three month periods, starting in January, 1971).

Police Planning

Whan a problem exists, then the planning process is involked. This involves a systematic analysis of the facts and existing conditions in order to determine what is relevant to the problem. A predetermined course of action is then developed which offers the optimal combination of effectiveness and economy in conjunction with the greatest potential for success.

Determining what is relevant is the primary aspect of good planning and requires the judgment of the police administrator. The police administrator relies upon his planning team or a technical assistance team to generate decision-making information, thereby providing him with a basis for decision-making. A decision involves a choice. If a police administrator is given no alternatives, no decision is possible. A good decision is dependent upon the following factors:
1. Recognition of the right problem.
2. Isolation of the problem to permit the collection of pertinent data.
3. Analysis of available alternatives.
4. Predicted consequences of each alternative.
5. Selection of the best solution.
6. Acceptance of the decision by the organization.

The Pilot City Team in Dayton will provide the criminal justice planning models which will enable the police administrator to select the best combination of policies, operations and resource allocation. The planning models will assist the police administrator in accomplishing the following:

1. Achieving maximum effectiveness within available budget.
2. Determining what resources are needed to maximize effectiveness based upon the present workload. (Time available "manpower"-time required "workload.")
3. Determining what will be needed to maximize effectiveness based upon the projected workload.

This provides the police administrator with the answer to three questions for which he is responsible:

1. What can be done with existing resources?
2. What resources do I presently need?
3. What resources will I need in the future?

The planning models will assist the police administrator in developing the in-house capability to plan for programs that are the best solution to real problems and then budget for those programs.

"A planning model is an abstract representation of an object, process, or system by use of symbols, functions, and relations. It describes the relationship between variable inputs and calculated outputs.

This analytical planning process develops and uses mathematical criteria, systems and tradeoff models.

These combined models *predict* the expected worth of each competing alternative in order to eliminate the undesirable ones. The choice candidates then undergo a trial demonstration to measure their actual worth. The use of models for eliminating some of the worthless changes is less costly and disruptive than test demonstrations of all feasible changes.

The planning models used to predict values, achievements and costs of alternatives are described as follows:

1. *Community System and Cost Model*—a descriptive model which estimates the achievements and costs of crime

prevention objectives resulting from a specified change in the social, economic and psychological characteristics of the community. These characteristics are represented in the model by demographic statistics and public attitudes.
2. *Criminal Justice System and Cost Model*—a combined description and network model which estimates the achievements and costs of criminal justice program objectives resulting from a specified change in the policies operations, and resources of the criminal justice system.
3. *Community Criteria Model*—a descriptive model which estimates the effectiveness, economy and worth resulting from a specified change in achievements and costs of crime prevention program objectives.
4. *Criminal Justice Criteria Model*—a descriptive model which estimates the effectiveness, economy and worth resulting from a specified change in achievement and costs of criminal justice program objectives.
5. *Tradeoff Model*—an optimization model which optimizes the system characteristics for each alternative based upon the criteria of maximum effectiveness, economy, or worth.

". . . While the base models are being used by sytems specialists, the analytical specialist makes successive model refinements which provide more-and-more-accurate predictions."

Problems Related to the Implementation of PPBS

The first problem that will be encountered by the police administrator who decides to implement the PPBS will be the development of a program structure. It is difficult to deal with a police organization in terms of measurable objectives. The Dayton Police Department's program structure is the result of a time consuming ordeal that involved an extensive period of revision. The police department initially expressed its program structure in terms of how to accomplish the police function rather than determining the objectives of the police function. The most notable example of this problem was with the objective the police have in regard to reducing the apprehension time of criminal offenders. This objective was first stated as reduced response time. The police then proceeded to ask themselves why they wanted to reduce the time required to respond to a crime scene. This thought process revealed that their real objective was to decrease apprehension time. If reducing response time did not accomplish their objective of decreasing apprehension time, they would not be concerned with reducing response time.

The development of a meaningful program structure requires the police administrator to ask continually "Why?" until he has defined the police function in terms of measurable objectives.

The second problem confronting a police administrator in implementing this sytem will be defining what is meant by the objective and what data will have to be collected to measure that objective. This requires a police administrator to review his present data collection system and adapt it to his program structure. The data collection system will then go through a period of revision. Problems will be confronted during this developmental stage, such as "What *is* apprehension time?" The Dayton Police Department decided apprehension time was the period of time which elapses after the police are notified that a crime has occurred until the perpetrator of that crime is apprehended. However, it could have been defined as the period of time which elapses from occurrence to apprehension. It is up to the police administrator to make this decision, although he should avoid assuming responsibility for objectives over which he does not have control.

The third problem of the police administrator will be the selection, training and control of a planning team capable of using the tools of modern management. This team is essential if the police administrator is going to have the information necessary to make decisions.

The fourth problem confronting the police administrator is limiting his own expectations. The PPBS is not a substitute for an administrator's judgment, but a supplement. The system cannot be implemented overnight. It requires hard work and time to be perfected. The Dayton system is about 75 percent complete; the last 25 percent will result from its practical application and the resulting revisions.

Summation

Lt. Col. O'Connor, of the Dayton Police Department, summed up the advantages of PPBS when he said that after thirty years of being a policeman he now knows what is expected of him. Police administrators know how to do their jobs, but have not had the time to relate "what is done" to "what is needed." This has permitted each citizen who picked up a telephone and requested police service to define the police job for the collective community at that moment. Since the police administrator has limited resources, it is possible that the collective community would prefer that the police remove a seriously injured citizen to a hospital as opposed to searching for a lost house key. This question is dependent upon public needs and resources (tax dollars) available to the police operation.

Topics for Discussion

1. Discuss the four categories of the criticisms of classic bureaucracies.
2. Discuss three of the most significant organizational problems created by bureaucracies.
3. Discuss the advantages of the democratic police model.
4. Discuss the spheres of influence that control and affect police planning.
5. Discuss the importance of budget justification.
6. Discuss the differences between PPBS and other budgeting methods.
7. Discuss the techniques of relating programs to objectives.

FOOTNOTES

Introduction

ANGELL, J.E. (1967) "The adequacy of the internal processing of complaints by police department." M.A. thesis. School of Police Administration and Public Safety, Michigan State University.

ARGYRIS, C. (1957) "The individual and organization: some problems of mutual adjustment." *Administrative Sci. Q.* (June): 1-24.

BARD, M. (1969) "Alternatives to traditional law enforcement." Presented before the 77th annual convention of the American Psychological Association. Washington, D.C.

BARNARD, C. (1968) *The Functions of the Executive,* Cambridge, Mass.: Harvard Univ. Press.

BENDIX, R. (1962) Max Weber. *An Intellectual Portrait.* Garden City, N.Y.: Doubleday.

BENNIS, W. (1966) Changing Organizations. New York: McGraw-Hill. — and P. SLATER (1964) "Democracy is inevitable." *Harvard Business Rev.* (March-April): 51-59.

BROWN, J.A.C. (1954) *The Social Psychology of Industry.* Baltimore: Penguin.

CAUDILL, H.M. (1963) *Night Comes to the Cumberlands.* Boston: Little, Brown.

CLOWARD, R. (1959) "Social control and anomie: a study of a prison community." Ph.D. dissertation. Columbia University.

COCH, L. and J.FRENCH (1948) "Overcoming resistance to change." *Human Relations* 1, 4: 512-532.

GERTH, H.H. and C.W. Mills [eds.] (1958) from Max Weber: *Essays in Sociology.* New York: Oxford Univ. Press.

IGLEBURGER, R. and R. WASSERMAN (1970) "The incorporation of community-base field training into a police recruit training curriculum." Presented at the Third National Symposium on Law Enforcement Science and Technology. Chicago.

KATZ, D. and R. KAHN (1960) "Leadership practices in relation to productivity and morale," in D. Cartwright and A. Zander (eds.) *Group Dynamics, Research and Theory.* New York: Harper & Row.

KIMBLE, J. (1969) "Daydreams, dogma and dinosaurs." *Police Chief* 36, 4: 12-15.

MASLOW, A.H. (1965) *Eupsychian Management.* Homewood, Ill.: Dorsey.

MYREN, R. (1960) "A crisis in police management." *J. of Criminal Law, Criminology and Police Sci.* 50, 6: 500-605.

PETER, L. and R. HULL (1969) *The Peter Principle.* New York: Morrow.
President's Commission on Law Enforcement and Administration of Justice (1967) Task Force Report: The Police, Washington, D.C.: Government Printing Office.
TULLOCK, G. (1965) *The Politics of Bureaucracy.* Washington: Public Affairs Press.
WEBER, M. (1958) "Bureaucracy," in H.H. Gerth and C.W. Mills (eds.) From Max Weber: *Essays in Sociology.* New York: Oxford Univ. Press. —(1947 "The essentials of bureaucratic organization: an ideal-type construction," in H. Henderson and T. Parsons (eds.) *The Theory of Social and Economic Organization.* New York: Oxford Univ. Press.
WILSON, O. (1963) *Police Administration.* New York: McGraw-Hill.

Organizing: A Democratic Model

[1] This approach to leadership is not new. According to Maslow (1965: 123), "The Blackfoot Indians tended not to have general leaders with general power... but rather different leaders for different functions. For instance, the leader in a war party was the one whom everyone thought to be the best person to lead a war party, and the one most respected or the leader in raising stock was the man best suited for that. So one person might be elected leader in one group and be very last in the second group." Caudill (1963: 39) points out that Confederate soldiers often elected their officers during the Civil War. In addition, it is common practice in small hospitals for surgical teams composed of the same members to shift the leadership role for different types of operations. Finally, there is considerable research which suggests many advantages to functional leadership (Katz and Kahn, 1960: 554-570; Coch and French, 1948: 512-532; and Cloward, 1959).

[2] Teams of five positions have been chosen because a total of approximately 25 officers would be required to staff five positions on a 24-hour-a-day basis when days off and holidays are considered. Since informal communications among officers is essential, the team must be kept as small as possible to facilitate face-to-face communications and solidify the team.

[3] The research indicates that placing such responsibility on employees results in higher worker satisfaction (Katz and Kahn, 1960).

Planning

[1] Germann, Day, Gallati, *Introduction to Law Enforcement and Criminal Justice* (Springfiled, Illinois: Charles C. Thomas, Revised Sixth Printing, 1968), pp. 28-29.

² O.W. Wilson, *Police Administration, op. cit.,* pp. 89-90; O.W. Wilson, *Police Planning, op. cit.,* pp. 14-17; Bristow and Gabard, *Decision Making in Police Administration* (Springfield, Illinois: Charles C. Thomas, 1961), pp. 14-79; Bristow and Gourley, *op. cit.,* p. 70: John P. Kenney, *Police Management Planning* (Springfield, Illinois: Charles C. Thomas, 1959), pp. 21-23.
³ For a discussion of these post-developmental processes, see Kenney, *op. cit.,* pp. 22-23.
⁴ Walter Gellhorn, "Police Review Boards: Hoax or Hope", *Columbia University Forum,* Summer (1966).
⁵ Cf. LeBreton and Henning, *Planning Theory* (Englewood Cliffs, N.J.: Prentice-Hall, Inc., 1961), pp. 117-118.
⁶ Kenney, *op. cit.,* p. 21.
⁷ *Journal of Criminal Law; Criminology and Police Science; Police; The Police Chief; Law and Order; The FBI Law Enforcement Bulletin, et. al.* For methods of practical use of journals, see Bristow and Gabard, *op. cit.,* pp. 32-33.
⁸ For a discussion of methods of library research see Bristow and Gabard, *op. cit.,* pp. 37-39.
⁹ V.A. Leonard, *Police Organization and Management* (Brooklyn: The Foundation Press, 1964), pp. 135-180.
¹⁰ LeBreton and Henning, *op. cit.,* pp. 120-121.
¹¹ *Ibid.*
¹² In the selection of the most adequate alternative Bristow and Gabard, *op. cit.,* pp. 69-74 recommend a cancelling technique where "pros" and "cons" are assigned weighed values. LeBreton and Henning, *op. cit.,* pp. 103-116, recommend a system of assigning probability values and making numerical comparisons.

SELECTED READINGS

Burgoyne, G.J., "PERT: A New Tool for Police Science," *California Law Enforcement Journal,* Vol. 4, No. 3, January, 1970, pp. 120-123. Briefly describes the development of PERT and its application in the Los Angeles Police Department. This technique was utilized in the budgeting process, construction of a command post, the scheduling of radio installations and a major criminal investigation.

Diamond, Harry, "Quality Control in Police Work," *The Police Chief,* Vol. 35, No. 2, February, 1968, pp. 40-44. Discusses the need for quality control of police operations through the inspection concept and the report review concept. Differentiates between line and staff inspections, and reviews trends in inspection.

Dolan, John F., "The PPB Concept," *The Police Chief,* Vol. 35, No. 7, July, 1968, pp. 28-31. Identifies the basic differences between line item budgeting, performance budgeting and PPB. Presents the distinctive characteristics of Planning, Programming and Budgeting Systems.

Hennessy, James J., "PPBS and Police Management," *The Police Chief,* Vol. 39, No. 7, July, 1972, pp. 62-67. Emphasizes that PPBS is a method of management which incorporates the management by objectives process. Discusses goals, objectives and cost/effectiveness.

Hoover, Larry T., "Planning-Programming-Budgeting Systems: Problems of Implementation for Police Management," *Journal of Police Science and Administration,* Vol. 2, No. 1, January, 1974, pp. 82-93. Argues that it is not possible to implement a genuine PPB system in law enforcement agencies because needed data to analyze the cost effectiveness of alternative police programs is lacking.

Igleburger, Robert M., John E. Angell, and Gary Pence, "Changing Urban Police: Practitioners' View," *Criminal Justice Monogram, Innovation in Law Enforcement,* Washington: U.S. Government Printing Office, June, 1973, pp. 76-114. Presents a case for police administrators to adopt a philosophy supporting a consumer orientation and insuring that their organizations have sufficient exposure and flexibility to align themselves with the needs of their clientele.

Key, Charles R. and Miles R. Warren, "Kansas City: Long-Range Planning Program," *The Police Chief,* Vol. 39, No. 5, May, 1972, pp. 72-75. Describes a police long-range planning study including three principle elements: (1) A Long-Range Planning and Grants Section, (2) A Computer-Operated Police Planning System, and (3) A Research and Development Task Force Concept.

Martin, John A., "Staffing and Manpower Analysis in Support of Planning, Programming, and Budgeting," *Police,* Vol. 14. No. 5. September-October, 1969, pp. 70-73. Discusses the technique of staffing and manpower analysis as a means of developing a staffing pattern that reflects the type and level of staffing needed to accomplish program objectives.

● **The study of this chapter will enable you to:**

1. Define organization development (OD).
2. Contrast organization development and the human relations movement.
3. List the six values which people must regard highly, before an OD effort can succeed in an organization.
4. Identify the most common obstacles to organization development.
5. List the four successive stages of an organization development effort.
6. Identify the distinctive features of task force management.
7. List the purposes of a team building workshop.
8. Identify the factors which are most critical as criteria for assessing the condition of "readiness" for team building.

5

Organizational Development

Introduction*

Although a literal interpretation of the words *organization development* could refer to a wide range of strategies for organization improvement, the phrase has come to take on some fairly specific meanings in the behavioral science literature and in practice. We say "fairly specific" because the boundaries are not entirely clear, perceptions of different authors and practitioners vary somewhat, and the field is evolving.

In the behavioral science, and perhaps ideal, sense of the term, *organization development is a long-range effort to improve an organization's problem-solving and renewal processes, particularly through a more effective and collaborative management of organization culture—with special emphasis on the culture of formal work teams—with the assistance of a change agent, or catalyst, and the use of the theory and technology of applied behavioral science, including action research.*

By *problem-solving processes* we mean the way an organization goes about diagnosing and making decisions about the opportunities

---

*Source: Wendall L. French and Cecil H. Bell, Jr., *Organization Development*, Englewood Cliffs, New Jersey: Prentice-Hall, Inc., 1973, pp. 15-20. Reprinted with permission of the publisher.

and challenges of its environment. For example, does it see its environment, and thus its mission, in terms of ten years ago, or is it continuously redefining its purposes and its methods in terms of the present and the future? Does the organization solve problems in such a way that it taps the creativity and commitment of a select few, or does it tap deeply into the resources, vitality, and common purposes of all organizational members?

The notion of improving problem-solving processes is interrelated with the matter of improving organizational "renewal processes," which is perhaps a broader concept. Lippitt combines these ideas in his definition of *organizational renewal* which he sees as

> the process of initiating, creating, and confronting needed changes so as to make it possible for organizations to become or remain viable, to adapt to new conditions, to solve problems, to learn from experiences. . . .[1]

Argyris stresses organizational renewal and revitalizing in his description of organization development:

> At the heart of organizational development is the concern for the vitalizing, energizing, actualizing, activating, and renewing of organizations through technical and human resources.[2]

Similarly, Gardner, in writing about organizational *self-renewal*, refers to the avoidance of organizational decay and senility; the regaining of vitality, creativity, and innovation; the furtherance of flexibility and adaptability; the establishment of conditions that encourage individual motivation, development and fulfillment; and "the process of bringing results of change into line with purposes."[3] Thus, along with ideas about improved problem-solving and renewal processes are the important notions of purpose and direction—all of which are central to organization development activities.

By the term *culture* in our definition we mean prevailing patterns of activities, interactions, norms, sentiments (including feelings),[4] beliefs, attitudes, values, and products.[5] By including products we include technology in our definition, although changes in technology tend to be secondary in organization development efforts. However, technology—if one includes procedures and methods along with equipment—is almost always influenced, and is an influence, in organization development activities.

Our use of the term *culture* includes the notion of the "informal system"—feelings, informal actions and interactions, group norms, and values. In some ways the informal system is the hidden or suppressed domain of organizational life—the covert part of the "organizational iceberg," as shown in Figure 5-1.[6] Traditionally, this hidden domain either is not examined at all or is only partially examined. Organization development efforts focus on both the formal and the informal systems, but once the OD program is legitimated through the formal system, the initial intervention strategy is usually through the informal system in the sense that attitudes and feelings are usually the first data to be confronted.[7]

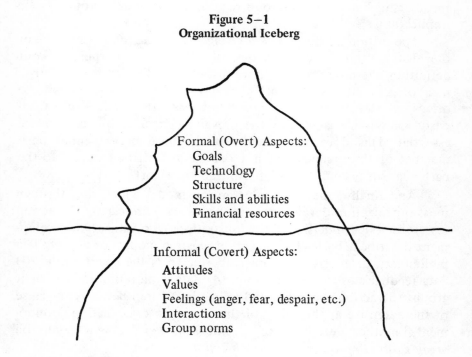

Figure 5–1
Organizational Iceberg

Formal (Overt) Aspects:
Goals
Technology
Structure
Skills and abilities
Financial resources

Informal (Covert) Aspects:
Attitudes
Values
Feelings (anger, fear, despair, etc.)
Interactions
Group norms

By *collaborative management* of the culture we mean a shared kind of management—not a hierarchically imposed kind. Who does what to whom is an important issue in organization development, and we want to stress that management of group culture must be "owned" as much by the subordinates as it is by the formal leader.

Our definition recognizes that the key unit in organization development activities is the ongoing work team, including both superior and subordinates. As we will elaborate upon in later chapters, this is different from more conventional ways of improving organizations. To give only one example, in most management development activities the focus is on the individual manager or supervisor—not on his work group. Traditionally he has participated in the learning experience in isolation from the dynamics of his work situation. Although we are emphasizing a focus on relatively permanent work groups to differentiate OD from traditional management development, in comprehensive OD programs extensive attention is also paid to temporary work teams, to overlapping team memberships, and to intergroup relations, as well as to total system implications.

The notion of the use of a *change agent,* or *catalyst,* as one of the distinguishing characteristics of OD has a purpose in our definition. We are somewhat pessimistic about the optimal effectiveness of OD efforts that are do-it-yourself programs. As will be discussed later, in the early phases, at least, the services of a third party who is not a part of the prevailing organization culture are essential. This does not mean that the third party cannot be a member of the organization but that he at least be external to the particular subsystem that is initiating an OD effort.

And finally, the basic intervention model which runs through most organization development efforts is *action research.* The action research model underlies all the illustrations of organization development described. Basically, the action research model consists of (1) a preliminary diagnosis, (2) data gathering from the client group, (3) data feedback to the client group, (4) data exploration by the client group, (5) action planning, and (6) action. Parenthetically, because of the extensive applicability of this model to organization development, another definition of *organization development* could be *organization improvement through action research.*

The above characteristics of organization development depart substantially from the features of traditional change programs, which Bennis categorizes as follows: "(1) exposition and propagation, (2) elite corps, (3) psychoanalytic insight, (4) staff, (5) scholarly consultations, and (6) circulation of ideas to the elite."[8] Bennis states that "exposition and propagation" are "possibly the most popular" and cites as illustrations the impact of the ideas of philosophers and scientists. The "elite corps" method is basically the infusion of scientists into key power and decision-making posts in organizations. "Psychoanalytic insight" as a change method is similar to the elite corps method but refers to effective change occurring through the medium of executives who have high self-insight and considerable "psychiatric wisdom" relative to subordinates. The "staff" strategy refers to the employment in organizations of social scientists who analyze situations and make policy recommendations. "Scholarly consultations" is a method of change involving "exploratory inquiry, scholarly understanding, scholarly confrontation, discovery of solutions, and, finally, scientific advice to the client." The sixth method described by Bennis is "circulation of ideas to the elite." One of the illustrations given is the Council of Correspondence, a chain letter which linked rebel leaders in the American Revolution.[9]

Organization development efforts depart substantially from these methods of organizational change. Of particular relevance are the two organizational consultation methods as categorized by Bennis, "staff" and "scholarly consultations." In both strategies an inside or an external expert studies a situation and makes recommendations; this is the traditional way of consulting. Organization development efforts are different. The OD consultant does *not* make recommendations in the traditional sense; he intervenes in the ongoing processes of the organization. For example, his end product is not a written report to top management, nor does he usually supply direct solutions to problems. He does, however, assist the client organization in the way it goes about solving problems. For example, he may be called upon to comment on the way a group is working together, or he may structure situations so as to highlight phenomena. As an illustration, if asked to assist in reducing conflict he may request two team members to role play each other and each other's point of view. But such interventions are in collaboration with the client group and are based on earned trust. Basically, he assists the client group in generating valid data and learning from them.

We see seven characteristics that we think differentiate organization development interventions from more traditional interventions:

1. An emphasis, although not exclusively so, on group and organizational processes in contrast to substantive content
2. An emphasis on the work team as the key unit for learning more effective modes of organizational behavior
3. An emphasis on the collaborative management of work team culture
4. An emphasis on the management of the culture of the total system and total system ramifications
5. The use of the *action research* model
6. The use of a behavioral scientist *change agent,* or catalyst
7. A view of the change effort as an ongoing process.

Another characteristic, number 8, a primary emphasis on human and social relationships, does not necessarily differentiate OD from other change efforts, but it is nevertheless an important feature.

Parameters of OD*

More people in business and government are searching for more meaningful, more effective, ways to get work done. As a result of their search, they are becoming optimistic about initiating creative and purposeful change in work organizations. They label their activities and concepts *Organization Development.* The search for new answers, however, also leads to more questions. Listed are many of the questions commonly asked about Organization Development. The answers which follow are tentative simply because the field is still emerging and the conclusions are not final.

Briefly, What is Organization Development?

Organization Development is a reorientation of man's thinking and behavior toward his work organizations. It applies the scientific method and its underlying values of open investigation and experi-

---

*Source: Lyman K. Randall, "Common Questions and Tentative Answers Regarding Organization Development," *California Management Review,* Vol. 13, No. 3, Spring, 1971, pp. 45-52. Reprinted with permission of the publisher.

mentation to individual and work group behaviors as they are directed toward the solution of work problems. It views both man and change optimistically. It applies a humanistic value system to work behaviors. It assumes people have the capability and motivation to grow through learning how to improve their own work climate, work processes and their resulting products. It accepts as inevitable the conflicts among the needs of individuals, work groups, and the organizations, but advocates openly confronting these conflicts using problem-solving strategies. Its goal is to maximize the use of organization resources in solving work problems through the optimal use of human potential.

Is Organization Development Simply the Human Relations Movement in a New Format?

Although related to earlier human relations work, OD differs from it in several ways. Many managers interpreted the message of human relations to be: "If morale is high, productivity will increase; and morale can be increased by "being nice to people.' " For these managers "being nice to people" eventually meant emphasizing what Frederick Herzberg later categorized as work hygiene factors: employee benefits, working conditions, facilities, administrative policies, and social relationships.[1]

Organization Development, on the other hand, concentrates on the accomplishment of work and the solving of work problems by people. The improvement of relationships between people and work groups is not an end in itself, as it often turned out to be in human relations. Instead, interpersonal and intergroup behaviors are the focus as they are relevant to the successful problem-solving efforts of the work unit.

Is Organizational Development Primarily Concerned With Restructuring Organizations?

No. Many people assume that the term, *organization development,* is closely related to the organization planning process with its emphasis on organization charts, and the like. Although the restructuring of an organizational unit could be one of the results of an OD effort, this activity might be only one consequence of OD.

On What Concepts is Organization Development Based?

The concepts fundamental to OD can be placed in two general categories: those that apply more often to work groups and large organizations and those that apply to the individual. Naturally, these concepts are interrelated.

OD Concepts Regarding Work Groups and Organizations

- *Systems.*—Organizations are laced together by systems and subsystems such as budgeting, purchasing, inventory and stores, and so on. All subunits or parts of a system are interrelated to the whole. When one part of a system is changed, the total system is affected. In OD, thinking in systems terms is necessary when planning the tactics and strategies of change.

- *Problem-Solving Interdependence.*—This concept is closely related to *systems.* When individuals and groups are working toward the solution of problems which affect other persons and groups, the problem is said to be interdependent rather than independent. In this situation people and groups have a common stake in the outcome and therefore need to have a voice in the solution. Treating interdependent problems as if they are independent usually leads to resistance to change and win/lose conflict between those who had a chance to have their say and those who did not.

- *Work Climate.*—Research indicates that work climate affects the kinds of results that individuals, groups, and total organizations produce; as atmospheric climate affects the quality of crops that a farm produces. Work climate is comprised of the values, attitudes, and underlying assumptions which determine how work gets done. Work climate is closely related to *OD Values* and *Theory Y* discussed later.[2]

- *Force Field.*—This theory states that any given behavior of a work group is held at a given level by two opposing sets of forces: driving forces which push the level of performance up to a certain point; restraining forces which prevent the behavior from rising beyond this same point. Using this concept we can see there are two basic options open to us if we wish to increase any behavior or set of behaviors such as, for example, work group productivity. We can add more driving forces to increase productivity or we can identify and remove some of the restraining forces which keep productivity from going any higher. The first option is more often used, but the second option often holds more promise for results since it removes a force that most individuals see as negative. Although OD focuses on both driving and restraining forces as determinants of job behavior, it often emphasizes the latter since restraining forces are more often overlooked in the analysis of work problems.

■ *Process and Tasks.*—Work groups and individual jobs exist to accomplish tasks needed for the organization to survive, remain healthy, and grow. Specific tasks are determined by goals of the group or corporation. Process, on the other hand, is what happens between individuals and groups as they work on their tasks. In the Broadway musical, "Zorba," *life* is described as "what happens to you as you are waiting to die." Similarly, process is what happens among people as they are working on tasks. Typically, managers focus on tasks much more frequently than they do on process although process problems usually cause below-standard task achievement. Perhaps the most difficult aspect of OD for many managers is learning to deal with process issues. Several specific factors distinguish *tasks* from *process*.

Task (the job to do)		Process (the "happenings" of getting tasks done)
Usually concrete and based on objectives	vs.	More nebulous and difficult to identify
More static and often repetitive	vs.	Very dynamic and fluid
Outside of ourselves	vs.	Involves "tuning-in" to what's going on inside ourselves
Much "There and then" time orientation	vs.	"Here and Now" time orientation
Usually dealt with intellectually	vs.	Involves much feeling and subjective perception

■ *Data Collection and Diagnosis.*— For years businessmen have been collecting data about such factors as markets, new product performance, and capital expenditures in order to diagnose problems that need to be resolved. These same men have only recently begun to learn to collect data about the human interaction process in their organizations. Problems involving individual and group perceptions, feelings, assumptions, and attitudes often provide the key for improving the accomplishment of work group tasks. Usually data are collected either by using anonymous questionnaires or by an outside resource person who interviews selected individuals. A summary of the data is then fed back to the group for discussion and problem-solving.

OD Concepts Regarding Individuals

■ *OD Values.* — These are values which people must regard highly before an OD effort can succeed in an organization. They are important ingredients in the work climate of an organization involved in OD. Many of these values have a concomitant skill or behavior which can be learned and practiced by individuals. These values and related skills include the following:

Trust and openness: This is perhaps the cornerstone of all OD work. Trust in interpersonal and intergroup relationships is essential if full and open communication is to occur. An open and nonmanipulative sharing of data is required for the effective solving of work problems. In most organizations or work groups, trust does not exist automatically. Traditional work orientations based on the manipulation of people have generated widespread distrust at all levels. This widespread distrust of supervisors, other work groups, and other employees is one of the initial problems often encountered in an OD effort. Building trust and openness throughout an organization is therefore one of the continuing goals of OD.

Leveling: "Tell it like it is!" OD work is usually called leveling. An individual with the skill and courage to share candidly with others meaningful information about how he thinks, reacts, and feels about work issues and co-workers is unfortunately the exception in today's organization. However, leveling skills are essential if OD is to succeed.

Feedback: People tend to make many assumptions about other people—how they feel, what they think, why they behave as they do—without ever checking out the accuracy of what they have assumed. Feedback is simply a communications skill for verifying or correcting these assumptions and thereby providing more accurate data on others as well as on ourselves. To be useful, however, feedback must be shared in a helpful, nonaccusatory manner. Feedback that begins, "The trouble with you is . . ." is generally destructive rather than helpful. Feedback can be given and received successfully only when relationships are based on trust and mutual respect. The result is a deeper and more accurate comprehension of what is going on in the process of accomplishing work.

Confronting conflict: OD values hold that conflict is a natural occurrence between people and work groups. Conflict issues should therefore be dealt with openly and problem-solved. Unfortunately traditional work orientations, perhaps influenced by the military, hold that conflict is negative and should be avoided, denied, or smoothed-over whenever possible. The result of this approach is the perpetuation of unresolved problems. Conflict confrontation skills include such factors as trust, leveling, and the ability to give and receive feedback.

Risk-taking: This refers to the ability of individuals to "stick their necks out" in meaningful ways. Examples include taking an unpopular stand in an important issue, conflicting with a superior on preferred solutions to a problem, trying to initiate action on a problem seemingly avoided by others, and sharing with co-workers personal feelings such as commitment, concern, anxiety, or caring.

Owning of personal experience: In today's organizations it is very easy to point the finger of blame or responsibility at others. In fact many, many people have spent years learning how to defend themselves from real or imagined accusors by denying responsibility, concern, and feeling regarding their work experiences. Personal experience also underlies an individual's ability to level about himself, give feedback to others, build interpersonal trust, take risks, and to confront conflict—all of which are essential to OD activities.

■ *Self-Actualization.*— This term is used by Maslow to describe man's highest order of motivational need. It includes his personal needs for learning, growth, achievement, competence, recognition, and the striving toward his fullest potential. These are the elements Herzberg has recently formulated as the Motivating Factors of people at work. Self-actualization is premised on an optimistic view of man. A self-actualizing man is an individual behaving at his most creative and productive level. An individual is most likely to behave in self-actualizing ways if the climate in which he works is characterized by openness, trust, confrontation, and so on, and he has challenging goals.

■ *Theory Y.*—In reviewing the findings of behavioral science research in industry several years ago, Douglas McGregor formulated a set of assumptions about man at work that seemed to be supported by the research. He called this formulation *Theory Y*.[3] It is built on the following premises:

- Work is as natural to man as play and rest.
- Man will use self-direction and control when he is committed to objectives.
- Man learns, under proper conditions, to accept and even to seek responsibility.
- Creative ability is widely dispersed among individuals.
- In industry, man's potential is only partially utilized.

McGregor then contrasted Theory Y with another set of assumptions about man at work. He labeled these *Theory X*, and observed that most organizations behaved as if they believed these assumptions rather than *Theory Y. Theory X* assumptions include:

- The average man dislikes and will avoid work.
- He must be forced, controlled, and directed to work.
- He prefers to be directed in his work since he has little ambition.
- He seeks only security.

Obviously, OD is based on *Theory Y* assumptions. Many of the work climate values and organizational practices which are change-targets for OD stem from *Theory X* assumptions.

- *T-Groups.* — T-Groups are specially designed learning experiences which focus on most of the OD concepts discussed above, i.e. trust, openness, process, leveling, feedback, risk-taking, owning experience, and so on. For many individuals, T-Groups provide unique opportunities to assess themselves regarding where they stand in relation to these values and skills. For this reason, some version of a T-Group is often used as a first step in launching an OD effort in an organization.[4]

How Do Many Organization Development Efforts Get Initiated?

Many OD activities begin when a manager becomes uneasy about the effectiveness of his own work group. Often he is not aware of the causes of ineffectiveness although he can usually point to several symptoms. These might include frequently missed deadlines, conflicts between employees that are never dealt with directly, low-level innovation in problem-solving, distrust between individuals or groups, the disowning of personal responsibility, et cetera. In other words, OD begins with a manager looking for more effective ways of accomplishing the goals of his group and the organization. He is usually a man who sees himself as reasonably competent but with more to learn.

Is it True That Organization Development Must Always Begin at the Top of the Organization?

A recent survey by the National Industrial Conference Board found that approximately half of the OD efforts studied had been initiated by corporate presidents or board chairmen. In the other organizations the early OD experimenting and risk-taking was done by managers further down the hierarchy.

"OD must begin at the top!," is sometimes used by a person who wants to deny any personal need or responsibility for change. In this case he is actually saying, "My work group is in fine shape, but it's those S.O.B.'s who always foul-up things!" Pointing to the top of the organization gets him off the hook nicely since there is little he can do personally to make top management aware of *their* need to change.

Eventaully OD must include top management. Organizational climate is formed at the top. Similarly the rewards for and restraints against risk-taking, leveling, and confronting conflict issues are established here. The overall quality of teamwork characteristic of the organization is heavily influenced by the behaviors of the top team.

Where Can an Individual Find an Organization Development to Fit His Needs?

Many managers are accustomed to thinking about their activities in terms of concrete programs that begin at a specific time and end by a preset date. It is therefore natural for them to expect OD to fit this same pattern. Unfortunately, it does not. By definition OD must begin where individuals and the work group are. This requires an initial collection of data from members of the work group followed by a basic diagnosis of restraints that seem to be interfering with the work group's problem-solving efforts and the achievement of its goals. Planning specific OD activities must evolve from this initial diagnosis.

Does OD Have a Beginning, Middle, and End Like Most Programs?

As mentioned above, many managers are accustomed to thinking in terms of programs which have clearly defined start-up and conclusion phases. If, however, an OD effort is successful, it will become "a way of organization life" and therefore will have no identifiable conclusion. This is often a difficult and frustrating point to understand. Just as *managing* is a means to an end, OD is also a process rather than an end product. One of its goals is to build into an organization the human dynamics essential for continuous self-renewing change. OD should lead a work group toward continuing purposeful adaptation and away from reactive change.

Several reasonably distinct phases of the OD process can be identified.[5] Initially a key manager must feel some heightened discomfort from internal or external pressures before an awareness of serious problems can occur. If none of his usual options for handling the problem prove effective, he may then look for new ways of attacking it. This is often the point where a behavioral science consultant is included in the work. The organization then goes through what is often called "an unfreezing process." Common psychological and communication restraints are purposely diminished. Because people and groups are now freer to communicate, considerably more relevant data and ideas become available for diagnosing and resolving the critical problems. As increasing degrees of trust and openness are built into the climate, more people become involved with and committed to creating the changes necessary to resolve the problem of the group. This greatly reduces the commonly encountered resistances to change. The next stage of change is characterized by increased experimentation and testing of new ways for working together and solving problems. If moderate to considerable success is experienced, the new work values and approaches to problems become the new work norms. This improved level of reintegrating work resources represents the final phase which makes possible a natural recycling of the whole process. At this point, managers in the group have learned to become OD managers.

Must a Manager Attend a T-Group or Management Training Laboratory Before He Can Undertake OD in His Group?

Although attending a T-Group is not a prerequisite for a manager interested in initiating an OD effort, it usually helps to prepare him for the experience in several ways. It enables him to experience what it is like to learn purposefully from his own experience. T-Group experience also helps him to learn more about OD concepts and related skills such as: process, "owning-up," trust, openness, leveling, giving helpful feedback, risk-taking, and confronting conflict. These cannot be learned except through experience, and T-Groups provide a short-cut to this kind of experience.

What if a Manager Cannot Afford the Time or Expense Demanded by an OD Effort?

This question is also based on the already-examined assumption that OD is some kind of new program that an organization carries on in addition to everything else. As we saw earlier, this assumption is false. Instead OD is a process through which work is accomplished more efficiently and effectively and without the human fallout and contamination often resulting from more traditional orientations toward work. Its payoffs are to be found in the saving of time and resources.

If OD is so Promising, Why Aren't More Groups or Organizations Practicing It?

OD is not more widely practiced at the present time for at least three reasons. First, individuals have only started to learn within the last twenty years how to create the process of change and growth within their own organizations. Before this, change was something that happened to them. OD is an attempt to integrate what we have learned in this area, and is a very recent idea. Another reason is that OD is a complex process and requires considerable time to spread throughout a total organization. Finally, OD has not always been an irrefutable success. People and organizations are still learning about OD.

Why is it Often Difficult to Get the Top Executives of an Organization Involved in OD?

There are many reasons. Many top executives have been successful in their careers, in part, because they have learned to live by the traditional values of the organization. It is difficult for them to contemplate modifying or discarding that for which they have been so well rewarded.

In some cases top executives are among the last to know about some of the serious problems which drain an inordinate amount of time and energy from the organization. As John Gardner recently noted "We have still not discovered how to counteract the process by which every organization filters the feedback on performance in order to screen out the things it doesn't want to face up to."[6]

Other executives may fail to see any interdependence between their own behavior and the problems they identify at lower levels. Some executives believe that they cannot afford the time to engage in OD efforts.

Finally, most executives are also human beings. They have the same anxieties, or perhaps even stronger ones, about more openly exposing themselves as vulnerable individuals. They have valid reasons to distrust others in the organization who often try to influence them by using distorted information and manipulation. They perhaps even sense some of the seething anger and frustration at lower levels of the organization and hesitate to put themselves in a position to have these feelings dumped on them.

If OD Encourages People to Express Their Feelings, Doesn't It Lead Toward Subjective Anarchy and Away From a Rational and Scientific Approach to Managing?

Feelings and subjective experiences are as relevant to the successes and failures of organizations as are budgets and statistical reports. However, until recently they were largely ignored since they were considered nonrational and therefore inappropriate. It is paradoxical that the more we learn about our own feelings, the more rational we become.

What are the Most Common Obstacles to OD?

OD efforts must overcome obstacles that commonly block any other management activity: misusing authority to bring about change; denying or avoiding conflict of central importance; disowning personal responsibility for initiating action or taking a stand on an issue; waiting for someone else to make the first move; resting on early or easy successes, instead of pushing on for higher levels of effectiveness; involving only a few people at the top in the planning of change rather than working toward widespread involvement; reacting to failure experiences by finding a scapegoat rather than searching for the real causes of failure; expecting to accomplish new levels of effectiveness without learning essential new concepts and skills from qualified experts; taking action without having clear goals in mind due to an initial lack of data. Fortunately, however, these are all specific problems to which OD efforts are particularly sensitive.

How Can You Evaluate the Pay-off of an OD Effort?

OD results are extremely difficult to evaluate in a strict, scientifically controlled design for several reasons. OD is a slowly evolving process involving numerous people over an extended period of time. Many variables, such as key personnel promotions and transfers, occur during the OD experimentation period and also have an impact on the results accomplished by an organization. Who is to say whether improvements are attributable to the OD effort or to the change in leadership? Both are likely influences toward change.

Robert Blake and Jane Mouton conducted an OD experiment several years ago at a Humble Oil Refinery which Louis Barnes and Larry Greiner from the Harvard Business School were asked to evaluate. Although Barnes and Greiner argued logically and persuasively that the experiment resulted in several millions of dollars of profit, they were unable to prove the direct cause and effect relationships.[7]

There are two different ways in which the results of experiments can be judged. One approach is called *quantitative validation*. It attempts to demonstrate statistically quantified relationships between experimental activities and the results achieved. To a limited extent OD efforts can be measured by this method. For example, an initial diagnosis might be made of a work group's perception of itself, using a written questionnaire. The OD activities which follow can then focus on the problems identified by the diagnosis. Later a similar written survey can again be conducted to determine if the earlier problems have been resolved. The results of the two surveys can be statistically compared.

A second approach is called *experiential validation*. It places importance on judgments individuals make about their own experiences. As an example, research shows that subordinates are reasonably accurate in discriminating between an effective and an ineffective boss. They base their judgments on their own private experiences with the world of bosses. OD results can more readily be judged through this same kind of experiential validation than by quantitative means. Most individuals in a work group that is involved in an OD effort can point to several critical incidents which would probably not have happened if OD had not been underway.

Aren't the Values on Which OD is Based Somewhat at Odds with Practical Organization Realities?

Yes, they are. One of the basic goals of OD is to change the traditional climate of organizations in order to utilize their resources more fully. Often the phrase, *practical realities,* is used by persons who feel the need to defend the status quo against the need for change. Being practical and realistic are two norms frequently advocated in the management-world of organizations. But who is to define what is practical and what is real? OD is built on the assumption that it is more scientifically valid to engage a large number of people at all levels of the organization in the investigation and discovery of what is both real and practical.

Is Organization Development an Attempt to Build an Industrial Utopia?

In many ways, yes. The goal of OD is to maximize organizational productivity through actualizing the potential of individuals and the work groups in which they are members. Certainly this is an idealistic objective, but does this make it an unreasonable one to strive toward?

The focus of OD is on the difference between what we are and what we are capable of becoming. Individual and organizational change are directed toward narrowing this gap. An executive involved in OD recently described the paradox involved in this type of goal: "We will continuously have to be dealing with resistance to change, including resistances within ourselves. People are not standing in line outside our doors asking to be freed up, liberated, and upended. Cultures are not saying: 'Change us; we can no longer cope; we are unstable.' Commitment to trying as hard and as well as we can to implement these [OD] values is not commitment to an easy, soft existence. On the other hand, the rewards we experience can be precious, real, and profound. They can have important meanings for us individually, for those with whom we work, and for our organizations. Ultimately, what we stand for can make for a better world—and we deeply know that this is what keeps us going."[8]

Conclusion

In his recent book, *The Revolution of Hope Toward a Humanized Technology*,[9] Erich Fromm asks: "Are we confronted with a tragic, insolvable dilemma? Must we produce sick people in order to have a healthy economy, or can we use our material resources, our inventions, our computers to serve the ends of man? Must individuals be passive and dependent in order to have strong and well-functioning organizations?"

Organization Development is an attempt to provide humanly affirmative answers to these difficult questions.

Organizational Team Building*

The Nickel Auction

Take a group of policemen from your agency. Next delegate three officers to bid on ordinary five-cent coins, the highest bidder of each coin takes possession. Auction off seven nickles, tally the cost per coin of each bidder, and then repeat the process with another seven coins, and again figure the cost for each bidder to "win" these ordinary nickels.

Now examine the process. One bidder may consistently bid only five cents and no more, his rationale being that this is the face value. The other bidders may have paid 19, 24, 30 cents or more per nickel. For them, "winning" was the greatest number of nickels regardless of cost.

But the trainer or consultant now introduces the key question: "Why must the nickel auction (and more significantly, organizational life) be dealt with only in a "win-lose" situation? Could not all of the bidders have "won" the nickels—and at a fraction of their face value?

Nothing in the rules prohibited the bidders from agreeing among themselves to alternate in the bidding so that awarding each

---

*Source: Melvin J. LeBaron and Fred D. Thibault, "Organizational Team Building: The Next Generation of Training," *California Law Enforcement Journal*, Vol. 9, No. 2, October 1974, pp. 71-81. Reprinted with permission of the authors.

coin rotated to the next person. Nothing in the rules prohibited a bid of one cent, with no second bid.

The message of this simple game is two-fold. First, we must devise methods of "win-win" among organizational members, not "win-lose." This boils down to collaboration, cooperation, and trust among fellow employees. Second, police organizations, like all of government, are terribly expensive. We must search for means to get a nickel's worth of use at a penny's cost.

This is one of the major goals of organizational development of which team building is a part.

Organization Development

Organization development—"O.D."—is a term describing an on-going process of continuous assessment of the organizational condition and the introduction of changes to revitalize the organization. It is an attempt to focus on the developmental needs of a total system, not only in terms of personnel, but also in terms of structures and functioning. O.D. is directed toward creating an environment for employee growth through involvement, an environment characterized by a strong willingness to deal openly with issues facing the organization and involvement nurturing a work-force of problem-solving and productive employees.

O.D. not only encompasses a loosely-defined field of modern public administration concepts, but more significantly, is a result of organizational values shifting more toward a value system of adaptation, effectiveness and involvement. O.D. is possible because both the organization and the individual have changed. Current organizational values governing behavior represent a transition away from traditions of the past. There is a renewed focus upon the individual's needs and desires, and greater efforts are being made to provide a fulfillment of these needs within the organization.

During recent years O.D. has become a particularly important concept for police organizations. Greater emphasis on professional education, minority recruitment, technological utilization, community acceptability, and organizational effectiveness has brought about a "new breed" of policemen, and with him has come a shift in organizational values away from the bureaucratic model of the past.

The central theme of the old model was that the policeman was a rational person who performed best in a formalized system in which roles and behavior norms were carefully spelled out. This model further assumed that the policeman's feelings are essentially irrational and must be prevented from interfering with his rational calculations and the organization must be designed in such a way to neutralize and control his feelings.

The "new breed" policeman has brought about new ways of doing things and is requiring things to be done which were never before required. He is becoming more mobile and has less loyalty to a particular department. Probably most important is his attitude that change must be made and he desires to be a part of that change. Some of his changing concepts are that power should be based on collaboration and reason, rather than on the traditional basis of threat and coercion; and that organizational values should be based on humanistic and democratic values replacing mechanistic ones. Perhaps the most important shift being brought on by the "new breed" policeman is one which recognizes that work should encourage personal growth, as well as to achieve organizational goals.

O.D. recognizes that in order for police organizations to cope with these changes new knowledge and skills must be developed. Particular attention must be given to increasing one's skills in dealing with people—a process which is initiated by a self-examination and understanding of:

1. One's assumptions of the needs of people.
2. Perceptual and real differences and similarities of people.
3. Ability to communicate and act effectively in groups.
4. Utilization of feedback for evaluation and personal growth.

Team Building

Team building is a sub-unit of O.D. that has gained popularity with many police agencies concerned about increasing effectiveness and improving the organizational climate. Through the University of Southern California's School of Public Administration, Center for Training and Development, over 50 police agencies have been exposed to team building training. Such efforts are replicated across the country by many major universities and private consultants.

Unlike many consultant packages, "team building programs" are largely tailored for the individual agency's needs and expectations. The concept that the police agency that contracts for team building *owns* its contents is sometimes difficult for the chief newly exposed to O.D. efforts to grasp.

Most initial contacts start with the police administrator requesting information about how he can improve the effectiveness of his department. In the early stages, the police administrator must feel or have an awareness that there is room for improvement in his organization, and O.D. efforts may offer a start to overcoming these deficiencies. This is known as the "diagnostic" stage.

When the outside trainer is consulted, an initial effort is made to assist the police administrator in recognizing and vocalizing what the department has going for it and what it may want to do something about. Frequently, the administrator will sense something is less than optimum without being able to pinpoint the causes. Dependent on the individual situation, the consultant may propose interviews with key organizational people, attend staff meetings, gather data from various sources, to better crystalize the issues. This is then fed back to the administrator in summary form while preserving the confidentiality of the individuals involved. The administrator is free to accept or reject the analysis.

If the administrator chooses to continue, the trainer may now list the various options open to the administrator, including the contents of a team building program, as well as membership in a workshop or series of workshops to be designed to meeting the individual needs and expectations of the department. It is significant to note that the police administrator is built into the planning process and each action step is subject to his rejection or acceptance.

Nature of Team Building Workshops

The contents vary considerably, and it is questionable if there is a representative workshop. However, in California, the Commission on Peace Officers Standards and Training has underwritten a semi-prescribed team building workshop which is the model used by police agencies in this state. Characteristics of this program are:

1. The program is planned and conducted for the department on-site, if desired, or at a convenient retreat location.
2. Each workshop is directed toward the unique organization and employee development needs of the department. While the workshop is planned specially for the department, most programs have evolved around three areas:
   a. providing greater emphasis on the personal development of the individual toward greater effectiveness as a team member.
   b. providing emphasis on the development of managerial-supervisory skills and attitudes.
   c. establishing a problem-solving vehicle for identifying and working through departmental problems.
3. The number of participants per program is usually limited to a minimum of 12 and maximum of 16. More than one workshop can be scheduled for a department.
4. Typical time period for a workshop is three days, to be scheduled at the convenience of the department.

A typical three-day agenda is:

Introductions and Workshop Orientation
"Rating Team Effectiveness"
— a synthesis of issues and concerns
FACTORS AFFECTING TEAM/GROUP EFFECTIVENESS
"Personal Need Systems"
— administration and interpretation of the FIRO-B
"Styles of Leadership"
— administration and interpretation of the "Styles of Leadership Survey"
"How to Choose a Leadership Pattern"
"Decision-Making"
— group exercise and discussion
"The Nature of Conflict"
— its generation and steps to resolution
— group exercise and discussion
"Personal Reactions Under Stress"

— positive and negative use of strengths
— administration and interpretation of the "Life Orientations Survey"
"Communications: Personal and Organizational Information-Sharing"
— one-way/two-way communication
— perceptual sets and biases (exercise)
"Personal Relations Survey"
'What Does It All Mean?"
— review of team analysis
Planning for Future Action and Activity
Evaluation
Adjournment

The purpose of team building workshops usually includes:

1. To set goals and/or priorities.
2. To analyze or allocate the way work is performed.
3. To examine the way a group is working: its processes, such as norms, decision-making communication.
4. To examine relationships among the people doing work.

Usually, police agencies restrict membership to their own personnel, known as a "family group," so that results can be directly carried back to the work setting for continued organizational improvement. Membership may be hierarchially horizontal, vertical or diagonal—so long as there are chain-of-command linkages to the top person in the department.

The police administrator gives a brief opening statement followed by the trainer's orientation. Depending on the group, some theory may be introduced. Lecture-style learning is at a minimum, as participatory training by all present is stressed.

The first day is generally devoted to organizational diagnosis ("Where are we and where do we want to be?") coupled to increased self awareness. The group, frequently through written instruments, learns to take a reading of itself both as a group and as individuals.

Members learn tolerance of others' views, and while withholding value judgment, they learn to feed back information accurately. They also learn to take ownership for the data they themselves generate.

Organizational Development

The second day is often devoted to conflict resolution, i.e., to face and resolve problems in an adult manner, rather than burying them. Team building efforts appear. A group of 15 may be broken into three sub-teams and given 30 minutes to design a "logo" representative of the organization's climate. The members suddenly learn to utilize each member's strengths, expertise, and interests to collaboratively produce a team effort. The old saying, "two heads (or four, or six) are better than one" takes on new significance.

Members learn how to listen more carefully, as well as gain an appreciation of other's positions. For example, trainees may be asked to play the roles of both parts in a scenario. First the role of a subordinate expressing a grievance to the chief, and then the same individual turning around and responding as the chief.

The final day can be summarized as commitment, action and closure. Closure may come in the form of action agendas, or specific programs and projects to be carried on back at the job. Trainees may undertake role negotiations or contract setting where supervisor and supervisee reach a consensus as to their goals and obligations, each having a clearer and more specific understanding of responsibilities and expectations. Commitments may also result with peers as to rights and responsibilities to each other—"quid pro quo."

The significance lies in the fact that the group largely motivates itself as to the direction it takes. Participation (or non-participation) is largely voluntary. Decisions are generally by consensus, not majority vote or an autocratic decision. Everyone's values, beliefs, and feelings are respected, yet visible progress is made.

Results of Team Building

What do police administrators who have undergone team building workshops have to say?

One police chief took over a department and found some indications of employees having overlapping duties. In addition, some structural changes were being made. Further, the chief wanted a higher problem-solving ability by his staff. Team building helped the department in correcting these deficiencies and moving in a common direction.

The first workshop was held for staff personnel, followed by additional sessions for line personnel, generally matched by supervisors and the actual officers they supervised.

The results? All but two of his total sworn complement have perceptually altered their behavior to their fellow employees in a more positive manner. There is more cooperation and far less competitiveness. Officers have set up committees to deal primarily with operational matters, not so much in new projects as improving existing processes. Twenty-five topics, subsequently condensed to eleven main headings with sub-points were dealt with by these committees. Membership on a committee is voluntary and determined by the interest of the officer, irrespective of rank. The life of committees has run from one day on up to a year. Only one project was a failure and it was inconsequential to the department.

A second chief commented that he had received more ideas from just one of the workshops than in five years as department head prior to the training exercise. (Conversely, it can be argued that team building made known to lower echelon personnel a desire by the chief for their ideas). This administrator's approach stresses the individuality available to a department.

Rather than set up a number of committees, trainees submitted cards with specific items they felt needed improvement. The vast majority of these suggestions were followed through. In addition, a training committee was set up on the premise that the individuals who know best the training they need are those to be trained. An annual complaint session is held "off campus" where anything goes and confidentiality is respected. Subordinates have also had input (but *not* capitulation) in promotability procedures, resulting in a perception of greater fairness and justice. Employees are regularly interviewed by the administrator and probed to suggest an organizational improvement. Some role negotiations also transpire. (These administrators also appeared acutely aware of a need to sense the environment that their agency serves to determine perceived effectiveness and ways for improvement.)

Most spectacular was linking all employees to a single objective. Excluding only the four top members of the department, everyone from the cadet and filing clerk on up to the lieutenant are now motivated to a common objective or reducing selected major crimes. If the task is successful, a bonus in salaries is linked to the percentage drop in selected crimes. Thus the entity eliminates the need for additional personnel while existing employees show a greater effectiveness, and increase the citizen's perception of safety in his community.

Coupled to the pay incentive plan, the department took a systems approach by some alteration in structure, brainstorming for new tactics and strategies in dealing with the criminal, plus a public awareness campaign of how the citizen can help himself from being a victim. While the experiment is still underway, early results show a substantial drop in targeted crimes. In addition, the department has introduced two independent variables to add the reliability: The city manager's office inspects the reports (plus command personnel rotation); and criteria of the district attorney's office is used in the initial classification of crime.

This administrator attributes team building as opening the door for these innovative projects.

A third police chief, also having a 100% department participation in team building workshops, developed a number of innovative projects. In addition, he developed a technique of assisting individuals in matching personal goals to organizational goals. Rather than stifling excess energy of employees away from the job, the chief individually counseled personnel and thus utilized individual interests and abilities to enhance the department's effectiveness. He also attributes team building as the spark that ignited this progress.

In the case of this department, members were asked to make evaluative comments about the value of the team building workshop some time after they had returned to the work environment. Areas of major emphasis upon which participants focused were:

1. The effect of the team bulding workshop upon their own particualr attitudes and behaviors.
2. The effect of the team building workshops upon their attitudes toward the department.
3. The participants' preconceived assumptions as to the need and productivity of the team building workshop.
4. The participants' attitude toward the future utility of team building workshops for their department.

All respondents indicated that the team building workshop had been contributing to their own individual growth and knowledge. An added benefit was seen as the insights the individuals had gained about their co-workers and administrators which had led to more satisfying relationships in the work environment.

Further, the introduction to the concepts and methods of participative management and team effort was credited with enhancing the member's skills in interpersonal communication and conflict resolution capabilities throughout the levels and structure of the organization. Many participants mentioned that they had noticed positive individual and organizational changes, heightened morale, greater confidence and respect, and a general dealing and overcoming of organizational inertia as a result of the team building workshops.

These favorable responses are in sharp contrast to what an individual would have given prior to attending a team building session. Many evaluators indicated they were, at the outset: apprehensive, skeptical of the goals and purposes, not looking forward to it, pessimistic of the benefit, and sure it would be a waste of time and money. However, their conclusions from the experience indicates they left with new commitments to improve the organization and a new-found enthusiasm toward their endeavors.

With the exception of a small few, the participant's general tone indicated that the team building workshops were very worthwhile, rewarding, enlightening enjoyable and educational. The few who had misgivings about the experience felt that there had been a problem with group members per se—i.e., low participation and minimal open expression, rather than a problem with the organization and coordination of the sessions.

Some representative comments from the members of this police department who responded to their team building experience are:

"I thoroughly enjoyed the seminar and it wasn't until after I went back to work that I discovered how it had affected and changed me. At first I thought this condition would only be temporary. However, this was not the case and I am still benefiting from the seminar."

"In summation, I can't help but feel that the experience was one of the best things that has happened to help unity and better communications within this department and strongly recommend continuing programs such as this one."

"One of the outstanding aspects of the Team Building Workshop was that of being able to talk directly with the administration with a feeling of confidence that your thoughts and complaints were being heard and considered. I feel this is a must for organizations such as police departments or other similar civil service occupations."

"The retreat allowed me to see myself from a different perspective resulting in a lot better insight into myself. I found that I could communicate at different levels of rank in a relaxed atmosphere. This in itself was very rewarding to me. I was able to get to know my fellow co-workers on a more personalized basis. In the past, it was only my close friends. I gained a better understanding of how this department is run and why certain things were done as they were. I felt like I belonged."

"The seminar also indicates to me that the department is attempting to create a better environment for its officers. This is done by instructing the participants in the art of communication and understanding."

"I also believe that the workshop taught everyone present the importance of establishing priorities and also enlightened us as to the major problems the chief encounters, particularly with respect to budget restrictions."

"I believe the knowledge received helped all officers attending to work together more as a team rather than as an individual."

"This team building workshop was to me an interesting experience. We had a chance to explore all the problems with "our" department. We also had a chance to solve some problems and set up the working conditions so that we may be able to solve the problems we still have and will have in the future."

"Being allowed, as a patrolman, to voice my opinion and participate in inter-department affairs was a very gratifying experience inducing me to aspire working at 110 percent efficiency to accomplish my job."

"Finally, the most important concept to me at this workshop, and I feel the most valuable to me, was self-evaluation. I learned things about myself that I absolutely did not know. I feel as a result of this workshop with all of its concepts, theories, and especially personal evaluation, I am a much better person and, I feel, policeman."

"I found that my own attitude changed as far as working with instead of against my fellow workers. I also observed a change in attitude of other officers plus more cooperation throughout the department."

"The chance to get away from rank structure and communicate with the brass on a more equal level enabled me to feel a more important part of the team, not just a number but a member of a team."

"The program did not provide any cut and dry answers to any of the problems of the individuals or to the problems confronting the department. The program did provide the tools through which the individuals could find a solution to these problems as they arise."

"The seminar provided directions to be taken that would lead to the realization of set goals. The members of the organization were pulled together to pool their efforts rather than act on an individual basis."

"I found the sessions themselves quite interesting. Although many of the participants were obviously exercising some restraint in their statements and actions, several of the more outgoing personalities did as I expected and carried the ball. I heard more favorable comments from other participants both during and after the sessions themselves. It is interesting since comments by the same persons before going to the workshop were primarily negative in nature. There seemed to be good general acceptance of the program by the vast majority of personnel who attended."

"It is my personal belief that this team building workshop was the most enlightening and rewarding experience in the seventeen years I have been on the department and should be continued on an annual basis."

"The team building workshop was the most enjoyable experience that I have ever been required to take part in as a member of this department. I feel that any department of this size that would want to undertake a program of this type could not do anything but benefit from the program."

"I attended the workshop for approximately two days before I even realized what it was about. At that time, I learned the benefits of true communication and feedback between myself and my fellow officers."

"I feel that every police department should attend a workshop of this nature for many reasons.
1. It develops a close relationship between fellow workers.
2. It develops a closer relationship between administrative personnel and employees.
3. It makes each individual aware of his own assets and liabilities, as well as his co-workers."

"I did not realize the high degree of distrust, misconceptions and lack of mutual understanding that existed within our agency. It was very gratifying to see faults brought out into the open and freely discussed, breaking down hindering barriers which had been built over the years due primarily to poor networks of communications. I felt rather ashamed to realize that I also was a contributing factor to this unhealthy situation and it also prompted me to develop more insight into my role in the organization.

As a result I have made definite modifications in my behavior patterns in relation to my job in a more positive direction. The seminar session has, in general, re-established confidence in my abilities to effectively perform my function within the organization in a more harmonious manner."

"I think that one of the best results of the seminar was to bring all of the men together and to see that we all want the same things—a better police department where we all work."

"In my opinion, this program was most beneficial and effective in bringing the department closer together and in solving many problems that had not been discussed in the past. I would highly recommend this program to all police agencies."

"I was impressed with the effectiveness of the team building workshop. Everyone who attended agreed that they were made aware of their weaknesses, and were given tools to make them more effective. There was more understanding of the effectiveness of inter-departmental communications. We were shown that by each of us becoming involved, working together, we could accomplish much more and become a part of the decision making process rather than just be a follower. By everyone participating, each of us learned that each individual is important and can effectively add to the entire group. Some of the tools we were made aware of were effective communications, improved feedback, group participation, free exchange of ideas, effective listening and the stimulation of individual initiative."

"I have noticed that since going to the workshop most of the personal conflicts people had, have been resolved or at least coped with."

"I think the most important part of the workshop with regard to my own personal need was the fact that I now can communicate with the members of the staff without feeling inferior as I learned that they are also police officers just as I am and the only difference is our jobs. There is no doubt that this alone will enable me to better myself and assist in bettering or at least help in bettering the department I work with."

"Initially I wondered what the team building workshop was all about, but later I found that it did exactly what its name implied—it helped promote team unity in work situations. I see my team working together more now than before and the sergeant is now a part of our team rather than just a management representative."

"The workshop has not only uncovered individual problems but also mutual problems within the entire structure of the department and disclosed methods of overcoming many of these problems. It has enlightened many as to their intended goals and has given many, myself included, the incentive and/or motivation to commit themselves towards reaching those goals."

Readiness for Team Building

The value of team building within a police agency is directly related to the state of "readiness" that exists within that organization. Experience gained from conducting over one hundred police department team building workshops brings attention to several factors which are most critical as criteria for assessing the condition of "readiness" within the department.

1. The first step toward "readiness" is an awareness on the part of the police administrator that his department can improve. From his viewpoint, he must be willing to share some participatory management, largely on operational matters that directly affect personnel. This is something which should be viewed as a "gaining" rather than a "giving-up" process. As one chief said, "No one forgets who the boss is"; nevertheless, it is possible to exchange some degree of internal democracy in return for the shedding of dysfunctional activities and increasing personnel commitment.
2. The police administrator must be responsible for the initial establishment of a team building effort and feel the organization can move forward with the program. He

cannot leave the initial responsibility to a training officer or administrative aide.
3. The top level decision-makers of the department must be willing to acknowledge the existence of important problems within the organization and must have a genuine desire to improve the organization climate.
4. There must be a willingness in the organization to acknowledge the existence, legitimacy and importance of interpersonal relationships (member's feelings about themselves and others) as a part of operating dynamics. This variable should be strongly felt by the administrative group in the department.
5. There must be individuals at the administrative and supervisory levels in the organization who are willing to sanction, and, at least to some degree, personally accept "openness" and mutual trust as desirable qualities for relationships within the department.
6. There must be a high degree of willingness by the administrative staff and the supervisors to share planning and decision making functions with subordinates.

Conclusion

Organizational team building is results oriented. Feedback to date, justifies and validates this claim but time and circumstance will be a better judge. Team building workshops are designed as the beginning of a process and not an event so it is too early to close the door on the evaluation process.

Long range results from police department team building is yet to come. As a result, there remains the possibility that the team building workshops have been more entertaining than permanent in their effect. One thing for sure, the process needs maintenance and reinforcement within the organization, if the renewed environment is to be given a chance of survival.

Team building workshops—if nurtured—may well represent the next generation of police training in offering the opportunity of opening doors to greater effectiveness which is being increasingly demanded.

OD – An Action Plan*

In recent years a new term—organization development (OD)—has sprung into vogue in professional business and administrative circles. Commonly, the term denotes both a process of planned change and a desired end result. It embraces many component programs and strategies, some or all of which may be combined into a total OD effort: sensitivity training, T-groups, confrontation meetings, data feedback, team-building, task force management, management by objectives, participative management, managerial grid, laboratory training, and group problem-solving.

Each of these OD techniques, irrespective of its procedural diversity, shares the ultimate goal of creating a self-renewing organization. Organizational development attempts to improve an organization's problem solving capabilities and ability to cope with changes in its external and internal environment.

Objectives or organizational development can be summarized as follows:[1]

1. To increase the level of trust and support among organizational members.
2. To increase the openness of communications laterally, vertically, and diagonally.
3. To increase the incidence of confrontation of organizational problems within groups and among groups.
4. To increase the level of personal enthusiasm and satisfaction in the organization.
5. To create an environment in which authority of assigned role is augmented by authority-based knowledge and skill.
6. To find synergistic solutions. (Synergistic solutions are creative solutions through which all parties gain more through cooperation than through conflict.)
7. To increase the level of self and group responsibility in planning and implementation.

---

*Source: Revis O. Robinson, "Organization Development: An Action Plan for the Ontario Police Department," *California Law Enforcement Journal,* Vol. 8, No. 4, April, 1974, pp. 177-183. Reprinted with permission of the author.

The above listed objectives focus on an obvious fact: OD invariably concentrates on the values, attitudes, relations, and organizational climate—the "people variable"—as the point of initial change. Thus it is understandable that an external change agent, usually a professional behaviorial scientist, is employed at least throughout the early phases of an OD program.

Purpose of the Study

At the time of this writing all staff supervisory personnel of the Ontario Police Department have attended two three-day OD seminars conducted through the University of Southern California. During these sessions the group was introduced to the concepts and assumptions of OD, and more critically, they *experienced and practiced* various OD techniques as mutual problem identification and solving, self-analysis instruments, and team-building exercises. Initial acceptance of the program was so high among all participants that it was decided not only to continue the program among the supervisory staff, but to expand it into an entire OD effort encompassing the whole department.

This report then is an offshoot of these seminars and is designed to formulate an organization development plan for the Ontario Police Department. There are a number of objectives and sub-objectives of this study which will be addressed in the following pages:

1. To define the nature and extent of an OD program for the department.
    1.1 Is the department ready for an expanded OD effort?
    1.2 What conditions are necessary for an expanded OD program?
    1.3 What specific groups of persons be exposed to what type of OD strategy?
    1.4 What is the proper time sequence for scheduling the OD programs?
2. To establish an in-house, continuing OD capability in the future.
    2.1 What types of refresher courses should be given to which groups?

    2.2   Should there be an in-house OD expert and if so who?
    2.3   What other changes should be made in the organization to complement and reinforce the OD efforts that have proceeded it?
3.   To determine a suitable methodology for reviewing the OD program for effectiveness at all levels.
    3.1   Review reasons for failures in other OD programs.
    3.2   Develop strategies for ensuring program acceptance among departmental participants.
    3.3   Identify symptoms of program failures and analyze underlying causes(s).

Need for the Study

Perhaps the best statement as to the need for an OD planning and goal setting program has been advanced by Bennis:

Obviously everyone cannot be in the OD program. This raises the question of priorities and choice. Can OD be isolated in certain components of the organization, leaving other components without it? Or should attempts be made to include segments of all subsystems of the client system in the initial stages of the program? In any case, a careful diagnosis needs to be undertaken in order to trace the most strategic circulation of effects throughout the total client system.[2]

The appropriateness of the Ontario Police Department to conduct this diagnosis is evident from another statement made by Bennis:

... The change agent (outside consultant) should attempt to involve the client system (Ontario Police Department) in planning and goal setting for the change program.[3]

Scope of the Study

Organization development efforts are typically divided into four successive stages:

1.   *Diagnosis*—a determination that there is some discrepancy between expected and desired results and actual results

and the conversion of the identified trouble into a defined problem capable of being impacted by an OD strategy.
2. *Action-Planning*—development of the plan of action which shows the most promise of altering the performance of the system in the desired direction.
3. *Action-Implementation*—translation of the selected plan into actual behavior.
4. *Evaluation*—comparison of the planned goals with actual results and diagnosing the variance and its causes.

This particular study will focus principally on the first two stages: diagnosis and action-planning. It will provide a general overview of the problems presently encountered within and between major departmental sub-systems (e.g., Investigation, Patrol, Services Bureaus) based upon personal observations of the writer and comments elicited from affected personnel in the concerned subsystems. In short, it will highlight areas of the department in need of and amenable to change through OD interventions.

Action-planning will be performed inasmuch as OD strategies and tactics most appropriate for system improvement will be identified and evaluated for their potential application.

The sections immediately following discuss these stages in the context of specific operational difficulties (diagnosis) and remedies (action-planning).

Organizational Diagnosis
1. *Group-to-Group Interface*

An excessive amount of dysfunctional energy is being spent by organizational units in criticizing and opposing the actions of other units. Much of this conflict can be traced simply to a lack of understanding of the other units operational needs, although sometimes it is obvious that each unit considers itself and its operations to have superceding importance. These conflicts manifest themselves in one form or another at the following organizational unit interfaces:
- Records Section — Patrol Division
- Patrol Division — Investigations Bureau
- Patrol Watch 1 — Watch 2 — Watch 3

2. *Individual-to-Group Interface*

There are really very few effective *teams* functioning in the department despite the fact that the organizational structure has created numerous fragmented work units. Most work effort is performed by individuals with a highly specialized assignment and who are exclusively concerned with fulfilling their job requirements. Oftentimes broader team goals have not been defined, or if they have, they are not clear, or shared by all team members.

Organizational processes such as decision-making and problem-solving are generally performed in typical bureaucratic fashion—by deference to the formal power structure through a method best described as "decision by authority rule". Participation in these processes, if provided at all, is probably considered tokenism or mere "window-dressing" to the individual employees.

3. *Individual-to-Organization Interface*

A relatively large number of persons are only minimally committed to the department's goals and only half-heartedly motivated toward diligently applying themselves to their jobs. Apparently individuals' needs for achievement, recognition, and self-development are not being satisfactorily fulfilled, or individual needs are in some ways conflicting with organizational demands.

4. *Individual-to-Individual Interface*

Disruptions at this interface usually surfaces as interpersonal conflicts between co-employees within the same work unit. Personality clashes, perceived differential treatment by supervisors, and role misunderstandings typify the types of incidents which may spark an open confrontation between two parties.

Action-planning

To effectively intervene these organizational problems a number of custom-tailored OD strategies are proposed in this section. The arrangement suggested is essentially a packaged OD program, integrating on a planned and multi-phase basis, a number of program efforts organization-wide.

1. *Team-Building and Personal Development*

It is recommended that five separate three-day sessions be provided to the following groups:

- Records Section
- Investigations Bureau
- Patrol Watch 1
- Patrol Watch 2
- Patrol Watch 3

This phase of the OD program is intended to focus primarily on the tasks of the teams and the working processes designed to achieve those tasks. Hopefully, the teams will learn to function more effectively while, incidentally, improving their individual performance.

These programs may begin by having each member complete a series of instruments which give his view of the quality and character of team action and individual efforts. This includes how he sees himself operating as a team member and his own barriers to excellence. What he sees other team members doing or not doing which furthers or rescues the effectiveness of the team is also part of this effort. The completion of the selected instruments provides a foundation from which the members can study and resolve team problems.

The first task of the groups is to ask the question, "What can I (we) do to be more effective on the job?" Because a group can get so involved with what causes problems, it may spend an inordinate amount of time diagnosing and analyzing, instead of solving their troubles. Therefore, the question has to be asked: "Considering what we are doing that causes us problems, what alernative actions or behavior would be more effective?" Another question to be asked is: "What do we need to do to achieve the alternatives we have chosen?"

The major activity in these meetings is the working through of the information produced. The groups solve those problems that can be fully dealt with at the meeting. They make action plans for dealing with those items which need to be handled by the team or some subpart of it after the meeting. They also develop mechanisms for handling items that have to be forwarded on to some other part of the organization. At the end of the meetings the teams usually have a list of follow-up activities, including meetings and a timetable of actions or an implementation schedule.

Two key elements in the proposed team building sessions are (1) the involvement of the groups' supervisor(s) and (2) the technical assistance of a trained third-party facilitator. Inclusion of the groups' supervisor(s) is necessary if a genuine "team" approach to problem-solving is to be initiated and sustained in the department. Further, since part of the problems identified may directly or indirectly involve the behaviorial intercourse between the supervisor and his team members, it is essential to confront these conflicts openly in a face-to-face environment.

The services of an outside professional trainer are needed to coordinate the sessions, to serve as a resource and facilitator, and to observe, collect, and feedback useful data to the group to help it reach its goals. This person(s) should be well-versed in small-group behavior and in techniques of behaviorial science intervention.

2. *Interteam Development*

The objective of organization development efforts at this interface is to achieve collaboration or integration between these groups of specialized employees so that they can make a coordinated effort toward total organizational goals.

To achieve this end, it is recommended that the manager and several representative members of each functional group in the department attend a three-day interteam development seminar.

The following bureaus, divisions, or units should be represented:
- Records Section
- Patrol Watch 1
- Patrol Watch 2
- Patrol Watch 3
- Detective Division
- Youth Division

Although a number of intervention strategies can be employed at this interface to improve group coordination, a method referred to as the "Differentiation Laboratory" by Lawrence and Lorsch may be appropriate. The design of this laboratory consists of dividing the participants into groups of six—one representative from each functional group. Members of each functional group are asked to talk about what is important in accomplishing their job and what satisfaction they derive from their work. They also are asked to discuss what other functional groups do to block accomplishing their objectives. When these subgroups reconvene in a general session, a discussion develops in which the participants work out problems between functional groups and develop a clearer understanding of their misconceptions and stereotypes of each other and how these inhibit communication and conflict resolution.

3. *Management Team Development—A Continuing Effort*

The initial series of three-day OD seminars for all supervisory, management, and administrative personnel inculcated the values and knowledge necessary to sustain an open, direct, problem-solving capability in the department. It is therefore recommended that similar once-a-month off-site meetings be continued as an integral part of the total OD program. In fact, one such meeting was held on November 2 as an extension of the formal OD training program conducted by USC. However, a significant number of shortcomings were observed at that session which indicate the need for a more structured approach to improve the group's effectiveness and make the most productive use of available time.

Insofar as possible a monthly agenda should be established by requiring that all persons desiring to discuss a certain problem or situation, submit it in writing to the Chief's secretary two days in advance of the meeting. The established agenda should be distributed to all attending persons at least one day prior to the meeting. In the event no agenda is established, no meeting should be held (highly unlikely. The length of the meeting should normally be no longer than half-day with a certain amount of time left open after coverage of the agenda items for free discussion. Leadership for the meetings should rotate among all participants every month. The same individual acting as coordinator should take minutes which briefly reflect the discussions that occurred, decisions reached, and most important, action items. Each problem discussed which cannot fully be resolved within the meeting or requires follow-up implementation should be assigned to one or more group members with a definite timetable for completion.

Immediately following each meeting but before the group disperses, a brief post-meeting session should be held to discuss the following items:

- A recapitulation of the actions taken and recommendations made during the meeting recently concluded.
- A clear statement of what is to be done (and by whom) before the next session.
- Any thoughts or contributions a member developed after adjournment.
- The question: What specifically does each member believe could be done to improve the effectiveness of the next meeting?

The setting for these monthly meetings should continue to be off the Police Department premises in order to preclude work distractions and psychologically encourage freer expression of personal thoughts and ideas.

Periodically, perhaps two or three times a year, an external OD consultant familiar with departmental management personnel (preferably one that has conducted previous OD seminars for the department) can be invited to observe the monthly management team meeting. The observer does not participate, nor is the session conducted in any special fashion for his benefit. After the meeting, however, the onlooker is asked to give feedback to the team on what he observed, especially inter-personal processes such as the degree of openness and leveling, conflict resolution techniques, mutual support, and goal orientation.

Evaluating the OD Program

The OD consultant as part of his involvement as observer/evaluator of the management team meetings will decide when professional outside consultation is no longer necessary, that is, when the department is capable of sustaining its own OD efforts. Most likely, the consultant will assess such factors as satisfaction, communications patterns, morale, and mutual trust and support in reaching his conclusion.

With a withdrawal of reliance on external consultants, the department must devote attention to identifying and strengthening internal change resources who can help with group and intergroup training activities, such as team-development programs, intergroup workshops, goal-setting programs, etc.

It is possible that the department's Training Sergeant can assume this new role with minimal specialized training in OD interventions. Another alternative would be to appoint an OD training committee from existing department members already possessing some knowledge of OD strategies and programs.

If needed, occasional direct linkages between department personnel and external consultants working as a "project team" can be formulated. For example, if the department desires to adopt a Management By Objectives (MBO) program the temporary services of an external consultant may be employed to provide technical input and implementation assistance. The advantage of this arrangement, of course, is that a balance is struck between the expertise of external consultants and the extensive knowledge of the organization possessed by the internal change agents.[4]

Measures to Increase Program Success

The single most crucial condition for ensuring success of the OD program is the commitment of the chief executive to the goals and methods of organization development. However, even ostensible advocation is insufficient by itself in determining whether or not the OD program leads to organizational change. The critical factor is the boss' behavior from day to day. If his leadership does not encompass the new spirit of tolerance and candor, the mood soon evaporates. His behavior thus undermines the program's goals and eventually causes it to fail.[5]

Another factor bearing upon the possible success or failure of the program is the timing with which the effort is expanded throughout the department. The OD program should not be expanded downward in the organization until all management personnel are able to reinforce their new behaviorial skills in their interactions with others. When they have internalized new behaviorial patterns, they will be better able to influence their subordinates.[6]

The action plan presented herein presumes that all management personnel have internalized the appropriate OD values and, furthermore, have demonstrated a willingness to change their organizational behavior accordingly. If this is not the case, the program will be doomed from the outset; it will be considered by many as the epitome of administrative hypocracy!

In undertaking any planned organizational change utilizing OD techniques, the voluntary commitment of the participants is a crucial factor in the success of the program.[7] Departmental personnel at the operational level (patrolmen, detectives, records clerks, etc.) must undertake the training in a completely voluntary spirit. It is highly doubtful that they will learn if this condition does not prevail.

Voluntary participation of all members can be solicited through sound "selling" methods which are calculated to explain the benefits of the program to all concerned. Trigger words which sometimes carry offensive overtones such as "sensitivity training" should be avoided, although persons should be told that they will be encouraged to freely and openly express themselves in the session. In this respect, employees must be guaranteed that whatever they say during the course of the seminars will not result in any reprisals by management personnel. In other words, job security of the participants must be protected at all costs. (It may be advisable for the OD consultant to conduct an informal and preliminary survey of participants at the operational level to determine the nature and extent of sensitive problems which they may be reticent to discuss openly at the sessions. Among other things, the responses will indicate whether a confidential written questionnaire should be employed to elicit problem statements within the course of the OD programs.)[8]

Future OD Program Components

Other programs which can develop as logical offshoots of the basic organization development program (and which build upon the behavioral expertise previously established), should be considered for future implementation within the department.

1. *Management by Objectives (MBO)*—Management by Objectives, also referred to as Management by Results is an approach to management planning and evaluation in which specific targets for a year, or for some lengths of time, are established for each manager, on the basis of the results which each must achieve if the overall objectives of the organization are to be realized. At the end of this period, the actual results achieved are measured against the original goals—that is, against the expected results which each manager knows he is responsible for achieving.

    Currently, Management by O/R serves many ends, chief among them being organizational planning, management control and coordination, personnel utilization, the evaluation of managerial effectiveness, and management training and development.[9]

    Although the MBO/R approach was originally adopted and is predominately practiced by private corporate enterprises, a growing body of evidence indicates that MBO can be successfully implemented in non-profit institutions (such as police organizations).[10]

    Experience also suggests that the organizational capabilities for a successful MBO/R process are essentially the same as those developed through OD. Therefore, one author strongly suggests that OD become the initial effort, and that MBO/R be phased-in later.[11]

2. *Task Force Management (TEM)*—As a component of its on-going OD program, the department should experiment with the concept known as Task Force Management. Interdivisional teams with temporary, but real authority, tackle major departmental problems under this operation. Major departmental objectives such as producing a policy and procedure manual, selecting and implementing a new management information system, or creating a braintrust to create and explore innovative ideas and programs could all be performed more effectively by the initiation of task forces.

    A distinctive feature of the task force is that its responsibility extends far beyond suggesting solutions or making recommendations. The true task force has operational responsibility for what it proposes. Its work is not done until it has implemented its recommendations or created the instruments to carry on its work.[12]

Another trademark of the task force is that it cuts across departmental boundaries (and barriers). It is an instrument of integration and collaboration; it ensures that important functions are performed with the assistance of others responsible for other functions.[13]

Postscript

The recommendations of this report may sound like jargon and theory-spouting to some. Charges of faddism may be raised by still others. However, the sincere intent of this report is to introduce modern management approaches into a heretofore traditional bureaucratic management structure. It recognizes the need for experimentation with new administrative theories and models, but it tempers this need with the understanding that these new technological tools are not ends in themselves, but simply means to produce systematic and ordered organizational improvements.

Topics for Discussion

1. Discuss the concepts regarding work groups and organizations on which OD is based.
2. Discuss the methods of evaluating OD efforts.
3. Discuss the characteristics of a team building workshop.
4. Discuss the diagnosis stage of an OD effort.
5. Discuss the action-planning stage of an OD effort.
6. Discuss the relationship of an OD effort to MBO.

FOOTNOTES

Introduction

[1] Gordon L. Lippitt, *Organization Renewal* (New York: Appleton-Century-Crofts, 1969), p. 1.

[2] Chris Argyris, *Management and Organizational Development: The Path from XA to YB* (New York: McGraw-Hill Book Company, 1971), p. ix.

[3] John W. Gardner, *Self-Renewal: The Individual and the Innovative Society* (New York: Harper & Row, Publishers, Harper Colophon Books, 1965), pp. 1-7.

[4] Whyte and Hamilton see sentiments as referring to "the mental and emotional reactions we have to people and physical objects" and as having three elements: "(1) An idea about something or somebody . . . , (2) emotional content or affect, (3) a tendency to recur upon presentation of the same symbols that have been associated with it in the past." William Foote Whyte and Edith Lentz Hamilton, *Action Research for Management* (Homewood, Ill.: Richard D. Irwin, Inc., 1965), p. 184.

[5] Kroeber and Kluckhohn cite 164 definitions of culture; our above definition is congruent with their synthesis: *"Culture consists of patterns, explicit and implicit, of and for behavior acquired and transmitted by symbols, constituting the distinctive achievement of human groups, including their embodiments in artifacts; the essential core of culture consists of traditional (i.e., historically derived and selected) ideas and especially their attached values; culture systems may, on the one hand, be considered as products of action, on the other as conditioning elements of further action."* See A.L. Kroeber an; Clyde Kluckhohn, *Culture: A Critical Review of Concepts and Definitions* (New York: Vintage Books, 1952), pp. 291, 357.

[6] This illustration is adapted from an address by Stanley N. Herman, TRW Systems Group, at an organization development conference sponsored jointly by the Industrial Relations Management Association of British Columbia and the NTL Institute for Applied Behavioral Science, Vancouver, B.C., Canada, 1970.

[7] Our use of the word *culture* includes Argyris's notion of the *living system:* "the way people actually behave, the way they actually think and feel, the way they actually deal with each other. It includes both the formal and informal activities." Chris Argyris, "Some Causes of Organizational Ineffectiveness within the Department of State," *Occasional Papers*, No. 2, Center for International Systems Research, U.S. Department of State, 1967, p. 2.

[8] Warren G. Bennis, "A New Role for the Behavioral Sciences: Effecting Organizational Change," *Administrative Science Quarterly*, 8 (September 1963), 130.

[9] *Ibid.*, pp. 130-134.

Parameters of OD

[1] For a fuller discussion of Herzberg's theory regarding hygiene factors vs. motivating factors, see "One More Time: How Do You Motivate Employees," *Harvard Business Review,* (Jan.-Feb. 1968).
[2] See George Litwin and Robert Stringer, *Motivation and Organization Climate* (Boston: Harvard University, 1968).
[3] Douglas McGregor, *The Human Side of Enterprise* (New York: McGraw-Hill, 1960).
[4] Chris Argyris, "T-Groups for Organizational Effectiveness," *Harvard Business Review* (March-April 1964).
[5] Larry Greiner, "Patterns of Organization Change." *Harvard Business Review* (May-June 1967).
[6] John W. Gardner, "We, the People," Millikan Award Address, California Institute of Technology, Noverber 21, 1968.
[7] Robert Blake and Jane Mouton; Louis Barnes and Larry Greiner, "Break Through in Organization Development," *Harvard Business Review* (Nov.-Dec. 1964).
[8] Robert Tannenbaum and Sheldon Davis, "Values, Man, and Organizations," *MIT Research Report* (Oct. 1967).
[9] Erich Fromm, *The Revolution of Hope Toward a Humanized Technology* (New York: Harper & Row, 1968).

OD—An Action Plan

[1] Wendell French, "Organizational Development Objectives, Assumptions and Strategies," *California Management Review,* Vol. XII, No. 2, Winter, 1969, p. 51.
[2] Warren G. Bennis, *Organization Development: Its Nature, Origins, and Prospects,* Reading, Massachusetts: Addison-Wesley Publishing Co., 1969, p. 51.
[3] *Ibid.*
[4] Paul R. Lawrence and Jay W. Lorsch, *Developing Organizations: Diagnosis and Action,* Reading, Massachusetts: Addison-Wesley Publishing Co., 1969, pp. 53-54.
[5] Chris Argyris, "The CEO's Behavior: Key to Organizational Development," *Harvard Business Review,* Vol. 51, No. 2, March-April 1973, pp. 59-60.
[6] *Ibid.,* p. 63.
[7] Bennis, p. 74.
[8] Dale D. McConkey, *How to Manage by Results,* New York: American Management Association, 1967, p. 15.
[9] *Ibid.,* p. 22.

[10] See, for example, Dale D. McConkey, "Applying Management by Objectives to Non-Profit Organization. *"Society for the Advancement of Management (SAM) Journal,* Vol. 38, January 1973, pp. 10-20; Rodney H. Brady, "MBO Goes to Work in the Public Sector," *Harvard Business Review,* Vol. 51, March-April 1973, pp. 65-74; and T.P. Kleber, "The Six Hardest Areas to Manage by Objectives," *Personnel Journal,* Vol. 51, August 1972, pp. 571-575.

[11] Arthur C. Beck, Jr. and Ellis D. Hillmar, "OD to MBO or MBO to OD: Does It Make a Difference?" *Personnel Journal,* Vol. 51, November 1972, pp. 827-834.

[12] Thomas L. Quick, *Your Role in Task Force Management,* Garden City, New York: Doubleday and Company, Inc., 1972, p. 55.

[13] *Ibid.*

SELECTED READINGS

Blumstein, Alfred, "Science and Technology," *The Police Chief,* Vol. 36, No. 12, December, 1969, pp. 56-61. In this article, the author describes the benefits that can be obtained from major advances in science and technology. Consideration is given to technological dilemmas, operations research and technical assistance.

Brianas, James G., "Behavioral Technology: A Challenge to Modern Management," *Public Personnel Management,* Vol. 2, No. 4, July-August, 1973, pp. 290-298. Provides a practical frame of reference from which a manager can proceed to a better understanding of current organizational problems and the directions for insuring meaningful change.

Clary, Thomas C., "Motivation Through Positive Stroking," *Public Personnel Management,* March-April, 1973, pp. 113-117. Identifies the "stroke" as a basic motivation for human survival and enjoyment of life. Describes positive stroking techniques and a consciously planned program which can be utilized by a manager.

Halal, William E., "Organizational Development in the Future," *California Management Review*, Vol. 16, No. 3, Spring, 1974, pp. 35-41. Presents a scenario for organizations of the future and describes the nature of future organizational development programs. Argues for a balance between science and art in the organizational development movement.

Hollis, Joseph W. and Frank H. Krause, "Effective Development of Change," *Public Personnel Management*, January-February, 1973, pp. 60-69. Describes a theoretical model based upon a systems approach utilizing the ideas and attitudes of people to effectuate change. The four major aspects to change are initiators, procedural developers, implementation and modifiers.

Meyer, John C., Jr., "Policing the Future," *The Police Chief*, Vol. 38, No. 7, July, 1971, pp. 14-17. Suggests that of all occupational and social groups, the police are subject to the effects of change quite unlike any other single group. The police must cope with the four essential aspects of "future shock": acceleration, transcience, novelty and diversity.

Sperling, Ken, "Getting OD to Really Work," *Innovation*, No. 26, 1971, pp. 38-45. Emphasizes that organization development goes beyond managing human resources or finding the right organizational structure. OD is a means of helping managers shape and attain goals.

Weisbord, Marvin R., Howard Lamb, and Allan Drexler, *Improving Police Department Management Through Problem-Solving Task Forces: A Case Study in Organization Development*, Reading, Massachusetts: Addison-Wesley Publishing Co., 1974. Case study of a medium size police force by social scientists, from the NTL Institute for Applied Behavioral Science when an organizational development program was instituted. The strategy was to build a more effective middle management.

● **The study of this chapter will enable you to:**

1. Identify the three major steps in managing.
2. List five attributes of a management by objectives system.
3. Write a short essay describing the major elements common to an MBO system.
4. Distinguish between goals and objectives.
5. List five criteria to be followed when writing objectives.
6. Differentiate between goal oriented and task oriented individuals.
7. Define key results analysis.
8. List the main sections of a key results analysis form.

6

Managing for Results

Introduction

Management by objectives is a systematic approach to management. It is not a set of rules, a special technique, or a specialized aspect of the process of managing.[1] Objectives have been utilized by managers since the biblical times but it was not until the early 1950's that Peter Drucker popularized objectives as a basis for a management system.[2] It is best understood, as a total philosophy of management permeating every element of an agency. In the application of MBO, major differences exist in the techniques utilized and in the degree of application.[3] Some law enforcement agencies utilize it as a personnel evaluation process. Still other agencies employ MBO as a sophisticated control mechanism while a few agencies utilize it as an overall management system.[4]

Even though applied in varying ways and to differing degrees, management by objectives, when successfully employed for a reasonable period of time, has resulted in improved communications, coordination, control, and as a key motivational factor. As pointed out by Dale D. McConkey, the impact of MBO on the management process and on the individual manager has been profound and dramatic.[5] He emphasized that the rather common and traditional definition of management usually defined as getting things done through people by planning, organizing, directing, and controlling was rendered obsolete by a recent study which concluded that management is comprised of three steps: establishing objectives, directing the attainment of objectives, and measuring results. The three major steps (Figure 6-1) in managing are divided into eleven elements as indicated by the management wheel, and the three major steps are strongly and positively oriented toward objectives. When management is viewed in this light, the functions such as planning, organizing, directing and controlling, become sub-functions of the three larger steps.[6] This is not to suggest that the functional approach to management created by Henri Fayol are not important managerial activities, but that they are clearly subordinate to the objectives of management and the achievement of those objectives

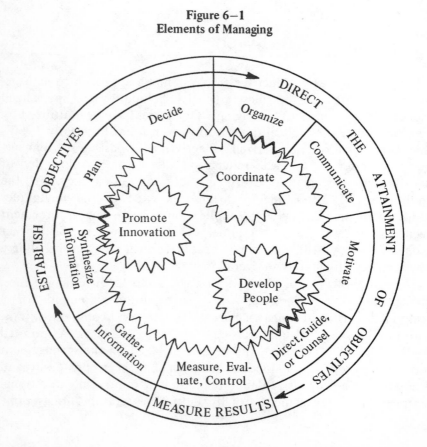

Figure 6–1
Elements of Managing

278 *Managing for Results*

by the organization. The assumption is that orientation toward objectives will insure successful managerial performance. This approach to management beyond the general principles of management, as viewed by the traditionalist, reflect the strong influence of the behavioral sciences and the systems approach to management. It results in a balancing of the organizational functions and the management functions within the frame of reference of objectives.

The system of management by objectives brings spirit and life to the hierarchy with the provision of a bureaucratic atmosphere emphasizing a balanced participative style. Such a style of management must accentuate to a maximum degree, the following attributes:

1. In-depth delegation.
2. Maximum participation in the objective setting and planning processes.
3. Managers permitted to make some mistakes.
4. Change encouraged and planned for.
5. Policies and procedures minimal and subject to change when necessary.
6. Controls tight but only the minimum imposed to keep the unit in control.
7. A meaningful reward system.
8. A high degree of self-management, self-discipline, and self-control from managers.

Experience has proven that MBO cannot be successfully foisted on the wrong management style. The entire management style and approach must be supportive of MBO or it will not reach its success potential.[7]

As a system, management by objectives is especially applicable to managerial positions and can be extended as far down as the first line supervisor. George S. Odiorne has emphasized that management by objectives has helped in overcoming many of the chronic problems of managing managers. He points out: (1) It provides a means of measuring the true contribution of managerial and professional personnel. (2) It defines the common goals of people and organizations and measures individual contributions to them, and it enhances the possibility of obtaining coordinated effort and teamwork without eliminating personal risk taking. (3) It provides solutions to the key problem of defining the major areas of responsibility for each person in the organization, including joint or shared responsibilities. (4) Its processes are geared to achieving results desired, both for the organization as a whole and for the individual contributors. (5) It eliminates the need for people to change their personalities, as well as for appraising people on the basis of their personality traits. (6) It provides a means of determining each managers span of control. (7) It aids in identifying potential for advancement and in finding promotable people.[8]

Management by objectives is viewed differently by individuals as well as agencies and consequently the definition of MBO varies accordingly. Humble defines MBO in such a way as to clearly emphasize the importance of planning[9] while Reddin defines it with an emphasis on effectiveness and a measurement of that effectiveness.[10] Odiorne's definition emphasizes the relationship of the individuals within the organization and he describes it as a "process whereby the superior and subordinate managers of an organization jointly identify its common goals, define each individuals major areas of responsibility in terms of the results expected of him, and use these measures as guides for operating the unit and assessing the contribution of each of its members."[11]

Management by Objectives/Results

While the emphasis on management by objectives varies with each definition, there are major elements common to an MBO system:

1. Objectives established for positions.
2. Use of joint objective setting.
3. Linking of objectives.
4. Emphasis on measurement and control.
5. Establishment of review and recycle system.
6. High superior involvement.
7. High staff support in early stages.[12]

Management by objectives, as a systematic approach to management, provides for a statement of organizational goals and related objectives. It prescribes a process for turning objectives into specific programs and identifies activities for organizational members as well as determining the resources required to achieve those activities. It calls for a continuing review of the organizational structure in long range planning. It helps to identify key areas in management and provides for a continuing clarification and refinement of the organizational goals and objectives relating to those goals. In management by objectives this is accomplished by a thorough analysis of the purposes of the organization and by a comprehensive analysis of internal and external factors influencing achievement and identifies the changes in those factors that can probably be expected in the future. It provides for an in-depth assessment of the relevancy of police goals as they relate to the organization, and finally, it provides a frame of reference for the identification of the planning and control systems through which this process is implemented. In short, management by objectives is a complete management and control system.

As a control system, however, management by objectives differs from any alternative forms inasmuch as it provides for superior and subordinate participation in the objective-setting process.[13] The standards by which the individual manager is expected to be held are not established by executive fiat. Subordinate managers participate in the establishment of criteria for measurement and control and this is done in consultation with their supervising manager.

When confronted with the fact that MBO was developed primarily in an environment of profit-oriented business organizations and that some have questioned its applicability to non-profit organizations, C.C. Leek has pointed out that while the achievement of profit is not the motivating force in the police service, it has, however, some primary objectives just as important to it and the society in general—the maintenance of law and order and the operation of a whole series of societally authorized rules designed specifically to make social living a much more rewarding and wholesome thing. It is his position that any improvement that can be brought about concerning the manner in which police forces deploy their labor to achieve primary objectives, must be to the good.[14]

Audit Check List

The following questions will serve as an audit check list for helping to determine the applicability of management by objectives to nonprofit organizations:
1. Does the organization have a mission to perform? Is there a valid reason for it to exist?
2. Does management have assets (money, people, plant and equipment) entrusted to it?
3. Is management accountable to some person or authority for a return on the assets?
4. Can priorities be established for accomplishing the mission?
5. Can the operation be planned?
6. Does management believe it must manage effectively even though the organization is a nonprofit one?
7. Can accountabilities of key personnel be pinpointed?
8. Can the efforts of all key personnel be coordinated into a whole?
9. Can necessary controls and feedback be established?
10. Is it possible to evaluate the performance of key personnel?
11. Is a system of positive and negative rewards possible?
12. Are the main functions of a manager (planning, organizing, directing, etc.) the same regardless of the type of organization?
13. Is management receptive to improved methods of operating?[15]

Affirmative answers to the above questions are directly correlated to the applicability of MBO to law enforcement agencies. The police enterprises in the United States clearly fall within this criteria and the limitations are minimal in terms of the application of MBO to this field.

Design

A design for implementing management by objectives (Figure 6-2) has been set forth by J.D. Batten providing for the completion of seven phases identifying the dynamics of the management process.

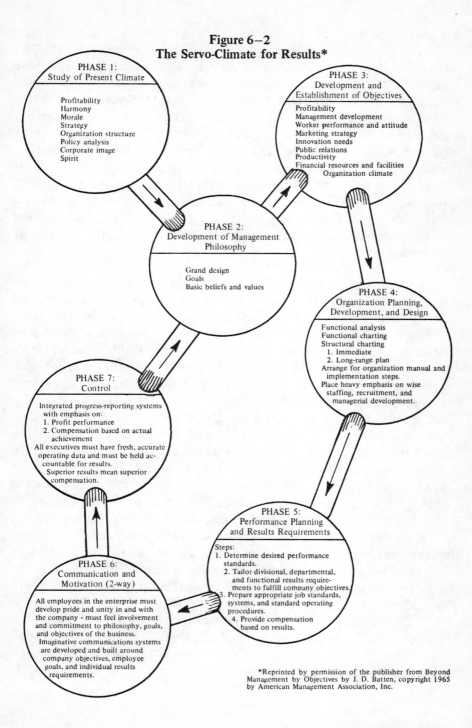

Figure 6-2
The Servo-Climate for Results*

*Reprinted by permission of the publisher from Beyond Management by Objectives by J. D. Batten, copyright 1965 by American Management Association, Inc.

Managing for Results 283

This design is implemented by a comprehensive analysis of the present organizational environment which then evolves into the creation of a total management philosophy that sets forth its basic beliefs and values as an organization and identifies the goals in terms of a grand design. The subsequent phases involve the actual development and establishment of objectives. It evolves through a detailed planning and design phase followed by a consideration of performance evaluation coupled with the dynamics of communication and motivation allowing for the conceptual underpinning of control. As this latter phase is completed, the loop closes and returns to phase 2, the development of a management philosophy, granting a continuing modification and updating of subsequent phases creating a servo-climate for results thus establishing a total system of management.[16]

In the study of the present climate of a law enforcement agency and its relationship to its basic beliefs and value system that become the foundation for the setting of organizational goals, the police policy-makers must take into account the interacting modifiers affecting the characteristics of the community.[17] One of these problems is crime as modified by the factors of politics, and social, economic and psychological conditions. In addition, there is a strong external environment influencing the police organization including other criminal justice agencies such as probation, parole, courts, district attorney, defense attorney and law enforcement agencies. Each of these criminal justice agencies affect the police performance by their own policies, procedures, resources, and operations. (Figure 6-3). Thirdly, an analysis of the internal state of the agency itself is essential, in order to pinpoint modifiers such as style, philosophy, mission, and organizational resources. All of these factors influence, to a varying degree, the creation of an optimal objective model that maximizes organizational effectiveness and allows for the attainment of objectives. Such a model provides a frame of reference for maximal organizational response as well as opportunity for self-actualization.

One of the most pressing and challenging duties of the police chief executive is establishing meaningful goals and objectives for all personnel of the police agency. However, the achievement of goals necessitates agency-wide cooperation.

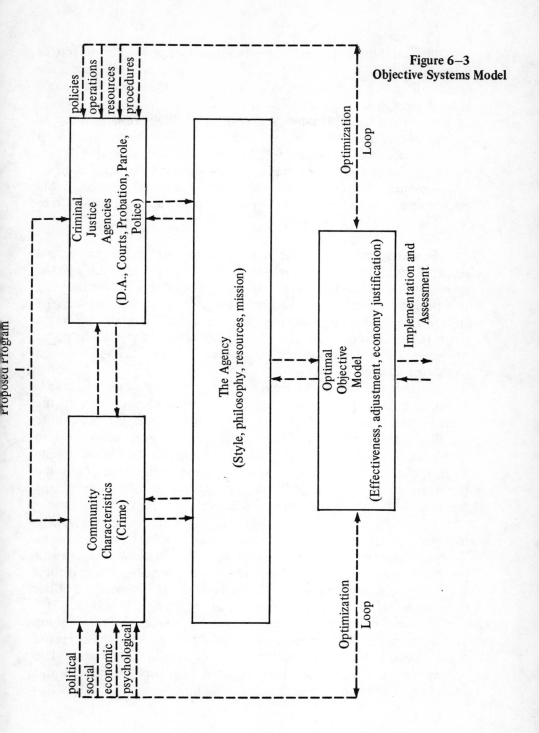

Figure 6–3
Objective Systems Model

Managing for Results

The establishment and accomplishment of goals and objectives follow such similar paths that the words "goal" and "objective" are often interchanged. Authorities generally agree that "goal" is a more general term than "objective."

> Goal—A statement of broad direction, general purpose or intent. A goal is general and timeless and is not concerned with a particular achievement within a specified time period.
>
> Objective—A desired accomplishment which can be measured within a given time frame and under specifiable conditions. The attainment of the objective advances the system toward a corresponding goal.

Fundamental to the establishment of all goals and objectives is a perception of the problems encountered or anticipated by the agency. Clear definition and careful analysis of the factors generating the problem may clearly indicate possible solutions and suggest appropriate goals and objectives. On the other hand, the failure to perceive or to understand a problem may lead to the establishment of goals or objectives that could be nonproductive or even counterproductive.[18]

Neither the police chief executive nor the unit commander can, alone, establish goals and objectives. All employees within the agency, particularly employees at the executive level, can contribute to the understanding of the problem. These employees have face-to-face contact with members of the community and are coping with problems in the field. They, in turn, will understand the problems to be met only through contact and discussion of the problems with members of the community.

Obtaining input from within the agency requires an atmosphere that encourages all employees, regardless of rank, to submit ideas. Such an atmosphere must be genuine; it must start at the top and permeate the entire agency. Employees should understand that their evaluation of the problem, their analysis of cause, and their suggestions of possible solutions are all needed. The Kansas City, Mo., Police Department has formalized this process by establishing task forces composed of personnel of all ranks, with emphasis on the lower ranks, to work out solutions to problems in certain areas. The agency has succeeded in creating the desired atmosphere, but it is too soon to evaluate the contribution of these task forces.[19]

Other agencies of government often are a good source of ideas and assistance, as are community and service organizations. In order to obtain a response that is representative of the community, the police agency should take care to solicit input from private as well as official sources. Additionally, and of utmost importance, is the requirement set out in the first standard of this report that the goals and objectives of the police department must be directed by the policies of the governing body that provide formal authority for the police function.

Community meetings can be valuable for providing private input, although in some cases there is a disappointingly small number of people involved in community activity. Furthermore, those responding have not always been representative of the community, and in some cases response from community meetings does not always reach the top levels of the police agency. The last is always a risk if command personnel do not attend the community meetings.

In establishing goals and objectives, the police chief executive is bound, of course, by constitutional limitations and by the laws under which he serves. Additionally, he must insure that the goals and objectives of the agency are consistent with the police role that he has defined. These goals and objectives, in turn, should provide constraints upon commanders of agency units is establishing unit goals and objectives.

Just as agency goals and objectives are necessary to coordinate and direct the efforts of agency personnel, so unit goals and objectives are essential to coordinate and direct efforts of unit personnel to fulfill the mission of the unit. Unit goals and objectives should be put into writing and should be reviewed by the chief executive. They should complement the agency's overall goals and objectives.

All goals and objectives, agencywide and unit, should be directly responsive to community needs. Normally, if problem definition and analysis have been adequate and alternative solutions carefully screened, responsiveness to community needs can be achieved.

The identification of goals for a police agency should not be taken lightly and can prove to be a time-consuming process when maximum input is sought from all of the external forces and from the internal organizational environment. The police management literature as set forth by O.W. Wilson and Dr. V.A. Leonard as well as the International City Management Association in its authoritative work entitled, "Municipal Police Administration" have listed various goals for law enforcement agencies but never in context of the total managerial system. The closest to this was Dr. Leonard who espoused that the dominance of an idea should permeate an organization and all efforts should be directed toward the attainment of that idea.

A recent study by the American Bar Association pointed out that the tasks to be performed by a law enforcement agency were not identified out of any community assessment of what it would like its police to achieve, but rather, generally reflect ad hoc judgments by the police of what they should be doing. The studies emphasized the need to determine more precisely what the police should be trying to achieve, how the police should react to various situations, and what priorities should govern police intervention. The American Bar Association study recognized that most police agencies are currently given responsibility, by design or default:

1. To identify criminal offenders and criminal activity and, where appropriate, to apprehend offenders and participate in subsequent court proceedings.
2. To reduce the opportunities for the commission of some crimes through preventive patrol and other measures.
3. To aid individuals who are in danger of physical harm.
4. To protect constitutional guarantees.
5. To facilitate the movement of people and vehicles.
6. To assist those who cannot care for themselves.
7. To resolve conflict.
8. To identify problems that are potentially serious law enforcement or governmental problems.
9. To create and maintain a feeling of security in the community.
10. To promote and preserve civil order.
11. To provide other services on an emergency basis.[20]

Goals

The above list of police responsibilities is presented as a guideline and should not be accepted as a panacea for all law enforcement agencies. It will certainly serve as a starting point for the extended discussion and involvement of all individuals concerned with establishing the goals for a law enforcement agency. A typical design which sets forth the basic police goals of a medium sized police department is listed below:

1. Provide for crime prevention programs involving and utilizing all available resources.
2. Provide for positive measures against established criminal activities.
3. Provide for expeditious and prudent apprehension of suspected offenders.
4. Provide thorough and appropriate police-related investigation.
5. Provide for emergency services that relate to the protection and preservation of life and property.
6. Provide for social order during times of unusual occurrences.
7. Provide services that contribute to the preservation of community health, safety and general public assistance.
8. Provide for effective coordination between agencies related to the criminal justice process.
9. Provide for effective and efficient departmental administration and for the career needs of departmental personnel.[21]

The development and establishment of objectives is one of the most critical steps in this management system process. As indicated previously, the goal statement is general and continuing in nature while an objective is quite specific and is designed for measureability and accomplishment. A goal provides direction whereas an objective identifies specifics and the means of accomplishment as well as being a measurement indicator. Each objective should be clearly linked with one of the goals and should emphasize output elements of a managerial decision. The following serves to illustrate a number of specific objectives listed under various goals:

● Goal 1—Provide for crime prevention programs involving and utilizing all available resources.

Objective: Reduce residential burglaries by _____ during Fiscal Year 19___ over previous fiscal years.

Managing for Results

Means: Implement Operation Identification in __homes in the community by the beginning of Fiscal Year 19__ through homeowners' groups, Chamber of Commerce, community programs and other civic organizations.

Measures: Measure the effectiveness of Operation Identification by the reduction of home burglaries in general and the rate of burglaries among homes participating in the program.

Recovery rate of property stolen from Operation Identification homes.

Means: Participate with city planners to review residential building codes to make recommendations regarding minimum standards for locks, windows and other security devices. This is the responsibility of the Planning and Research Section.

Measures: Evaluate the burglary rate in statistical reporting areas that represent the housing developments incorporating the new security standards.

Means: Coordinate with school officials to identify potential truants and control truancy. Use Community Programs Technicians for this program.

Measures: A reduction in daytime burglaries.

<u>*Objective*</u>: Reduce commercial burglaries by __during fiscal Year 19__.

Means: Develop a program for merchants identifying common factors leading to community commercial burglaries and techniques available to combat those factors. Use Community Programs Technician and Police Services Technician Trainee for this program.

Prepare and gather supplementary brochures and visual aides for use in presentations to business groups.

Provide trained PSTs to tour business establishments and make recommendations regarding burglary prevention techniques.

Provide follow-up checks with commercial representatives to determine what preventive actions have been taken by those businesses previously inspected.

Measures: Amount of decrease in burglaries committed at businesses adopting recommended crime prevention measures, and for those merchants who did not follow recommendations.

Significant percent of those businesses receiving a prevention inspection that actually implement recommended crime prevention measures.

Objective: Reduce the number of repeated distrubance of the peace calls during the calendar year of 19___.

Means: Implement a thirty-hour crisis intervention program for all patrol personnel between February 1, 19___ and May 31, 19___.

Develop a referral system of Bay Area agencies or individuals that provide various public services. This system is to be utilized in crisis intervention referrals by all members of the department.

Implement crisis intervention as a patrol function by June 1, 19___.

Develop a follow-up procedure on referred cases with the P.S.T. IV Referral Technician assigned to the Community Relations Section.

Measures: A statistically significant decrease in repeat disturbance of the peace calls between June 1, 19___ and December 31, 19___ as compared to similar previous time periods.

Record the number of referrals and follow-ups made during the same reporting period.

● Goal 2—Provide for positive measures against established criminal activities.

Objective: Increase burglary arrests by _____ over a sixty-day period.

Means: Organize special enforcement units to stake out high burglary areas identified on the basis of statistical data and police intelligence.

Coordinate all burglary reports with all other forms of police intelligence (Field Interrogation cards, suspicious circumstances reports, etc.) to develop comprehensive burglary dossiers to be provided to Selective Enforcement Unit and other operations personnel.

Provide for extensive interviews of all burglar arrestees by SEU personnel, to develop data on MOs location of stolen property, identification of additional suspects, etc.

Measures: Increase of burglars arrested over sixty-day period as compared to previous time periods.

<u>Objective:</u> Increase the percent of stolen property recovered by ___ over a sixty-day period.

Means: Utilize a Selective Enforcement Unit (SEU) to provide surveillance on known fences, pawn shops and flea markets.

Use Investigative Technicians to compile comprehensive lists of stolen property emphasizing those items readily identifiable (i.e., serial or ID number, complete descriptions, uniqueness of items, etc.) for dissemination to SEU (Selective Enforcement Unit).

Provide extensive interviews of all suspected thieves arrested by SEU personnel.

Use Investigative Technicians to provide follow-up on all thefts emphasizing accurate property descriptions.

Measures: Increase of recovered stolen property over sixty-day period as compared to previous time periods.

The specific kind of special enforcement should depend upon an analysis of criminal statistics. This data should be fed into a Management Information System on a daily basis and analyzed monthly. The results of this analysis should determine the offensive actions of the Department.

● Goal 3—Provide for the expeditious and prudent apprehension of suspected offenders.

Objective: Decrease the incidence of "resisting arrest" by_____ in FY 19_-_ as compared to FY 19_-_ and 19_-_.

Means: Implement a twenty-five hour training course through squad meetings (one-half hour twice a week) to run between January 1 and June 31, 19__. Course material to include practical training in laws, methods, and techniques of arrest. (Methods would include training in planning arrests, timing arrests, etc.)

Implement an ongoing physical fitness/self-defense training program to be attended by all Operations field personnel.

Implement a training course concerning the psychology of arrest and the principals of persuasive speaking to run concurrently with the practical training course listed in paragraph # 1, Means, Goal 3 above.

Measures: Percent decrease in charges of "resisting arrest".

Decrease in officer injuries in the commission of an arrest.

Decrease in citizen complaints evolving from arrests.

Objective: Bring about 20 percent increase in arrests regarding valid "in progress" crimes. Comparison periods to be FY 19__-__ with FY 19__-__.

Means: Develop a system of strategic manpower deployment based upon intelligence data and MIS data aimed at providing personnel in high crime areas.

Develop systems of closing avenues of escape by response techniques established by a special study force and included in written form and squad room training. Systems to be enforced by responsible unit or responding supervisor.

Analyze information gathering techniques utilized by desk personnel and data dispatch procedures to seek improved means, through training, of dispatching personnel to crime scenes.

Measures: Decrease in response time to crime scenes due to improved manpower deployment SEU activity, response techniques, and data dispatch procedures.

Increase of at-the-scene arrests.

Increase of on-view arrests.

● Goal 4—Provide thorough and appropriate police-related investigations.

<u>*Objective*</u>*:* Significantly increase the development of investigative leads resulting in arrest or probable identification of suspects through use of criminal intelligence files.

Means: Develop and implement a total system capable of effectively gathering, collating, storing and disseminating criminal intelligence by July 1, 19 .

Establish an intelligence unit under the Investigative Coordinator utilizing Investigative Technicians having received data systems training. Theirs will be the responsibility of developing a prototype intelligence system. Assistance will be afforded through an advisory task force comprised of a representative cross-section of departmental personnel.

Establish through the above named groups a report technique to be utilized by field personnel in preparing intelligence reports.

Implement briefing–debriefing meetings between shift personnel with an intelligence monitor to record pertinent data.

Establish an intelligence file system which will make total intelligence packages (suspect histories, acquaintances, vehicles, MOs, cross-references, etc.) readily available to police personnel.

Contact neighboring communities and agencies to tap their criminal intelligence for inclusion in departmental files and reports.

Measures: Increase in clearance rates resulting from utilization of criminal intelligence files during FY 19__-__ as compared to FY 19__ and FY 19__.

Objective: Significantly improve investigative cost effectiveness by decreasing the ratio of investigative hours expended to cases cleared.

Means: Establish criterion for determining case expenditures that will provide for the greatest degree of successful case dispositions and manpower utilization. Investigative standards to be implemented by July 1, 19__.

Analyze case variables to determine those elements conducive to a successful investigation.

Develop minimum standards for case investigations including the level of investigatable elements available, working caseload, time allotted in relation to investigatable elements, etc.

Implement levels of investigation utilizing Investigators, special unit personnel, and nonsworn Investigative Technicians.

Measures: Decrease in man/hours expended on noninvestigatable cases.

Increase in cases cleared in ratio to intelligence manpower expended.

● Goal 5—Provide for emergency services that relate to the protection and preservation of life and property.

Objective: Increase employees' capability in first-aid and safety activities.

Means: Develop and implement a twenty-six hour first aid and safety course to be presented in fifty-two consecutive one-half hour sessions annually beginning the week of July 1, 19__. Components of this course will include:

1. Utilization of City swim lagoon and lifeguards to instruct departmental patrol personnel in lifesaving and water safety.
2. Utilization of PG&E safety personnel to instruct departmental patrol personnel in electrical safety.
3. Utilization of departmental accredited instructors to provide refresher first aid, applicable to field unit use (i.e., choking, bleeding, respiration, shock).

4. Utilization of fire department personnel to instruct in rescue techniques.

Measures: Pre-instruction and post-instruction exams to ascertain the percent increase in knowledge and ability gained through the training program.

Objective: Decrease by a statistically significant degree the response time to emergency calls while decreasing in the same manner personnel injuries and property damage sustained as a result of emergency responses. Comparison to be FY 19__-__ with previous years.

Means: Develop and implement by April 1, 19__ a sixteen hour drivers training program designed to provide qualifications for all sworn personnel by July 1, 19__. The program will utilize departmental instructors certified by the California Highway Patrol Driver Instructor six-week course.

Deploy shift personnel in a strategic manner based on statistical data while continuously reviewing with personnel the most direct routes to various city locations.

Train Information Technicians in data gathering and radio procedures to direct field personnel utilizing the most complete location available (i.e., _____ address, the Smith Manufacturing Company or the Lexington Garden Apartments, east complex, southeast side).

Measures: Decreased response time (statistically significant).

Reduced police vehicle damage and personnel injury (statistically significant).

Reduced incidents of personnel injury and property damage due to emergency responses (statistically significant).

● Goal 6—Provide for social order during times of unusual occurrences.

Objective: Increase the level of departmental readiness to respond to the scenes of unusual occurrences.

Means: Develop and maintain a local emergency

activation program that will have deployed all available personnel within a two-hour period at proper locations, equipped and informed as to their responsibilities in the operation. All personnel will know the purpose and procedure of this program.

Squad room training sessions reviewing personnel emergency action, reporting stations and equipment check-off list.

A field exercise simulating the call up and deployment to specified locations and the performance of specific duties.

Operational guidelines and directives updated at least once annually and disseminated to all personnel involved.

Measures: The time required to call out and deploy fully equipped personnel to a simulated disaster area.

Written tests concerning assembly point, code designations, equipment location, and identification procedures for equipment release, and individual's role and responsibility within the operation.

Objective: Increase departmental capability for controlling mass disorder threatening life and property.

Means: Organize and actively train an unusual occurrence force expert in crowd control, rescue techniques and the suppression of a planned and prolonged armed attack.

Coordinate with adjoining law enforcement agencies and emergency forces (i.e. fire, ambulance) to establish unusual occurrence procedures and responsibilities.

Prepare advanced procedures for making and processing mass arrests consulting with the court system to determine areas of special concern.

Determine equipment responsibilities and provide all necessary means for the personnel to meet these responsibilities.

Measures: Degree of readiness of unusual occurrence force as determined through testing and drills.

Percent of necessary equipment available.

Increased understanding among all personnel, of procedures to be followed regarding personnel deployment, supervision, individual responsibilities, and equipment issue.

Understanding determined through testing and drills.

● Goal 7—Provide such services that contribute to the preservation of community health and safety and general public assistance.

Objective: Increase formal action taken on abandoned property over FY 19_ - _ compared to FY 19_.

Means: Assignment of Police Service Technicians to seeking out and reporting parties responsible for the abandonment of property such as automobiles and refrigerators.

Measures: Decreased incidence of abandoned property. Increase in reports.

Objective: Correct all hazardous conditions in public streets.

Means: Assign responsibility to Police Service Technicians to spot and immediately report any hazardous condition and its exact location for corrective action by the appropriate city department.

Measure: Percent increase in hazardous conditions corrected.

Objective: Increase student awareness in traffic and personal safety.

Means: Develop a city-wide school safety program appropriate to the various grade levels for presentation by September 1, 19_.

Utilize Police Service Technicians and Police Officers in presenting programs to all city school children.

Utilize outside firms and sources for equipment, materials, expertise, and instructors for presenting specific areas of safety to students.

Measures: Passing test scores on all material presented to insure retention of important points within the program and to allow certification.

● **Goal 8**—Provide for effective coordination between agencies related to the criminal justice system.

Objective: Increase communications between the court system and the police department.

Means: Creation of a Court Liaison Officer whose primary responsibility is the coordination of data and relations between the court system and the police department. (The court system includes the judges, and staff, District Attorney's office, Public Defender's office and Marshal's office.)

Development of in-house training on the court system for police personnel with sessions involving court system personnel.

Exchange work program allowing members of the court system and police department to accompany one another in the course of their work day.

Measures: Before and after substantive examinations designed to measure increases in understanding of court systems operation.

Objective: Increase in "intelligence" exchange between local offices of the criminal justice system and the police department.

Means: Representative meetings between staff and line personnel of the various segments of the criminal justice system to ascertain specific intelligence needs. This program will be coordinated by an Investigative Coordinator, Investigation Unit.

Development of specific report techniques, data gathering systems, reproduction and routing procedures that will generate the required intelligence for appropriate dissemination.

● **Goal 9**—Provide for effective and efficient departmental administration and for the career needs of departmental personnel.

Objective: Review and refinement of departmental goals and objectives.

Means: Establish a task force representative of departmental structure to analyze existing goals and evaluate the validity and progress of objectives. It would be the responsibility of this force to insure departmental understanding of all goals and objectives pursued by this organization.

Measures: Test personnel understanding of departmental goals and objectives.

Increased personnel involvement in the goal/objective establishment process.

Objective: The equitable and meaningful performance evaluation of all departmental personnel.

Means: A task force representative of the department to analyze current personnel evaluation processes and proposed alternative. It would be their responsibility to ensure that evaluation measures would accurately reflect the worker's performance and provide real meaning to the worker as well as to the needs of the department.

Personal communications to all persons evaluated and to all persons evaluating setting forth in straightforward terms the intent, means and standards utilized in the evaluation process.

Continual involvement by all levels of departmental personnel in study and revision of personnel evaluation procedures.

Measures: Increased standardization of personnel evaluations.

Increased common understanding of the evaluation process by all levels of departmental personnel.

Increased training programs based upon meaningful personnel evaluations.[22]

Writing Objectives

The example cited above clearly illustrates the simplicity of well-stated goals as well as clearly identifying the relationship to an objective. The writing of objectives is not an easy task—the following criteria should be applied to each objective:

1. It starts with the word "to," followed by an action verb.
2. It specifies a single key result to be accomplished.

3. It specifies a target date for its accomplishment.
4. It specifies maximum cost factors.
5. It is as specific and quantitative (and hence measurable and verifiable) as possible.
6. It specifies only the "what" and "when"; it avoids venturing into the "why" and "how."
7. It relates directly to the accountable manager's roles and missions and to higher-level roles, missions, and objectives.
8. It is readily understandable by those who will be contributing to its attainment.
9. It is realistic and attainable, but still represents a significant challenge.
10. It provides maximum payoff on the required investment in time and resources, as compared with other objectives being considered.
11. It is consistent with the resources available or anticipated.
12. It avoids or minimizes dual accountability for achievement when joint effort is required.
13. It is consistent with basic company and organizational policies and practices.
14. It is willingly agreed to by both superior and subordinate, without undue pressure or coercion.
15. It is recorded in writing, with a copy kept and periodically referred to by both superior and subordinate.
16. It is communicated not only in writing, but also in face-to-face discussions between the accountable manager and those subordinates who will be contributing to its attainment.[23]

Unlike a public works agency (that can set specific tangible objectives) as well as other non-profit agencies, the police departments perform many activities that have somewhat intangible results. The difficulty of quantification and measureability is no excuse for not making use of objectives wherever possible. The test of what constitutes a sound objective as identified by Reddin includes: measureable, specific, realistic and attainable, and time-bounded.[24]

Acknowledging the difficulty of identifying objectives which must be matched by some method of measurement, Odiorne has identified criteria which can be utilized when attempting to measure objectives by some method other than raw data.[25] He suggests that objectives can be measured in terms of gains, risks, expected values, utility, or other criteria that will indicate change, or indicate the degree to which action has been effective.[26]

Positive participation is essential to the success of a management by objectives systems. This was emphasized by Douglas McGregor who (in describing how management by objectives should work) stated: "Genuine commitment is seldom achieved when objectives are externally imposed. Passive acceptance is the most that can be expected—indifference or resistance are the most likely consequences."[27] However, it must be pointed out that there is some controversy over the question of how much autonomy the subordinate should have in the creation of his own managerial objectives as well as for those of the unit in which he works. Participation is essential to success when the managers of a law enforcement agency anticipate participation in the decision making process and the creation of goals and objectives. In a police agency, job content is the source of motivation and encompasses the factors that can produce effective job performance for that agency. Once an individual has committed himself to organizational goals and objectives, his potential is tremendous, and it is the joining of the organizational objectives with that of the managers objectives which provides the basis for a synergistic organization. Personal meaning and motivation must be provided in the work environment, and this is best accomplished through job content which allows for the creation of a goal-oriented, rather than a task-oriented individual. Charles L. Hughes in his excellent book on goal-setting, emphasizes the importance of creating a work environment to develop individuals who are goal-seekers rather than task-accomplishers. He identifies goal-seekers as individuals who exhibit confidence and are action minded. It is a type of individual who has a strong need to tackle tough goals and achieve them, not just well, but with a high degree of excellence. They set long- and short-range goals for themselves and plan their job content activities accordingly, and as part of this type of accomplishment they seek feed-back and knowledge of results. According to Hughes, (see fig. 6-4) the environment that develops goal-seekers is that in which the supervisors take responsibility for results, and where goal-setting is more likely to occur when work is described in terms of opportunities to

**Figure 6–4
Goal Oriented Vs. Task Oriented***

GOAL-ORIENTED INDIVIDUAL	TASK-ORIENTED INDIVIDUAL
1. Seeks feedback and knowledge of results. Wants evaluation of own performance. Wants concrete feedback.	1. Avoids feedback and evaluation. Seeks approval rather than performance evaluation.
2. Considers money a standard of achievement rather than an incentive to work harder.	2. Is directly influence in job performance by money incentives. Work varies accordingly.
3. Seeks personal responsibility for work if goal achievement is possible.	3. Avoids personal responsibility regardless of opportunities for success.
4. Performs best on jobs that can be improved. Prefers opportunity for creativity.	4. Prefers routine nonimprovable jobs. Obtains no satisfaction from creativity.
5. Seeks goals with moderate risks.	5. Seeks goals with either very low or very high risks.
6. Obtains achievement satisfaction from solving difficult problems.	6. Obtains satisfaction not from problem solving so much as from finishing a task.
7. Has high drive and physical energy directed toward goals.	7. May or may not have high drive. Energies are not goal-directed.
8. Initiates actions. Perceives suggestions as interference.	8. Follows others' directions. Receptive to suggestions.
9. Adjusts level of aspiration to realities of success and failure.	9. Maintains high or low level of aspiration regardless of success or failure.

*Reprinted by permission of the publisher from Goal Setting by Charles L. Hughes, copyright 1965 by American Management Association, Inc.

set goals instead of simply establishing performance tasks. He suggests that the employee should be exposed to competition in his own managerial level and a climate should be created which allows for maximum feed-back and continuous evaluation. Managers in such a climate should be sensitive to people and their work environment and strive for a balance between the long-range goals of the individual and those of the organization. Hughes perceives this process as a means of achieving greater organizational mental health through the intrinsically therapeutic or preventative potential of work itself.[28]

The success of management by objectives in a law enforcement agency depends on the ability of the police chief executive to create an organizational environment of maximum responsiveness to the community and the employees of the organization. The total planning process, geared to the management by objectives system, should be focused on the future, and the organizational climate should be outward-looking with a strong people-orientation geared to the accomplishment of results. MBO can only be effective if it is practiced in a participative style encouraging maximum teamwork that creates the servo-climate for results.

MBO: The Human Side*

This paper therefore considers how police labour once obtained can be rendered more efficient and effective. It concerns in the main the appraisal of individuals and their contribution to the purposes of the organization. However, M by O is not content merely to appraise the individual's present performance but also provides for organizationally inter-related tactics for improving the individual's contribution to the organization's purposes.

Appraisal

When first formulated, management by objectives as an overall organizational strategy was no doubt intended to supplant earlier strategies for appraising the performance of individuals within the

---

*Source: Robert Goodall, "Management by Objectives, A Conceptual Application for the Police, 2—The Human Side," *Police Journal,* Vol. XLVIII, No. 3, July, 1974, pp. 251-261. Reprinted with permission of the publisher.

organization. At that time most appraisal schemes focused on personality or potential appraisal.

While there may be Force differences as to the content of the appraisal form, the broad categories used have been in the past somewhat universal and have included such factors as initiative and diligence, co-operation and attitude, judgment, reliability, supervisory ability or the acceptance of responsibility, appearance and finally quality of work. This list is neither exhaustive nor exclusive. The major criticism of personality or potential appraisal was that several of the factors included in such appraisal required value judgments on behalf of superiors and therefore were subjective. Other factors could be relatively objectively assessed whether against predetermined standards or not. It was gradually realized that this form of appraisal went hand in hand with nepotism or favouritism with the result that frequently overall organizational performance suffered when particular individuals, who were not necessarily well qualified or suited, were pushed on up the organizational hierarchy. The system was generally self-generating.

A few of the more progressive managements realizing the drawbacks of this early system instituted systems of performance appraisal which either went hand in hand with systems of personality appraisal or replaced them entirely. If anything, the M by O scheme has swung to the polar extreme with the focus being almost entirely on the individual's performance rather than his personality. It is my own personal view that the two systems supplement each other and should therefore co-exist within the organization. The main points of difference between the two systems of personality appraisal and performance appraisal are that personality appraisal tends to view all individuals within a particular rank as having certain universal personality characteristics facing similar job tasks and likely to meet the similar problems in the performance of those task. To outline this, some organizations drew up job specifications or profiles of the work shown in Figure 6-5.

The M by O system of performance appraisal on the other hand tends to focus on the dissimilarities of individuals within a particular rank. In this sense, the individual, having regard to the circumstances of his situation which is determined, among other things, by the geographical, demographical, technological, infra-structural and sociological characteristics of his beat, section, sub-division or

Figure 6-5
Job Profile

Job Title: SUB-DIVISIONAL OFFICER
Rank: Chief Inspector
Superior: Deputy Divisional Commander
Immediate Subordinates: Duty Officers and Shift Inspectors

DUTIES AND RESPONSIBILITIES

1. Subject to the requirements of the Divisional Commander or his Deputy, directs and controls police activities in ... Sub-Division of ... Division.
2. With the assistance of immediate subordinates, ensures that other personnel (police) in his command effectively discharge the duties and obligations falling upon them.
3. Ensures that subordinate personnel are kept in brief with Force policy and procedures.
4. Maintains watching brief of hours of duty, overtime, extraneous duties arranged by immediate subordinates, and in accordance with instructions of the Divisional Commander or his Deputy.
5. In accordance with Force Policy on Major Incidents, ensures the effective implementation of major incident drill occurring within his jurisdiction.
6. Liaises with specialist departments of the Force and with external agencies, such as Fire, Ambulance, Air Sea Rescue, Bomb Disposal, on all matters requiring specialist attention.
7. Supervises the activities of immediate subordinates, completes appraisal forms in respect of them, and counsels them regarding the effective discharge of their duties.
8. Superintends arrangements for the Burgh court and attends to any matters arising therfrom which require the further attention of police.
9. Liaises with local notaries, councillors, local authority officials on all police matters arising within his jurisdiction.
10. Ensures that all press releases are attended by the appropriate Press Liaison Officer, in accordance with Force Policy.

EXTRANEOUS DUTIES

11. Ex-officio chairman of sub-divisional branch of Force Social and Recreational Club.
12. Refers to Force Welfare Officer, all matters believed requiring of his attention.

OVERALL RESPONSIBILITIES

13. The Sub-Divisional Officer has overall responsibility for his Sub-Division throughout a period of 24 hours, in terms of activities of personnel and in terms of buildings, vehicles and equipment in his command.
14. He is required to submit annually on the prescribed dates his budgetory and manpower returns and estimates in accordance with Force Policy.

whatever, is assessed according to predetermined standards of performance, standards which were not imposed arbitrarily on him by the organization but with regard to which he had contributed in a counselling session with his superior and which he personally had agreed to endeavour to obtain. It is necessary to emphasize that this scheme is a *management* by objectives approach to appraisal and therefore it is intended to focus only on those ranks which are of a management or in some instances a supervisory nature. It is possible that through time there could evolve a system of policing by objectives covering all ranks within the Service. However, even the first step, that of instituting a system of management by objectives, would be a major task. For the purposes of this paper the jobs we are interested in have some or all of the following characteristics: the job holder must have a clearly defined area of responsibility, geographical or functional, such as command of a division, subdivision, section or department. The job holder must have overall responsibility for subordinates acting on his behalf or on his instruction; this responsibility must extend through a given time period. The job holder must have responsibility for the material and equipment resources in his command at a given time. The limits of his authority must be clearly defined and must be in direct relationship to his area of responsibility. Finally, any extraneous duties or obligations which fall upon the job holder must be detailed.

Under the M by O scheme, the job specification or profile is superseded by a key results analysis (K R A). A specimen form is attached in Figure 6-6. This form is normally divided into six main sections—*the job title,* naming the holder and detailing the contribution of the job to the overall efficiency of the organization; secondly, the *position* of the job holder *in the authority structure* of the organization, that is to whom and for whom directly responsible; thirdly, *the scope of the individual's job,* his total responsibilities in terms of geographical area and operational resources, whether material, equipment or financial; fourthly, the *personal activities* of the job holder, that is duties and obligations which he personally attends to and which are not delegated on his behalf; fifthly the *limits* of the job holder's *authority;* and sixth and finally, the job holder's *key tasks.* This latter element is probably the most important singular aspect of the K R A and will be considered separately and at length.

Figure 6-6
Key Results Analysis

1. MAIN PURPOSE OF JOB (and its contribution to the efficiency of the organization)
To ensure that each man in his command efficiently and effectively fulfils the primary functions of the service and discharges the duties prescribed by Force Procedure, through direct personal supervision and with the aid of immediate subordinates.

DIVISION: SUB-DIVISION:

2. *Position in Organization*
Rank: Chief Inspector   Number: 1001
JOB TITLE: Sub-Divisional Officer
Immediate Superior: Deputy Divisional Commander
Immediate Subordinates:

3. *Scope of Job* (indicate total responsibilities in terms of men, material, facilities and equipment)
  i. Area of Command
  ii. Number of men in command
     ... Inspectors
     ... Sergeants
     ... Constables
     ... Civilians including Cadets
     ... Policewomen
     ... Force Task or Support Group
     ... Special Constables, etc.
     ... Personnel Carriers
     ... Prison Escort Vehicle
     ... Sub-Divisional Vehicles
     ... Panda Cars
     ... Major Incident and Subsidiary Equipment
  iii. Surplus Personnel
  iv. Operational Material Resources

4. KEY TASKS
  i. Description of Task
    a. To obtain an overall reduction in % rate of increase in crime p.a. of 4%
    b. To obtain similar reduction in road accidents
    c. Aspects of personnel management
    d. Occurrences and Industrial Accidents

  ii. Level of Performance
    a. 1% cumulative reduction in each of 3 monthly periods commencing January April, July and October
    b. As above

  iii. Data Used for Measuring Performance
Monthly Crime Statistics prepared by Divisional Statistician

Monthly Road Accident Returns of Traffic Dept. Statistician

  iv. Comments
    a. Increased ground coverage
    b. Liaise with Crime Prevention Department
    c. Training of Subordinates

    a. Liaise with Traffic Supt.
    b. Identification of Black Spots and reporting of potential hazards

N.B.: These categories are likely to differ from man to man and from area to area and are worked out by job holder and his superior and if necessary approved at higher level.

5. PERSONAL ACTIVITIES: Duties actually performed and NOT delegated

6. Limits of AUTHORITY

Key Tasks

As mentioned earlier, it was formerly assumed that persons of the same rank could be assessed in accordance with a uniform set of criteria. Such assessment does not however take into consideration the differing circumstances of jobs within a particular rank, at least not in a formal way. Key results analysis from the outset considers that different officers are faced with different situations, that while 80 *percent* of the content of jobs in a particular rank may be broadly similar, 20 *percent* may be radically different. That is not all, different persons tend to perceive the same situation in different ways. Key results analysis focuses on the individual's perception of his own situation and considers also his superior's perception of the same situation, this is most important. It has been suggested that the essential definition of key tasks are those duties which *must* be performed if the main purpose of this job is to be achieved, and that these tasks should be concerned with results or outputs rather than with activities, though there are instances where tasks are concerned with activities only. It is felt by some that defining key tasks is probably one of the most contentious areas when considering the application of management by objectives to the Police Service. I agree with this view but I do not necessarily accept that it renders M by O inapplicable to the Police Service. It poses an essential and exciting challenge.

When compiling the K R A the subordinate is asked to prepare a list of the key aspects of his job, those aspects which command his immediate and/or continuing attention. It is accepted that the individual may be considering his personal position in isolation and with little regard to the needs of the organization as a whole. Because of his subordinate position, he may have no knowledge of the overall needs of the organization though under the M by O scheme this would be unlikely as it is essential that the purposes of the organization be known to all the individuals within. In order that the needs of the organization as a whole be brought into line with the needs of the subordinate, the superior also reviews the situation this time from a wider perspective and prepares what he considers to be the key characteristics of the subordinate's job. When the superior and the subordinate then meet in a counselling situation, they endeavour to reach agreement on the factors of most importance.

These are then listed under key tasks. Now it may be considered that this procedure can undermine the authority of the superior. This is not necessarily so, for in the event of a disparity of opinion between the subordinate and his superior, the criterion applies that the needs of the organization take precedence over the needs of individuals within the organization.

Performance Standards

Assuming that agreement is reached on the key tasks and a description of them has been prepared, the M by O scheme requires that certain minimum standards of performance be set. It is important to note that in this section of the K R A we are not merely interested in the individual's performance at his job, that is the personal activities he performs. We are interested more in his ability to command. It must be borne in mind that we are appraising persons who are in their own right managers, and that while he agrees to the standards set, he must consider the capabilities of the men in his command, the strength of his resources, his abilities as a manager, and so on. Having agreed the standards, he must encourage, counsel or otherwise procure his subordinates, whether by personal example or otherwise, to increase their performance in order to bring about the standards acceptable to the organization as a whole. There are a number of different strategies or tactics open to the police manager to achieve the standards to which he has agreed, either by re-deployment of personnel and resources or by liaison with other force units, specialists, and so on. It is up to the unit commander to choose which line of action he personally is to pursue. It is he who will be held accountable for the overall performance of the unit at a later date. It might be considered that there is a danger where the individual, in attempting to meet the agreed performance standards, neglects the other aspects: the other 80 *percent* of his total job. For this purpose there may be instituted a periodic monthly or bi-monthly informal review. However, it must be noted that any aspect of an individual's total job consistently neglected would, in an organization such as the police, soon come apparent to all parties concerned; the man, his subordinates, his superiors, and in extreme cases the public.

In formulating key tasks and setting standards of performance, we are interested in what is happening within the geographical area for which the officer is responsible. Is there unnatural wastage of personnel; are complaints against the police in that area increasing or decreasing; are the number of crimes or occurrences of road accidents, reported or made known to the police, increasing or decreasing? If so, what is the individual doing about it; how is he motivating, deploying, superintending his subordinates, and many many other factors. Any or all of the above-mentioned aspects; wastage, complaints, crime, occurrences, road accident, or whatever may be considered by the superior and the subordinate to constitute a key task, a task to which the subordinate must give special attention.

We are then faced with the question of whether performance standards can be set for the police manager. There is a view, and I would not necessarily disagree with it, that within the Police Service there are many jobs for which performance standards can not be set. In some instances we have to recognize that in many cases we are seeking goals which can only be described and that to impose quantification upon these goals before they are shown to be capable of quantification could make the measures of performance meaningless and may serve to distract attention from the real purposes of the job. However, there exist within the realm of police work many areas which can produce quantifiable standards. For example, in the realm of crime, road accidents and occurrences, most if not all forces produce statistics detailing percentage increases or decreases in reported crime, road accidents, occurrences or whatever; and I would submit that in the case of operational job tasks, such statistics as are at present produced can be used as a basis for quantification of performance standards.

Assume, for example, that a given sub-divisional officer is faced with the situation that crime, road accidents and occurrences in his sub-division are showing percentage rates of increase of 7 *percent* per annum per crime, 5 *percent* per annum for road accidents, and 8 *percent* per annum for other occurrences. A statement of performance standards would be to obtain for example 3 *percent* reduction in the annual percentage rate of increase in each of the categories, comprising a 1.5 *percent* decrease in each of the periods of six months commencing respectively January 1 and July 1 in the year over which performance is to be appraised. If we were to further assume that the standards which were set were achieved over that year, then we would expect the Chief Constable's Annual Report on that sub-division to show that crime had increased over that year by 4 *percent,* road accidents by 2 *percent* and other occurrences by 5 *percent.*

Data Used in Measuring Performance

The key results analysis form should also detail the sources of data which will be used for measuring the individual's performance. Use could be made of the operational statistics currently produced by most forces, though some consider that this data is of doubtful value certainly in judging success or otherwise of achieving the standards set; alternatively, there may be a need for the police management to review the systems of data collection currently used and produce new methods of data collection relevant to the system of management by objectives. What is important is that the job holder and his superior know where to go for the information against which performance will be assessed to date.

Comments

Generally, the K R A form includes a section for comments, under which are listed suggestions for job improvement and possible limitations on job performance. The individual who is having trouble in motivating his subordinates or with excessive manpower wastage for example, may have need for a course of specialist training in man-management. On the other hand, if a particular area is faced with a high incidence of crime to the exclusion of many other things, this may indicate the need for the specialist assistance of the Crime Prevention Department, or the need for the help of the Task Force or Support Group. At the same time, this section under "comments" might include suggestions for job improvement of the individual, such as planned experience in other aspects of the organization's purpose as an aid to further progression and within the authority structure of the police organization. But, whatever is included under this column would again be agreed by both the superior and the subordinate. We have therefore under the category of key tasks in the K R A form; description of the task to which special attention must be given, the level or standard of performance which has to be obtained, the data which is to be used to measure the performance, and comments on likely limitations and performance and/or suggestions for personal improvement of the individual.

What of the man who does not meet this targets, whose performance falls short of that agreed in the couselling session? In the short run, there may be good reason for this: short term exigencies of personal duties, a number of major unforeseen crises, circumstances beyond the man's control resulting from senior policy-making decisions for which the job holder can not and should not be held accountable. In the long run however, failure to meet targets persistently may indicate either unrealistic target setting or a need for intra or extra-organizational training in the sphere in which performance is lacking, or a need for the individual to be transferred to some other job for which he is more suited, or in the extreme case which is virtually unknown to any person to whom I have spoken, that is a need to demote the individual to his former rank, a rank in which he was presumably an effective contributor. What ever action is taken, M by O requires that organizational considerations have to take precedence over individual considerations however painful they might be to the individual.

Key Results Analysis and Other Aspects of the Personnel Function

There is a two-way relationship between K R A and other aspects of the personnel function as shown in Figure 6-6. In some cases the relationship is a direct one, as in the case of promotion, training, development (both organizational and individual) and management succession, while in other cases the relationship is less direct, such as with welfare, transfers and recruitment.

Management by objectives is concerned not merely with the development of some individuals within the police organization but with all persons. This is best illustrated by an example, *vis.*, it is possible that the K R A form for a patrol inspector might include, under the heading Key Tasks, to ensure that each member of his section receives adequate training in the various roles which he has to play, including development in other aspects of the police function. We would then expect the patrol inspector, having regard to the constraints imposed, to critically examine the contribution and the potential of each person in his section and to make recommendations under Part 2 of Figure 6-6. Now it may be that he feels that some members of his section would benefit from a spell as an aide to the C.I.D. or to the Traffic Department or the Crime Prevention Department or the Road Safety Department, or whatever other department of the force he considers suitable. Alternatively, he may feel that there is a need for re-training of a number of members of his shift or section in the general purposes of the police organization with a view to bringing them up to date with aspects of the law or procedural duties; and he may feel that other members of his section are adequately trained and perform well in their present job role and that, as they have the necessary qualifications, they are suitable for promotion.

The patrol inspector would then be expected to put his various recommendations, regarding each of the personnel in his section, on paper for consideration by higher authorities in the Service. Now assuming that not one but a number of patrol inspectors recognize there to be a similar need in training, for example, in latest Crime Prevention techniques, we would expect the higher authorities to pick this up and put forward a recommendation to the Training Officer to establish a course in this deficient need. Again assuming that one of the key tasks of the Training Officer is to attend with expediency any requests for a training course in whatever subject is required, we would expect the training officer to establish a course on Crime Prevention techniques and make the necessary arrangements with the various Divisional Commanders to have the officers released for such a course.

Many persons may feel that this sort of procedure is currently carried on within the police organization at present and that there is not a need to alter the same. However, the difference between this procedure and that currently in operation is that the decision on which officers require training or development is taken at a much lower level in the police organization, namely by the patrol inspector. It is he who is in touch with the various personnel in his section and it is he who knows by what training they would benefit. Therefore, it seems feasible under the management by objectives scheme that he should initiate the training or development procedure.

Conclusions

There can be little doubt about the necessity for a standard system of appraisal with a view to the pending developments in 1974/75. In order that each individual has equality of opportunity and consideration by the heads of the newly amalgamated forces, objectivity in appraisal so far as it can be achieved must be sought. Initially, potential appraisals, such as those suggested by the committee on staff appraisal for the Police Service in Scotland, may suffice for the lower ranks in the Service. However, because of their essentially subjective nature they may not be entirely suitable, particularly for the promoted ranks. A measure of objectivity can be obtained within the management structure in the Service using the procedures outlined in this paper notwithstanding certain obvious drawbacks. However, where potential appraisal and performance appraisal are used in conjunction, the heads of the amalgamated forces concerned would be furnished with a more complete, though perhaps still wanting, work picture of the various officers in his force. This truth has been attested in other organizations to which the M by O scheme has been applied, and I believe it can be realized in the Police Service also.

With the advent of regionalization, now is probably the best time to consider implementing a scheme like management by objectives. As I have argued, management by objectives is a complete organizational system, which if systematically applied could provide the necessary cohesion for the larger regionalized units. The policies of the regionalized units would be known, plans would be formulated to achieve those policies, individuals would know how they are to contribute to the plans, and steps would be taken to ensure that each individual received the training which he or she would require in order to perform those duties.

Topics for Discussion

1. Discuss the advantages and the disadvantages of MBO.
2. Discuss the values of participative management.
3. Discuss the measurability of police objectives.
4. Discuss the relationship of the budgeting process to MBO.
5. Discuss the need for top management support when implementing an MBO system.
5. Discuss the philosophy of MBO.

FOOTNOTES

Introduction

[1] George S. Odiorne, *Mangement by Objectives* (New York: Pitman Publishing Corporation, 1971), p. 54.
[2] Peter F. Drucker, *The Practice of Management* (New York: Harper and Brothers, 1954).
[3] Dale D. McConkey, "Applying Management by Objectives to Non-Profit Organizations," *Advanced Management Journal,* Vol. 38, No. 1, January, 1973, p. 11.
[4] Robert Goodall, "Management by Objectives, A Conceptual Application for the Police—1," *The Police Journal,* Vol. XLVIII, No. 2, April, 1974, p. 180.
[5] Dale D. McConkey, "MBO—Twenty Years Later, Where do we Stand?" *Business Horizons,* Vol. XVI, No. 4, August, 1973, p. 27.
[6] Dale D. McConkey, "MBO—Twenty Years Later, Where do we Stand?" *Business Horizons,* Vol. XVI, No. 4, August, 1973, p. 28. Reprinted by permission of the publisher.
[7] *Ibid.*
[8] Odiorne, *op. cit.*, p. 55.
[9] John W. Humble, *Management by Objectives in Action* (London: McGraw-Hill, 1970), p. 3.
[10] W.J. Reddin, *Effective Management by Objectives* (New York: McGraw-Hill, 1971), p. 19.
[11] Odiorne, *op. cit.*, pp. 55-56.
[12] Reddin, *op. cit.*, p. 13. Reprinted by permisison of the publisher.
[13] Goodall, *op. cit.*, p. 180.

[14] C.C. Leek, "Management by Objectives," *The Police Journal,* Vol. XLIV, No. 3, July-September, 1971, p. 214.
[15] McConkey, "Applying Management by Objectives to Non-Profit Organizations," S.A.M. *Advanced Management Journal,* Vol. 38, No. 1, January 1973, p. 12. Reprinted by permisison of the publisher.
[16] J.D. Batten, *Beyond Management by Objectives* (New York: American Management Association, 1966), p. 54. Reprinted by permisison of the publisher. Copyright, 1966, by American Management Association.
[17] Goodall, *op. cit.,* p. 181.
[18] National Advisory Commission on Criminal Justice Standards and Goals, *Police* (Washington: U.S. Government Printing Office, 1973), pp. 49-50.
[19] *Ibid.,* p. 50.
[20] American Bar Association, *The Urban Police Function* (New York: American Bar Association, March, 1972), p. 53.
[21] Social Development Corporation, *Use of Manpower in a City Police Force* (Bethesda: Social Development Corporation, January, 1973), pp. 18-19.
[22] *Ibid.,* pp. 21-31.
[23] George L. Morrisey, *Management by Objectives and Results* (Reading, Massachusetts: Addison-Wesley Publishing Company, 1970), p. 62.
[24] Reddin, *op. cit.,* p. 89.
[25] George S. Odiorne, *Management Decisions by Objectives* (Englewood Cliffs, N.J., 1969), p. 27.
[26] *Ibid.,* p. 33.
[27] Odiorne, Mangement by Objectives, *op cit.,* p. 142.
[28] Charles L. Hughes, *Goal Setting* (New York: American Management Association, 1965), p. 38. Reprinted by permission of the publisher. Copyright, 1965 by the American Management Association.

SELECTED READINGS

Brady, Rodney H., "MBO goes to work in the public sector," *Harvard Business Review,* March-April, 1973, pp. 65-74. Discusses the application of management by objectives to the Department of Health, Education and Welfare. Problems are described that must be overcome before MBO can be effective in the public sector.

Goodall, Robert, "Management by Objectives, A Conceptual Application for the Police—1," *Police Journal,* Vol. XLVII, No. 2, April, 1974, pp. 178-186. Presents the case for applying MBO to the police as a comprehensive management system providing for systematic management, timely forecasting and planning, and a network of management control tempered with participative management.

Griffin, Gerald R., "Goal Setting for Police Organizations," *The Police Chief,* Vol. XLI, No. 5, May, 1974, pp. 32-34 and 74. Suggests that a program of goal setting can solve many police problems. Discusses the setting of goals and the fact that they should be positive, definite and measurable. Reviews the advantages of goal setting and cautions to be considered.

Leek, C.C., "Management by Objectives," *Police Journal,* Vol. XLIV, No. 3, July-September, 1971, pp. 212-223. Outlines the principles of management by objectives with an emphasis on its advantages to the organization. Reviews the benefits of MBO and includes a detailed description of performance review.

Peart, Leo E., "A Management by Objective Plan," *The Police Chief,* Vol. 38, No. 6, June 1971, pp. 34-40. Describes the theory and practice of management by objectives and shows how it can be developed and applied to a police organization. Discusses the six phases in applying the concept.

Pudinski, Walter, "Mangement by Results in the California Highway Patrol," *California Law Enforcement Journal,* Vol. 8, No. 4, April, 1974, pp. 194-198. Reviews the application of Management by Results in the highway patrol in four areas: (1) integration of the concept of participative management, (2) education and training of managers, (3) implementation of an OD program, and (4) development of a long range planning process.

Pudinski, Walter, "Managing for Results," *The Police Chief,* Vol. XL, No. 1, January, 1973, pp. 38-39 and 64. Presents in general terms the value of combining management by results with participative management. Supports the position that responsibility, achievement, recognition and growth opportunities are factors that motivate people to better performance.

The study of this chapter will enable you to:

1. List the three elements perpetrating the myth of the magical leader.
2. Differentiate between magical leadership and diagnostic leadership.
3. Identify four of the more significant demands of the leadership role.
4. Write a short essay describing the synergistic results of a manager's heredity, environment and self-generated effort.
5. List four suggestions that serve as guidelines when dealing with a work group.
6. Identify the reasons why a manager should be an astute organizational analyst.
7. Write a short essay describing the continuum of leadership behavior.
8. List the three forces that a manager should consider when deciding how to manage.
9. Identify four internal forces that affect the manager.

7

Leadership Styles

Introduction*

One of the most troublesome and perplexing dimensions of any leader's role concerns his "ability to motivate his people." Conversations with executives and others who fill leadership roles in a wide range of organizational contexts support the assumption that the "motivator-leader" model is a highly prized concept. The notion is also found in the literature on leadership, which literature, when "programmed" into training schools and executive development plans becomes promulgated as a *leadership ethic* which many leaders take personally and seriously. An oversimplified reduction of this ethic would be the following: "Good leaders have some personal quality which (as if by magic) equips them to inspirationally motivate their followers."

There are, in the writer's view, two broad reaching and important reasons for the widespread acceptance of this leadership ethic and its many variants.

The first is a cultural or anthropological pattern which specifies and sanctions the myth. The concept that the Good Leader does

---

*Source: J.W. Lawrie, "Leadership and Magical Thinking," *Personnel Journal*, Vol. 49, No. 9, September, 1970, pp. 750-756. Reprinted by permission of the publisher.

something "to motivate his followers" has deep roots in Western culture. Historical analyses, folklore and some brands of political science (from Plutarch to Schlessinger) tend to reinforce the mystique that there is a fleeting and elusive quality "in" the character or personality of some leaders which—as if by magic—equips them to motivate their followers. The "charisma" of such diverse people as Hitler on the one hand, and Jesus Christ on the other is offered as an "explanation" of their "power". Who has not heard of the magic of the FDR fireside chat? Who does not wonder at the adoring throngs at Gandhi's feet? Who cannot thrill at the image of an eight foot tall Charlemagne at the head of his adoring armies? This is exciting and heady stuff, and as children we learn and play out the glamor of the myth.

In an important sense these heroic-leader myths then become our conscious models for what it means to be a "leader of men." We all know Prince Valiant has a magical "singing sword" which *only he* can wield, and this kind of magical quality can become an automatic prototype in our thinking about what it means to be a leader.

Now this myth, like most myths, has a purpose and a function. Its function is to legitimize a certain style of leadership and a certain kind of followership *when the leaders and the followers are widely separated from each other.* It is politically and probably socially useful to create and sustain the adoration of the Indian untouchable for a Gandhi: it is solidifying for the black-lunged coal miner to idolize John L. Lewis; it is valuable to the corporation if its middle managers idolize "our founder" and hang his picture in their offices.

Thus, when leaders and followers are not in personal contact with each other, when they don't really "rub up against each other" in the everyday job of accomplishing the organization's tasks, "charismatic-leadership-at-a-distance" may recruit and sustain a kind of loyalty. In the face-to-face situation, however, the myth model does not work.

There is another root to our acceptance of the magical leader myth. This is a psychological dynamic that, as in most cases, complements or gives rise to the cultural manifestation.

The myth that the good leader is somehow a magical motivator is probably a conscious representation of something that originates at quite deep levels of our personal psyche. There is a part of us that "remembers", however dimly, the powerful magic of a Father who could, *at his own whim,* "make" us feel loved, good and valuable. That same father had the power to "make" us feel small, unimportant and rejected. Now the power of this experience lies in the fact that the child feels he is either valuable or useless *because* of what the Father does and *not* because of anything he does himself. He "believes" the following: "If my Father thinks I'm good, I am . . . If my Father thinks I'm bad, I am." What "I am" seems to depend entirely upon the thoughts of someone other than me. The Father can "make" the son feel good by telling him "You're a good boy." The Father can "turn" him into a bad boy by simply calling him bad. If that isn't magic, nothing is.

Add to this the fact that the Father is the first and, therefore, the most psychologically powerful leader any boy ever has, and you get a feeling for the strength of this early interaction. It can be demonstrated that such early experiences do not simply disappear with time. They can stick with us as psychological vestiges and are unconsciously carried forward as a "frame of reference" into which subsequent experiences are automatically fitted.

It happens, therefore, that two early and powerful types of training experiences join forces to pre-dispose us to psychologically embrace the magical motivator myth. What happens in the home reinforces the cultural myth, and the myth is the conscious representation of the dynamic relationship within the home. With two such formidable and mutually reinforcing influences operating at once, it is no wonder we come to take it for granted that good leaders have, in a magical sense, the ability to motivate us . . . "to turn us on."

In short, the myth of the magical leader can be seen to develop as follows:

1. We are born into a culture that, for political and social reasons, contains a folklore that cherishes a mythology of magical leadership. We incorporate this while young.
2. We, due to our own magical style of thinking when young, associate the first leader, the Father, with magic.
3. These two early influences are mutually reinforcing and blend into each other to become the basis of an unconscious frame of reference which we bring to leader-follower relationships in later life.

This early-earned frame of reference can become translated and even codified into a leadership ethic, thereby perpetuating the values implicit in the myth. Ultimately, the ethic generates a set of norms which are statements about what a "good" leader "should be." These norms, when taken together, paint a word picture of the stereotypic magical leader:

1. The leader should be tall and strong.
2. He should be "masculine."
3. He should be "firm but fair."
4. He should arbitrate disputes among his followers and make judgments to settle their squabbles.
5. He should reward good followers and punish bad followers.
6. He should tell subordinates what to do and how to do it and when to do it.
7. He should have "presence," particularly when he speaks in public.

These norms, and others like them, come to form the basis for a set of expectations that surround leader roles in organizations. They carry with them the assumption that if a leader fulfills these expectations, he will, by charismatic alchemy, have the magic "to motivate" his followers.

If this sketch of a deep, powerful and prolonged conditioning process does justice to the facts, it would follow that leader and followers bring an implicit model or set of expectations to their roles in organizations. Supposing for the moment that the sketch is reasonable, what are the consequences for (a) the followers (b) the leader and (c) their organizations?

Consequences of the Myth: Followers

The most powerful consequence for the follower who brings the myth to his organizational role is that he must *come to hate the leader*. This is an ironic twist of fate for the follower because at the conscious level he seeks, indeed he needs, to "love" the leader, to idolize him. How might his "love-to-hate" transformation take place?

Followers can enter into an organizational leader-follower relationship carrying with them, at the unconscious or automatic level, the wish to reconstitute or mimic the relationship obtained between themselves and the "first leader", the Father. They may even be consciously reminded of the warmth and security they felt with their Father when they speak of the "Old Man." It is probably not accidental that the boss carries this kind of a designation.

But the relationship with the "old man" also has a deeper meaning; it is a kind of daily "instant replay" of relations that raise and re-raise the powerful psychological issue of independence *vs.* dependence. In the case of followers who may have unconsciously opted for dependence, they may wish to reconstitute the early feeling of safety that prevailed between themselves and the first "old man."

It can happen, therefore, that followers automatically or unconsciously bring to their organizational roles a set of "feeling habits" that come to be *unknowingly* anchored on the leader. Furthermore, it has been the writer's experience that this transference tends to operate in direct proportion to the organizational power and status of the leader: The "higher" the leader in the organization, the greater the likelihood that the magical leader myth will be brought to bear upon him. We are reasonably content to allow a foreman to be a "regular Joe", but our corporation president has to be someone we can "look up to." Again, the language of the wish is probably not accidental.

Thus, we may *want* the leader to be magical and, of course, in some ways, he seems to be. He works in better quarters than we do, people seek him out, he makes more money than we do, and in the organization (as well as the culture at large) he tends to have the legitimized authority to "make" us feel good or bad by controlling organizational rewards and punishments. It turns out then that we wanted him to be magical and many of our organizational arrangements serve to operationalize the wish.

So, he seems to have a magic, and we may hope he does. In due time, however, reality raises its punishing head. As we grow and get to know him, the leader "lets us down." If we relate to him on a close basis (as distinguished from political followership) *he must fail in our eyes* because we've unconsciously given him an impossible task. He *cannot* be equal to our magical fantasies . . . *no one ever could*.

Now, if I "believe" in someone as a magical leader and he ultimately "betrays" me, I have at least three ways to deal with my anger that arises out of the "betrayal":

(a) I can understand that my ideas about him were unrealistic and childish. I had given him an impossible job description.
(b) I can "discover" that he was evil all along and that he deceived me. This way it's really not my fault that I believed in him because he was "so clever."
(c) I can create or "find" evidence that someone or something "took" his magic away or changed it somehow.

The first of these resolutions, the one involving self-blame, is probably the least likely to be used. We give up the myth only reluctantly, if ever, because the acceptance of our unrealistic projections as "the culprit" is very difficult. No, we would rather find the culprit outside ourselves. Therefore, it usually turns out that we embrace either "his magic got changed" (assumedly magically) or "he deceived me" or some combination of these alternatives.

If we feel better with the "something sapped his magic" alternative, we can usually find the culprit in his wife, his associates, "changing times," or his health. Many is the "unacceptable" leader who is "excused" because of factors over which he had no control. This is demonology at its most primitive, leading to pity for the leader.

However, if we "discover" that he was "evil" all along and was merely projecting a "charismatic image," the playlet can take another direction with a different outcome. If he cleverly "conned" us, we are completely justified in venting our anger. At one extreme, slowdown apathy and non-involvement, and at the other, outright violence can be observed as the treatment that deceptive leaders "deserve." After all, "he had it coming."

These are possible outcomes when the leader fails to fulfill our wishes, as he ultimately must. But it can also happen that he seems momentarily to be magical. In this case the sequence of events is different, but its outcome is the same: hate and anger.

Suppose we "make" the leader a "Father-god" by projecting our wishes to reconstitute the loved-but-controlled relationship of earlier times. Suppose further that the leader, responding to his *own* wishes, puts on the mantle we offer. He then magically becomes the charismatic but (here's the rub) controlling motivator. It follows that we must "become as a little child" and this entails psychological cost. Indeed, if as Argyris and Maslow have postulated, the thrust of our development is in the direction of independence and self-appraisal, the role of becoming "as a little child" will be felt as psychologically suffocating.

With the passage of time, therefore, we will chafe under the magical motivator. We come to re-feel the pain of the adolescent: We want the leader to control (to "motivate") us and to tell us how to act; but when he does, we have to hate him for it because it feels like we're being smothered. By a twist in our defenses we therefore turn the leader into someone who is "holding me back."

Thus, with time, the result of the magical motivator myth being projected upon face-to-face organizational leaders turns out to be a subtle but powerful transformation of "love" to hate. If the leader lives up to our fantasy in the short run, he comes to be perceived as "dominating"; if he fails to live up to our expectation in the long run, he has "betrayed our trust." These are observable consequences of followers unconsciously playing out their wishes that derive from the powerful magical-motivator–leader myth.

Consequences of the Myth: The Leader

By far, the most damaged person in the myth can be the leader. The leader who takes the charismatic model seriously is in for personal and organizational trouble. But it's easy to see how he is seduced.

His wishes, his organizational charter and the symbols of his office, when taken in combination, all serve to thrust him (sometimes reluctantly) into the role of "Big Daddy." Being a leader is heady wine and the accoutrements of office, which includes the "love" of his followers, may serve to foster and accentuate his own residual wishes to fulfill the magical motivator myth. The short-run reinforcements for playing the magical game can be powerful.

However, when the leader bumps his nose on organizational realities, doubt and anxiety come to haunt him. He comes to feel in some dark corner of his gut that he is not "making it." He's more tired than usual, he may drink more, he finds reasons for being out of the office.

In coping with his feelings, the leader has a set of possibilities that complement those of the followers. He can reason:

(a) "My expectations for myself are unrealistic: It is childish to try to magically motivate my people."
(b) "I really am a motivator but my followers are unresponsive. They don't make'em the way they used to."
(c) "I really am a motivator but the organization has changed in such a way that my style is no longer appropriate as it once was."

Again, the least acceptable alternative is the recognition that his wishes are mythical and he cannot fulfill them. Like the follower, he will tend to look outside himself and is most likely to lay the "blame" on his immediate subordinates. *He* now feels betrayed or "conned" and this legitimizes the application of organizational "muscle." He tends to look for schemes and programs and methods that will *punish* his subordinates into being "motivated." A close look at some of the "management development" and "executive evaluation" plans that are currently fashionable reveals an underlying wish on the part of the boss to "get back at" subordinates who are "not motivated." (Ironically, in the operation of such plans, the most punishing thing that a President can say about one of his Vice Presidents is "he cannot motivate his people.")

If the leader opts for the idea that he is somehow "out of phase" with the organization, he will either hang on or look for a new company. If he hangs on, he becomes more and more lonely and guilty as time goes on because he wishes he could "still motivate my people." If he leaves, he carries the myth with him in the hope it will come true, only to be disappointed in his next assignment.

In either case the leader has unknowingly taken upon himself a responsibility to become that which is impossible because he has been seduced by the motivator myth and the "love" of his followers. Organizations, at every level, contain such leaders, and again the risks inherent in the motivator myth are greater as one ascends the organization hierarchy.

Consequences of the Myth: The Organization

There are consequences for organizations that may, albeit unknowingly, subscribe to the magical leader myth.

Many companies are fertile ground for the implantation of the myth because they are in a fast changing state of flux. One thing that technology brings is organizational change and organizational ambiguity. In the process of responding to demands in the market, to innovation and to competition, organizational ambiguity and corporate chaos increase.

There is nothing wrong with flexibility in organizational relationships, but when the heat is on and people are wondering how to respond, the likelihood of their projecting the magical motivator myth onto the leader is increased because of the ambiguity of the situation. In most companies these days "the heat is on."

What are the symptoms of an organization that incorporates the magical leader myth?

One problem is the selection of leaders. Many companies are forced to pass over their own people and "go outside" because they somehow sense that the men "in house" are too well known to be perceived as magical. Secondly, the energy waste and in-fighting that can take place in companies trying to name a top leader can be tremendously exaggerated by the myth. After all, if you're trying to find a god to become your president, it's a very serious matter. But if, more realistically, you seek a man with certain specifiable human skills, it's not such a life and death issue.

A second indicator in organizations that embrace the magical leader fiction is what may be called "motivation programitis." Since motivation is perceived as something that the leader *does to his followers,* it follows that new and different ways "to motivate" people have to be constantly developed. Fancy incentive systems, management development techniques and leadership training institutes are generated and become obsolete at a rate that rivals Detroit's model change-over. To test the impact of his motivational fadism, sit through a "manpower planning" session, and write down the *implicit criteria* that surround the discussion, and the fanciful wish-fulfillment of it all will make you sick. No one could possibly fulfill the expectations that are half-voiced in these sessions wherein the "magical-motivator" ethic of leadership becomes codified by priestlike staff men.

A dead "give-away" indicator that the myth is operating in an organization is heavily centralized decision-making. If the leaders and the followers have internalized the myth, it is very unlikely that they will be able to tolerate sharing in the decision process. Sharing seems to somehow minimize the magic of the leader and it may, therefore, be psychologically costly. People will defend against sharing decisions if it threatens to take the magic out of their relationship.

Finally, an organizational error that derives from the myth is the assumption that leaders can be expected, because of their "magic," to operate equivalently well regardless of the situation. If what makes for leadership is some charismatic quality of the leader, it follows that the nature of the task and the kind of subordinates are relatively less important. So leaders are transferred from function to function with the expectation that they will "do fine" because they "know how to motivate." This misconception has been known to backfire.

Diagnostic Leadership: An Alternative

The question becomes "Is there a better way; a more realistic conception?" Data and theory from psychology suggest that we seem to be moving toward an appreciation of a new model of face-to-face leadership. The concept is not fully formulated in a strict theoretical style, but there seems to be enough common thinking to justify the following description of the diagnostic leader model.

1. The diagnostic leader firmly eschews the notion that motivation is something that he must charismatically put "into" or "onto" his followers. He knows he cannot "make" them "want to do their jobs." He knows that motivation is a noun and not a verb. He has endured many talks delivered by other leaders and has come to learn that the effect of such "inspiration" is about as long lasting as the speech itself.

    On the contrary, the diagnostic leader begins with the empirically demonstrable fact that to be alive is to be motivated. As some wag put it: "The unmotivated man is a good definition of a corpse." He recognizes in the same sense that *solubility* is a quality of *sugar*; *motivation* is a quality of *human life*. This recognition, if it occurs at a relatively deep level of personal insight, serves to free the diagnostic leader from the unrealistic burden of being a magical manipulator of his followers.

2. The diagnostic leader further recognizes that the motives that are "in" his followers will show a general regularity, but also a high degree of variability over time. He does not believe he ever has any of his subordinates' "number" in a motivational sense because he recognizes this variability. Specifically, he is likely to adopt the position that there is some pattern in motives that can cause them to be:
   (a) to some extent shared; to some extent unique to each follower.
   (b) to some extent similar to his own; to some extent far different from his own.
   (c) to some extent conscious and maturing; to some extent childlike and unconsciously automatic.
   (d) to some extent long lasting and enduring; to some extent growing, dynamic and changing.

   An appreciation of (a) makes him very leery of "programs" that are "guaranteed to motivate" the work force. An appreciation of (b) protects him from assuming that his subordinates want (or ought to want) what he wants. Many is the executive who wakes up to the shocking realization that his people may not want what he wants them to want. An awareness of (c) and (d) above allows the leader to recognize that there will be times when his followers seem not to know what they really want or need, and that in due time this condition may change.

3. The diagnostic leader also realizes that, given the right environment, his people will be able to *signal* their conscious motives to him. It follows from this that his real skill as a leader will be the ability to learn to "read" the signals over time and over situations.

   Following this principle, he will come to put management tools such as "participative management," "management by objectives" and "sensitivity training" into their proper perspective. He will see these tools as procedures for setting up an environment within which follower motives can be more readily signaled to him. As in any other area of technology, he will expect changes in this area.

Furthermore, his recognition that followers can be expected to change, will cause him to continuously assess their motives. The repeated application of his "signal-reading" skills will help him to diagnose the *changes* that take place in the normal developmental flow of subordinates' motives.

4. The diagnostic leader is also aware that he is not the corporate social worker who "gives people what they want to keep 'em happy." He realizes that some of the motives of his people will be inappropriate to his organization and/or childish. When he sees that this is the case, he does not try to punish the motivation "out of" his subordinate. Rather, he presents the reality of the situation by saying that a particular need cannot realistically be expected to be satisfied in the present environment. The diagnostic leader does not hesitate to tell subordinates when their motives are unrealistic; he confronts subordinates with organizational and interpersonal realities.

5. Finally, the diagnostic leader sees his role as an *ecological manipulator*. Rather than becoming a people manipulator (which never works in the long run) he, in George Odiorne's words, "makes things happen." He concentrates upon the psychological and physical *environment* within which his people's motives are expressed. Following Warren Bennis, the diagnostic leader implicitly adopts an "agrarian" stance toward his role. Like the farmer, he knows he cannot "make" an acorn grow into a fir tree. But he does all he can to provide an ecology within which an acorn can, in the existential sense, maximize its potential. Thus, he manipulates the organizational and interpersonal environment in such a way that the realistic motives of his people have an ecological opportunity to be expressed rather than be smothered in the magical pollution of wishful thinking.

The notion of diagnostic leadership also carries implications for followers and organizations.

Organizations that take the diagnostic leadership position seriously will seek to evaluate and reward leadership in "developmental" as well as profit and service terms. Leaders will be measured in terms of the growth of their subordinates, since qualitative leaps in the contribution of subordinates will mark the successful leader.

The diagnostically oriented organization will place a positive value on innovation and experimentation. As a realistic response to changes in its own "system relations" with the outside world, the organization will seek to free its members from the inertia of the model of classical bureaucracy. Much of the "work itself" movement formulated by Herzberg and largely operationalized by the Bell System is built on the notion that development of subordinates often occurs at the expense of the organization charts and manuals.

Organizations that come to value diagnostic leadership can be expected to move beyond "programs" of diagnostic skills such as management by objectives. Their leaders will come to be evaluated in terms of the degree to which they are facile and innovative "match makers." They will evaluate leaders in terms of their ability to (a) diagnose follower motives and (b) match these with environmental manipulations and changes.

Within this kind of an organizational context followers will be expected to act differently. If the concept is made even partially operational, followers can expect to be held responsible to become more openly sharing about what they *really want and need*. If one of the pieces of input required by the diagnostic leader is "signals of follower motivations," the follower has a duty to become more inward looking and self analytical so that he can more clearly specify his motives and goals. He cannot "hold back" if he expects his role to be geared to that of the diagnostic leader. When this happens he comes to *share the responsibility* for making his organization "a great place to work."

Managerial Leadership*

It is a rare person who would not rather be an "alpha fish" than go through his career as a lowly member of the school. From time immemorial people have been divided into the leaders and the led, but just how an individual attains leadership status defies precise analysis. For instance, everyone at West Point receives basically the same education in essentially the same environment. Fifteen years later, some seem to be able to lead almost any group, whereas others would have difficulty leading a band of hungry orphans to a hamburger stand. In business the situation is no different.

Complicating the confusion, the experts have been overly imaginative in describing the various kinds of leaders—terms such as democratic, free-rein, custodial, task-oriented, psychologically distant or close, accommodative, and mediative abound. Added to this is the nettlesome fact that each authority, partisan to his own nomenclature, tends to be hostile to others', and the contradictions leave the busy practicing manager uncertain, if not at sea. Actually, leadership is more an art than management itself. There is no pattern of training and experience that will guarantee success in a given company or department; those who are recognized as leaders have followed different paths to their positions.

The executive who wants practical guidance to attain leadership effectiveness has several options open to him. He can, of course, give ear to facile prescriptions that promise much with little effort, but formulas of a "do this—don't do that" genre are usually preludes to failure. He can, on the other hand, delve into painstaking research, but few managers have either the talent or the time for it, and are likely, anyway, to be put off by the jargon and the cautions with which such research is replete. A third approach entails reading the biographies of those who have been outstanding leaders. One can always learn from the experiences of others, of course, but it can't be assumed that what has worked for one person in his situation will be equally effective for another in his.

---

*Source: James J. Cribbin, "The Protean Managerial Leader," *Personnel,* Vol. 49, No. 2, March-April, 1972, pp. 9-15. Reprinted by permission of the publisher, American Management Associations, Inc.

Therefore, more realistically, the executive can formulate his own hypotheses about appropriate leadership behavior in his particular milieu and test these hypotheses in his day-to-day activities, using the feedback from his subordinates to refine his original concepts and make them more congruent with corporate realities. The challenge to the manager is not to adopt what has worked for someone else, but rather to adapt and adjust his leadership behavior to certain crucial variables, which include at least the following: the manager's concept of the leadership process, his own personality, the "personality" of the group he leads, the situation in which leadership is exercised, and the organizational "givens" that define relevant and irrelevant behavior. These points call for detailed discussion.

The Manager's Concept of the Leadership Process

Newly appointed managers at times find it difficult to accept the fact that promotion entitles people only to position power—nothing more. The organization gives them a "mission," coupled with a grant of reward-punishment power that is theoretically adequate to ensure that subordinate performance is in line with corporate requirements. Leadership has to do with personal power. It is a process of social influence. One becomes a leader when others allow him (note the verb) to influence their thinking, their attitudes, and their behavior, when others accept him and see him as an effective medium for satisfying some of their more important job-related needs and aspirations. Thus, management status is given by the higher echelons, but followership is given by the powerless.

That is to say, with promotion comes the authority of the position, which the manager is not reluctant to use when appropriate and is not unduly concerned about. With leadership come the authority of interpersonal impact and influence and the authority of competence. In other words, leadership is a sharing process, brought into play by the individual who possesses the requisite expertise. The manager-leader must realize that his task is far more than developing a charisma that mesmerizes subordinates into obedience. Among the more significant demands of his role are these:

- To build a work team characterized by mutual acceptance and respect, which produce greater coordination and cooperation.
- To relate to each individual, as far as possible, in such a manner as to draw from him the plus factor that spells the difference between mediocrity and excellence.
- To energize and renew the work group so that it is determined to improve on past performance.
- To provide for the stability and maintenance of the group so that, in Aristotelian terms, it can function in a structured cosmos rather than in a chaos.
- To represent the work group's views to higher echelons and to see that there are no unfair incursions by other organizational units.
- As a result of all the foregoing, to enable the work group to measure up to corporate or departmental expectations and thus earn recognition, at the same time helping each individual to secure some "psychic income" from the work that he does.

The Manager's Personality

No manager—or any other person—can be all things to all men. Everyone has certain strengths that are individualistically his and everyone his personal limitations, the synergistic result of three variables: heredity, with the potential and limitations it imposes; environment, with the opportunities and constraints it connotes; and self-generated effort, which involves the effectiveness with which he has capitalized on his assets and coped with his deficiencies.

What can the manager do about all this? If he cannot be master of his fate, he certainly can avoid being the helpless reactor to mindless forces. Avoiding self-flagellating introspection, the would-be leader can have a candid confrontation with himself, enumerating his strong points and formulating strategies to capitalize on them more fully, and owning up to his remediable limitations, in order to plan a practical program for eliminating them. Finally, he can face up to his irremediable deficiencies, but with a decision not to waste his energy in seeking what can never be.

It is naive nonsense to suggest that a manager be "psychologically close" or "psychologically distant," as some advise. It may also be dangerous to make a fetish of consultative or participative management, excellent though these approaches may be in themselves; in fact, excessive use of participation may be an outward sign of the manager's inner inability to make up his own mind and to listen to his own drummer. Despite all the palaver, the truth is that some executives are most effective when they are more directive than some textbooks would recommend, even though they are shrewd enough to allow opportunities for subordinates to influence them. Still others find it difficult to be stringent and exacting; their affiliation needs loom too large on their psychological horizons.

One manager may be adept at caring for day-to-day matters but weak when it comes to innovation. Another may be a zealot in getting a new program off the launching pad but a bull in a china shop when required to manage a smooth-flowing operation. One may be conceptually oriented, a second person-oriented, and a third task-achievement-oriented. What is important is that the manager know what pans out best for him in working with and through his people to attain organizational goals.

Executives with strong egos and forceful personalities will always be inclined to be dominant, despite the fact that more introverted people may consider them domineering. And it is equally likely that less flamboyant types will play a valuable catalytic role, despite the fact that more action-oriented people may consider them colorless or passive. Shakespeare was on target when he suggested, "To thine own self be true." Being true to one's self implies the admission that there are some kinds of behavior that the manager will take to naturally and well, others that he will engage in only with a certain amount of self-discipline, and still others that he will handle with little or no adroitness, because they are simply not for him.

The Personality of the Work Group

If the manager's leadership behavior must be congruent with the expectations of his superiors, it is no less important that it be compatible with those of his people. Rightly or wrongly, every work group has its own idea of the kind of behavior the manager should exhibit, its own idea of the leadership style to which it will respond, even though neither is articulated as such. Belligerent employees may misinterpret courtesy and consideration as signs of weakness and indecision, whereas a well-disposed group may be alienated by the executive who is overly forceful or compelling. The following suggestions may serve as guidelines in dealing with the work group.

- The manager should try to "vacuum" his mind of his prejudgments and prejudices concerning the work group. Many managers make self-fulfilling prophecies, viewing their subordinates through the filters of negative attitudes; then they subconsciously are attuned to behavior that substantiates these attitudes and draw the erroneous conclusion that they were right in the first place.
- The manager should be sensitive to the social pressures, rewards, and punishments that the group can bring to bear on its members. Extensive research indicates that, given a choice of receiving rewards from management or from their peers, most of the rank and file prefer to be rewarded by the work group. A cohesive work group has centripetal fields of force that no manager can eradicate on the basis of mere position power.
- Einstein once remarked that when confronted with a problem, one should not seek to impose his will on it, but politely walk around it, and in time it would reveal itself to him. The manger needs no better advice. If he practices "people thinking," rather than "thing thinking,", he will soon learn to observe his people in much the same manner that a scientist observes any phenomenon that he would understand—impersonally and objectively.
- The manager should be careful of what he "sees" when he deals with employees. Perception is one of the trickiest problems in the whole of psychology. The way subordinates see themselves is determined by the way they think they are seen by other valued individuals and groups. The executive might bear in mind the comment of Goethe that if one wishes a man to continue as he is, all he need do is to think of him as he is, but if he wishes that man to become as he should and ought to be, then he must think of him as he should and ought to become.

The social world of any person is a vast ink blot; each interprets it according to his unique needs. Such being the case, it ill behooves the manager in his strain for consistency to indulge in selective social blindness. (Anyone who thinks that the avoidance of social blindness is easy might recall that Freud's most famous biographer and associate noted that the master was a poor judge of people).

- Again following the model of the scientist, the manager can experiment prudently with various approaches to his work group and the individuals within it. Learning by trial and error is the most clumsy way in the world to learn. All real learning is the fruit of keen observation, careful experimentation, sensitivity to feedback and the results attained, and a willingness to change in light of this feedback. The manager who complains that no one tells him anything is juvenile. If he kept his antennae up to catch what his people are sending his way, he would discover that he had ample clues to guide him.

- One of the limitations of even the best psychology books is that they concentrate on the necessity of learning about the needs of others, saying relatively little about the equally important necessity for the manager to let his own needs be known in dealing with his subordinates. Learning about the needs of employees, while revealing nothing about the manager's needs, can be a not-too-subtle form of manipulation. This is not to say that the manager should be totally revealing. What is essential is that the manger send clear signals to the troops so that both they and he comprehend their mutual expectations without wasting time and energy in second-guessing or playing cards close to the vest.

- The manager does not really have a team dedicated to organizational objectives, despite all the nonsense about teamwork that is printed. What he has is a group of individuals, each intent on satisfying his own needs and aspirations while contributing to organizational objectives. Each, for better or for worse, sees the organization, the manager, and the work group as vehicles for attaining his personal ends. Accordingly, the assignment of the manager-leader is one of orchestration.

Leonard Bernstein has three fundamental tasks. He must be faithful to the composer's intent and purpose, but he has the right to introduce his interpretation of the score. He must so relate to each musician that he is stimulated to play better than his mother thought possible, allowing ample opportunities when appropriate for each to shine in a solo capacity. He must blend the contributions of each so that the total performance will be polished and distinguished, careful that this or that small group of instrumentalists does not dominate at the expense of other equally important segments of the total orchestra. For the manager, the parallel here is obvious.

The Situation in Which Leadership is Exercised

All management is situational in character; so, too, is all leadership. Emergencies demand positive, forceful, perhaps even unilateral or harsh action. Situations in which the work group is achieving corporate objectives efficiently may call for a kind of gyroscopic leadership to ensure that subordinate effort stays on target. Situations that deal with the unfamiliar, for which there is neither precedent nor past experience, may require a highly consultative or participative approach, since no one really knows what should be done. In this case, it is not so much a matter of making people feel good or of arguing that many heads are better than one; rather, it is a matter of maintaining that there are few really poor ideas, although there are many incomplete ideas. Consultation and participation often enable the group to fill out the incomplete ideas and make them more realistic.

Situational analysis is hard work because of the built-in risk of error. For instance, when a company has come upon hard times, the manager may have several options open to him. He may institute a cost-reduction program and involve the group by encouraging its members to become more cost-conscious and production-minded. He may decide to improve the work methods, to stop doing the unproductive and launch a program of improvement and innovation.

He may trim the organizational fat by workforce reduction. He may shift key people about to imbue a new spirit of urgency to do better. Or he may resort to any combination of these tactics.

The danger may lurk in two phrases—personal preference and trained incapacities. It is as natural for a finance officer to think first of tighter cost controls as it is for the engineering director to think first of improved design and production. The term trained incapacities refers to the inability of a manger to see a problem from a viewpoint other than that of his own expertise. A Nobel laureate observed that new discoveries were elusive not because of ignorance; many people possessed basically the same competence and at times even the same data. But, whereas the many made certain assumptions, the prize winner as often as not acted on different premises. (The well-known history of Sears Roebuck versus Montgomery Ward illustrates the truth in this observation.) Prize winning aside, the manager who allows his personal preferences or trained incapacities to distract him from the realities of the situation is rushing into difficulties.

The Organizational "Givens"

To the outsider, a company or department is a technical system of men, machines, methods, and money, all geared to profit. In an orderly and static way things get done according to an overall plan with suitable controls. To the insider, the organization is an organism; it is a social system. It is the arena where the organizers meet the organized, with all the interactions that this posture connotes. It is the setting in which individuals and subgroups are determined to fulfill their career plans and reach parochial goals. It is where authority and status are distributed while power is grasped and influence earned, where hopes flourish and ambitions are thwarted. In any company, cooperation lives cheek by jowl with competition. If friendships abide, jealousies abound. Side by side are found enthusiasms and frustrations, loyalties and antagonisms, aspirations and fears, human strengths and human frailties. Team spirit and petty discord wax and wane. In any firm one finds the whole gamut of human reactions, even when an ingrained sense of decorum politely cloaks it with propriety and protocol.

The manager who would be a leader must be an astute organizational analyst. Each organization has its peculiar traditions, mores, and tribal rituals, its individual philosophy, climate, and tempo. Each has its particular sacred cows, sensitive nerve endings, and taboos. Each has defined, overtly or covertly, the parameters of desired, permissible, tolerated, and offensive behavior. Each has evolved accepted ways of getting the work done and accustomed ways of interacting. It is idle for the manager to pass moral judgment on such realities or to rail against these "givens." If he is to succeed, he must gain insight into the dynamics of how a particular firm or department functions, at times in the face of what is said or written in organizational documents. Armed with this understanding, he will be better able to decide important questions such as these:

Where are the rocks and shoals on which I may founder? In what areas can I act autonomously? Where should I proceed only after consultation with key people or groups? When should I withhold action until I have built a strong support base? What approach, tactics, and strategies are likely to make my leadership effective without inviting backlash? At what pace and tempo should I move?

This painstaking analysis is undramatic work, but failure to do it thoroughly may ultimately show the manager that he is not really a leader—that his good intentions and well-laid plans have come a cropper through his own social ineptitude and psychological obtuseness.

How to Choose a Leadership Pattern*

- ◊ "I put most problems into my group's hands and leave it to them to carry the ball from there. I serve merely as a catalyst, mirroring back the people's thoughts and feelings so that they can better understand them."
- ◊ It's foolish to make decisions oneself on matters that affect people. I always talk things over with my subordinates, but I make it clear to them that I'm the one who has to have the final say."
- ◊ "Once I have decided on a course of action, I do my best to sell my ideas to my employees."
- ◊ "I'm being paid to lead. If I let a lot of other people make the decisions I should be making, then I'm not worth my salt."
- ◊ "I believe in getting things done. I can't waste time calling meetings. Someone has to call the shots around here, and I think it should be me."

Each of these statements represents a point of view about "good leadership." Considerable experience, factual data, and theoretical principles could be cited to support each statement, even though they seem to be inconsistent when placed together. Such contradictions point up the dilemma in which the modern manager frequently finds himself.

New Problem

The problem of how the modern manager can be "democratic" in his relations with subordinates and at the same time maintain the necessary authority and control in the organization for which he is responsible has come into focus increasingly in recent years.

Earlier in the century this problem was not so acutely felt. The successful executive was generally pictured as possessing intelligence,

---

*Source: Robert Tannenbaum and Warren H. Schmidt, "How to Choose a Leadership Pattern," *Harvard Business Review,* May-June 1973, pp. 162-180. Copyright 1973 by the President and Fellows of Harvard College. All rights reserved.

imagination, initiative, the capacity to make rapid (and generally wise) decisions, and the ability to inspire subordinates. People tended to think of the world as being divided into "leaders" and "followers."

New focus: Gradually, however, from the social sciences emerged the concept of "group dynamics" with its focus on *members* of the group rather than solely on the leader. Research efforts of social scientists underscored the importance of employee involvement and participation in decision making. Evidence began to challenge the efficiency of highly directive leadership, and increasing attention was paid to problems of motivation and human relations.

Through training laboratories in group development that sprang up across the country, many of the newer notions of leadership began to exert an impact. These training laboratories were carefully designed to give people a firsthand experience in full participation and decision making. The designated "leaders" deliberately attempted to reduce their own power and to make group members as responsible as possible for setting their own goals and methods within the laboratory experience.

It was perhaps inevitable that some of the people who attended the training laboratories regarded this kind of leadership as being truly "democratic" and went home with the determination to build fully participative decision making into their own organizations. Whenever their bosses made a decision without convening a staff meeting, they tended to perceive this as authoritarian behavior. The true symbol of democratic leadership to some was the meeting—and the less directed from the top, the more democratic it was.

Some of the more enthusiastic alumni of these training laboratories began to get the habit of categorizing leader behavior as "democratic" or "authoritarian." The boss who made too many decisions himself was thought of as an authoritarian, and his directive behavior was often attributed solely to his personality.

New need: The net result of the research findings and of the human relations training based upon them has been to call into question the stereotype of an effective leader. Consequently, the modern manager often finds himself in an uncomfortable state of mind.

Often he is not quite sure how to behave; there are times when he is torn between exerting "strong" leadership and "permissive" leadership. Sometimes new knowledge pushes him in one direction ("I should really get the group to help make this decision"), but at the same time his experience pushes him in another direction ("I really understand the problem better than the group and therefore I should make the decision"). He is not sure when a group decision is really appropriate or when holding a staff meeting serves merely as a device for avoiding his own decision-making responsibility.

The purpose of our article is to suggest a framework which managers may find useful in grappling with this dilemma. First, we shall look at the different patterns of leadership behavior that the manager can choose from in relating himself to his subordinates. Then, we shall turn to some of the questions suggested by this range of patterns. For instance, how important is it for a manger's subordinates to know what type of leadership he is using in a situation? What factors should he consider in deciding on a leadership pattern? What difference do his long-run objectives make as compared to his immediate objectives?

Range of Behavior

Exhibit 7-1 presents the continuum or range of possible leadership behavior available to a manager. Each type of action is related to the degree of authority used by the boss and to the amount of freedom available to his subordinates in reaching decisions. The actions seen on the extreme left characterize the manager who maintains a high degree of control while those seen on the extreme right characterize the manager who releases a high degree of control. Neither extreme is absolute; authority and freedom are never without their limitations.

Now let us look more closely at each of the behavior points occurring along this continuum.

The manager makes the decision and announces it.

In this case the boss identifies a problem, considers alternative solutions, chooses one of them, and then reports this decision to his subordinates for implementation. He may or may not give consideration to what he believes his subordiantes will think or feel about his decision; in any case, he provides no opportunity for them

Exhibit 7–1
Continuum of Leadership Behavior

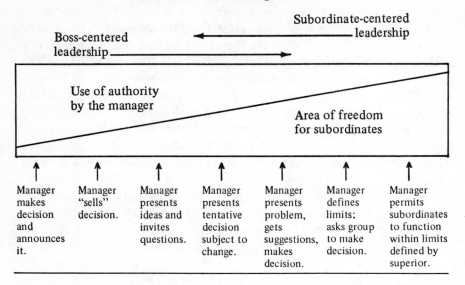

to participate directly in the decision-making process. Coercion may or may not be used or implied.

The manager "sells" his decision.

Here the manager, as before, takes responsibility for identifying the problem and arriving at a decision. However, rather than simply announcing it, he takes the additional step of persuading his subordinates to accept it. In doing so, he recognizes the possibility of some resistance among those who will be faced with the decision, and seeks to reduce this resistance by indicating, for example, what the employees have to gain from his decision.

The manager presents his ideas, invites questions.

Here the boss who has arrived at a decision and who seeks acceptance of his ideas provides an opportunity for his subordinates to get a fuller explanation of his thinking and his intentions. After presenting the ideas, he invites questions so that his associates can better understand what he is trying to accomplish. This "give and take" also enables the manger and the subordinates to explore more fully the implications of the decision.

The manager presents a tentative decision subject to change.

This kind of behavior permits the subordinate to exert some influence on the decision. The initiative for identifying and diagnosing the problem remains with the boss. Before meeting with his staff, he has thought the problem through and arrived at a decision—but only a tentative one. Before finalizing it, he presents his proposed solution for the reaction of those who will be affected by it. He says in effect, "I'd like to hear what you have to say about this plan that I have developed. I'll appreciate your frank reactions, but will reserve for myself the final decision."

The manager presents the problem, gets suggestions, and then makes his decision.

Up to this point the boss has come before the group with a solution of his own. Not so in this case. The subordinates now get the first chance to suggest solutions. The manager's initial role involves identifying the problem. He might, for example, say something of this sort: "We are faced with a number of complaints from newspapers and the general public on our service policy. What is wrong here? What ideas do you have for coming to grips with this problem?"

The function of the group becomes one of increasing the manager's repertory of possible solutions to the problem. The purpose is to capitalize on the knowledge and experience of those who are on the "firing line." From the expanded list of alternatives developed by the manager and his subordinates, the manager then selects the solution that he regards as most promising.[1]

The manager defines the limits and requests the group to make a decision.

At this point the manager passes to the group (possibly including himself as a member) the right to make decisions. Before doing so, however, he defines the problem to be solved and the boundaries within which the decision must be made.

An example might be the handling of a parking problem at a plant. The boss decides that this is something that should be worked on by the people involved, so he calls them together and points up the existence of the problem. Then he tells them:

> "There is the open field just north of the main plant which has been designated for additional employee parking. We can build underground or surface multilevel facilities as long as the cost does not exceed $100,000. Within these limits we are free to work out whatever solution makes sense to us. After we decide on a specific plan, the company will spend the available money in whatever way we indicate."

The manager permits the group to make decisions within prescribed limits.

This represents an extreme degree of group freedom only occasionally encountered in formal organizations, as, for instance, in many research groups. Here the team of managers or engineers undertakes the identification and diagnosis of the problem, develops alternative procedures for solving it, and decides on one or more of these alternative solutions. The only limits directly imposed on the group by the organization are those specified by the superior of the team's boss. If the boss participates in the decision-making process, he attempts to do so with no more authority than any other member of the group. He commits himself in advance to assist in implementing whatever decision the group makes.

Key Questions

As the continuum in *Exhibit 7-1* demonstrates, there are a number of alternative ways in which a manager can relate himself to the group or individuals he is supervising. At the extreme left of the range, the emphasis is on the manager—on what *he* is interested in, how *he* sees things, how *he* feels about them. As we move toward the subordinate-centered end of the continuum, however, the focus is increasingly on the subordinates—on what *they* are interested in, how *they* look at things, how *they* feel about them.

When business leadership is regarded in this way, a number of questions arise. Let us take four of especial importance:

Can a boss ever relinquish his responsibility by delegating it to someone else?

Our view is that the manager must expect to be held responsible by his superior for the quality of the decisions made, even though operationally these decisions may have been made on a group basis. He should, therefore, be ready to accept whatever risk is involved whenever he delegates decision-making power to his subordinates. Delegation is not a way of "passing the buck." Also, it should be emphasized that the amount of freedom the boss gives to his subordinates cannot be greater than the freedom which he himself has been given by his own superior.

Should the manager participate with his subordinates once he has delegated responsibility to them?

The manager should carefully think over this question and decide on his role prior to involving the subordinate group. He should ask if his presence will inhibit or facilitate the problem-solving process. There may be some instances when he should leave the group to let it solve the problem for itself. Typically, however, the boss has useful ideas to contribute, and should function as an additional member of the group. In the latter instance, it is important that he indicate clearly to the group that he sees himself in a *member* role rather than in an authority role.

How important is it for the group to recognize what kind of leadership behavior the boss is using?

It makes a great deal of difference. Many relationship problems between boss and subordinate occur because the boss fails to make clear how he plans to use his authority. If, for example, he actually intends to make a certain decision himself, but the subordinate group gets the impression that he has delegated this authority, considerable confusion and resentment are likely to follow. Problems may also occur when the boss uses a "democratic" facade to conceal the fact that he has already made a decision which he hopes the group will accept as its own. The attempt to "make them think it was their idea in the first place" is a risky one. We believe that it is highly important for the manager to be honest and clear in describing what authority he is keeping and what role he is asking his subordinates to assume in solving a particular problem.

Can you tell how "democratic" a manger is by the number of decisions his subordiantes make?

The sheer number of decision is not an accurate index of the amount of freedom that a subordinate group enjoys. More important is the *significance* of the decisions which the boss entrusts to his subordinates. Obviously a decision on how to arrange desks is of an entirely different order from a decision involving the introduction of new electronic data-processing equipment. Even though the widest possible limits are given in dealing with the first issue, the group will sense no particular degree of responsibility. For a boss to permit the group to decide equipment policy, even within rather narrow limits, would reflect a greater degree of confidence in them on his part.

Deciding How to Lead

Now let us turn from the types of leadership which are possible in a company situation to the question of what types are *practical* and *desirable*. What factors or forces should a manager consider in deciding how to manage? Three are of particular importance:

- Forces in the manager.
- Forces in the subordinates.
- Forces in the situation.

We should like briefly to describe these elements and indicate how they might influence a manager's action in a decision-making situation.[2] The strength of each of them will, of course, vary from instance to instance, but the manager who is sensitive to them can better assess the problems which face him and determine which mode of leadership behavior is most appropriate for him.

Forces in the manger: The manager's behavior in any given instance will be influenced greatly by the many forces operating within his own personality. He will, of course, perceive his leadership problems in a unique way on the basis of his background, knowledge, and experience. Among the important internal forces affecting him will be the following:

1. *His value system:* How strongly does he feel that individuals should have a share in making the decisions which affect them? Or, how convinced is he that the official who is paid to assume responsibility should personally carry the burden of decision making? The strength of his convictions on questions like these will tend to move the manager to one end or the other of the continuum shown in *Exhibit 7-1*. His behavior will also be influenced by the relative importance that he attaches to organizational efficiency, personal growth of subordinates, and company profits.[3]
2. *His confidence in his subordinates.* Managers differ greatly in the amount of trust they have in other people generally, and this carries over to the particular employees they supervise at a given time. In viewing his particular group of subordinates, the manager is likely to consider their knowledge and competence with respect to the problem. A central question he might ask himself is: "Who is best qualified to deal with this problem?" Often he may, justifiably or not, have more confidence in his own capabilities than in those of his subordinates.
3. *His own leadership inclinations.* There are some managers who seem to function more comfortably and naturally as highly directive leaders. Resolving problems and issuing orders come easily to them. Other managers seem to operate more comfortably in a team role, where they are continually sharing many of their functions with their subordinates.
4. *His feelings of security in an uncertain situation.* The manager who releases control over the decision-making

process thereby reduces the predictability of the outcome. Some managers have a greater need than others for predictability and stability in their environment. This "tolerance for ambiguity" is being viewed increasingly by psychologists as a key variable in a person's manner of dealing with problems.

The manager brings these and other highly personal variables to each situation he faces. If he can see them as forces which, consciously or unconsciously, influence his behavior, he can better understand what makes him prefer to act in a given way. And understanding this, he can often make himself more effective.

Forces in the subordinate: Before deciding how to lead a certain group, the manager will also want to consider a number of forces affecting his subordinates' behavior. He will want to remember that each employee, like himself, is influenced by many personality variables. In addition, each subordinate has a set of expectations about how the boss should act in relation to him (the phrase "expected behavior" is one we hear more and more often these days at discussions of leadership and teaching). The better the manager understands these factors, the more accurately he can determine what kind of behavior on his part will enable his subordinates to act most effectively.

Generally speaking, the manager can permit his subordinates greater freedom if the following essential conditions exist:

- If the subordinates have a readiness to assume responsibility for decision making. (Some see additional responsibility as a tribute to their ability; others see it as "passing the buck.")
- If they have a relatively high tolerance for ambiguity. (Some employees prefer to have clear-cut directives given to them; others prefer a wider area of freedom.)
- If they are interested in the problem and feel that it is important.
- If they understand and identify with the goals of the organization.
- If they have the necessary knowledge and experience to deal with the problem.
- If they have learned to expect to share in decision making. (Persons who have come to expect strong leadership and

are then suddenly confronted with the request to share more fully in decision making are often upset by this new experience. On the other hand, persons who have enjoyed a considerable amount of freedom resent the boss who begins to make all the decisions himself.)

The manager will probably tend to make fuller use of his own authority if the above conditions do *not* exist; at times there may be no realistic alternative to running a "one-man show."

The restrictive effect of many of the forces will, of course, be greatly modified by the general feeling of confidence which subordinates have in the boss. Where they have learned to respect and trust him, he is free to vary his behavior. He will feel certain that he will not be perceived as an authoritarian boss on those occasions when he makes decisions by himself. Similarly, he will not be seen as using staff meetings to avoid his decision-making responsibility. In a climate of mutual confidence and respect, people tend to feel less threatened by deviations from normal practice, which in turn makes possible a higher degree of flexibility in the whole relationship.

Forces in the situation: In addition to the forces which exist in the manager himself and in his subordinates, certain characteristics of the general situation will also affect the manager's behavior. Among the more critical environmental pressures that surround him are those which stem from the organization, the work group, the nature of the problem, and the pressures of time. Let us look briefly at each of these:

Type of organization—Like individuals, organizations have values and traditions which inevitably influence the behavior of the people who work in them. The manager who is a newcomer to a company quickly discovers that certain kinds of behavior are approved while other are not. He also discovers that to deviate radically from what is generally accepted is likely to create problems for him.

These values and traditions are communicated in numerous ways—through job descriptions, policy pronouncements, and public statements by top executives. Some organizations, for example, hold to the notion that the desirable executive is one who is dynamic, imaginative, decisive, and persuasive. Other organizations put more emphasis upon the importance of the executives ability to work effectively with people—his human relations skills. The fact that his superiors have a defined concept of what the good executive should be will very likely push the manager toward one end or the other of the behavioral range.

In addition to the above, the amount of employee participation is influenced by such variables as the size of the working units, their geographical distribution, and the degree of inter- and intra-organizational security required to attain company goals. For example, the wide geographical dispersion of an organization may preclude a practical system of participative decision making, even though this would otherwise be desirable. Similarly, the size of the working units or the need for keeping plans confidential may make it necessary for the boss to exercise more control than would otherwise be the case. Factors like these may limit considerably the manager's ability to function flexibly on the continuum.

Group effectiveness—Before turning decision-making responsibility over to a subordinate group, the boss should consider how effectively its members work together as a unit.

One of the relevant factors here is the experience the group has had in working together. It can generally be expected that a group which has functioned for some time will have developed habits of cooperation and thus be able to tackle a problem more effectively than a new group. It can also be expected that a group of people with similar backgrounds and interests will work more quickly and easily than people with dissimilar backgrounds, because the communication problems are likely to be less complex.

The degree of confidence that the members have in their ability to solve problems as a group is also a key consideration. Finally, such group variables as cohesiveness, permissiveness, mutual acceptance, and commonality of purpose will exert subtle but powerful influence on the group's functioning.

The problem itself—The nature of the problem may determine what degree of authority should be delegated by the manager to his subordinates. Obviously he will ask himself whether they have the kind of knowledge which is needed. It is possible to do them a real disservice by assigning a problem that their experience does not equip them to handle.

Since the problems faced in large or growing industries increasingly require knowledge of specialists from many different fields, it might be inferred that the more complex a problem, the more anxious a manager will be to get some assistance in solving it. However, this is not always the case. There will be times when the very complexity of the problem calls for one person to work it out. For example, if the manager has most of the background and factual data relevant to a given issue, it may be easier for him to think it through himself than to take the time to fill in his staff on all the pertinent background information.

The key question to ask, of course, is: "Have I heard the ideas of everyone who has the necessary knowledge to make a significant contribution to the solution of this problem?"

The pressure of time—This is perhaps the most clearly felt pressure on the manager (in spite of the fact that it may sometimes be imagined). The more that he feels the need for an immediate decision, the more difficult it is to involve other people. In organizations which are in a constant state of "crisis" and "crash programming" one is likely to find managers personally using a high degree of authority with relatively little delegation to subordinates. When the time pressure is less intense, however, it becomes much more possible to bring subordinates in on the decision-making process.

These, then, are the principal forces that impinge on the manager in any given instance and that tend to determine his tactical behavior in relation to his subordinates. In each case his behavior ideally will be that which makes possible the most effective attainment of his immediate goal within the limits facing him.

Long-run Strategy

As the manager works with his organization on the problems that come up day by day, his choice of a leadership pattern is usually limited. He must take account of the forces just described and, within the restrictions they impose on him, do the best that he can. But as he looks ahead months or even years, he can shift his thinking from tactics to large-scale strategy. No longer need he be fettered by all of the forces mentioned, for he can view many of them as variables over which he has some control. He can, for example, gain new insights or skills for himself, supply training for individual subordinates, and provide participative experiences for his employee group.

In trying to bring about a change in these variables, however, he is faced with a challenging question: At which point along the continuum *should* he act?

Attaining objectives: The answer depends largely on what he wants to accomplish. Let us suppose that he is interested in the same objectives that most modern managers seek to attain when they can shift their attention from the pressure of immediate assignments:

1. To raise the level of employee motivation.
2. To increase the readiness of subordintes to accept change.
3. To improve the quality of all managerial decisions.
4. To develop teamwork and morale.
5. To further the individual development of employees.

In recent years the manager has been deluged with a flow of advice on how best to achieve these longer-run objectives. It is little wonder that he is often both bewildered and annoyed. However, there are some guidelines which he can usefully follow in making a decision.

Most research and much of the experience of recent years give a strong factual basis to the theory that a fairly high degree of subordinate-centered behavior is associated with the accomplishment of the five purposes mentioned.[4] This does not mean that a manager should always leave all decisions to his assistants. To provide the individual or the group with greater freedom than they are ready for at any given time may very well tend to generate anxieties and therefore inhibit rather than facilitate the attainment of desired objectives. But this should not keep the manager from making a continuing effort to confront his subordinates with the challenge of freedom.

Conclusions

In summary, there are two implications in the basic thesis that we have been developing. The first is that the successful leader is one who is keenly aware of those forces which are most relevant to his behavior at any given time. He accurately understands himself, the individuals and group he is dealing with, and the company and broader social environment in which he operates. And certainly he is able to assess the present readiness for growth of his subordinates.

But this sensitivity or understanding is not enough, which brings us to the second implication. The successful leader is one who is able to behave appropriately in the light of these perceptions. If direction is in order, he is able to direct; if considerable participative freedom is called for, he is able to provide such freedom.

Thus, the successful manager of men can be primarily characterized neither as a strong leader nor as a permissive one. Rather, he is one who maintains a high batting average in accurately assessing the forces that determine what his most appropriate behavior at any given time should be and in actually being able to behave accordingly. Being both insightful and flexible, he is less likely to see the problems of leadership as a dilemma.

Topics For Discussion

1. Discuss the concept of a stereotype magical leader.
2. Discuss the effect on the organization of the magical leadership concept.
3. Discuss leadership as a sharing process.
4. Discuss the concepts of democratic leadership.
5. Discuss the conditions that must exist if a leader is to permit his subordinates greater freedom.
6. Discuss the impact of leadership on the attainment of objectives.

FOOTNOTES

How to Choose a Leadership Pattern

[1] For a fuller explanation of this approach, see Leo Moore, "Too Much Management, Too Little Change," *HBR* January-February 1956, p. 41.

[2] See also Robert Tannenbaum and Fred Massarik, "Participation by Subordinates in the Managerial Decision-Making Process," *Canadian Journal of Economics and Political Science,* August 1950, p. 413.

[3] See Chris Argyris, "Top Management Dilemma: Company Needs vs. Individual Development," *Personnel,* September 1955, pp. 123-134.

[4] For example, see Warren H. Schmidt and Paul C. Buchanan, *Techniques that Produce Teamwork* (New London, Arthur C. Croft Publications, 1954); and Morris S. Viteles, *Motivation and Morale in Industry* (New York, W.W. Norton & Company, Inc., 1953).

SELECTED READINGS

Bass, Bernard M., *Leadership, Psychology and Organizational Behavior,* New York: Harper and Row, 1960. Presents observations about leadership theory revolving around the concept that leadership involves influencing the group.

Bennis, Warren G., "Post-Bureaucratic Leadership," *Trans-action,* June-July, 1969. Reviews the problems facing organizations in the future pointing out that in addition to substantive competence and comprehension of both social and technical systems the new leader will have to possess interpersonal skills.

Bonner, John T., Jr., "Leadership for Lawmen," *FBI Law Enforcement Bulletin,* Vol. 42, No. 12, December, 1973, pp. 7-9 and 15. The major thesis of this article is that there are three major attributes of successful leadership—competence, courage and compassion.

Gibb, Jack R., "Dynamics of Leadership and Communication," *Current Issues in Higher Education,* 1967. Describes the elements of defensive leadership and presents, as an alternative, participative leadership. Reviews the implications of this model for ethical behavior.

Peoples, Edward E., "Measuring the Qualities of Police Leadership," *The Police Chief,* Vol. XLI, No. 5, May, 1974, pp. 30-31. Presents an analysis of experiments by the author to measure the behavioral needs relative to the leadership styles of over 400 police managers.

Spencer, Gilmore and Keith Jewell, "Police Leadership: A Research Study," *The Police Chief,* Vol. 40, No. 3, March, 1973, pp. 40-45. This study highlights the importance of self-confidence—along with ambition, ability to think in management terms, and good human relations skills—in considering police officers for higher levels of responsibility.

The study of this chapter will enable you to:

1. Compare the types of team policing in current usage.
2. Describe the type of team building at the management level.
3. Define team policing.
4. Write a short essay describing the characteristics of the colleague model.
5. Describe the functions of a top management team.
6. Identify two inherent difficulties in team policing.
7. List the four elements of a total team policing system.
8. List the five basic principles of management.
9. Define job enlargement.

8

Team Policing

Introduction*

Are we locked into a system? I look around and see the hierarchical system of organization "belching smoke" and defying our police agencies to change. The public administration organizational model espoused for so long by *Municipal Police Administration* by the International City Managers Association, and the late, great O.W. Wilson's *Police Administration* seems to decry change as do the International Association of Chiefs of Police and the California Commission on Peace Officer Standards and Training; two of the outstanding professional organizations; in their surveys of police departments. "The system has worked, thus, why meddle with it?" is the question oft repeated. And, may I add the syndrome of "never do anything for the first time."

I submit that change for the sake of change is not a good reason to meddle with our organizations but the time has come for us to be much more questioning and analytical about the way we are doing things. The world is changing ever so rapidly around us and we can only continue to hurt ourselves if we don't change with it. It is quite apparent that we continue to lose ground in our fight against crime

---

*Source: John P. Kenney, "Team Policing: A Model for Change," A Paper presented to the National Symposium on Urban Police Practices, Quanico, Virginia, November 12, 1972, pp. 1-16. Reprinted by permission of the author.

and indications are that the level of police services we are providing our communities has not substantially improved over the past several decades.

The question is, can changes in our approaches to organization improve our operations? I submit that they can and in those few progressive and innovative police and sheriffs' departments which are experimenting with organizational change, the results are encouraging. In my presentation today I shall deal with team policing, one model for organizational change.

Team policing is not new. In essence most of our smaller police agencies use a team approach in that all officers including the chief of police or sheriff are involved in practically all activities and must work closely together. The late August Vollmer in designing his organization for the Berkeley, California, Police Department made his field officers responsible for all activities on their beats and designated his detective, juvenile, and traffic specialists as support personnel to aid and assist each officer, as the needs arose, in team fashion. And almost every agency has assembled teams to deal with special situations or events such as a major investigation or a narcotics raid.

More recently, however, the approaches to team policing are of a different nature. Significant organizational changes have taken place and teams have been formalized and made an integral part of agency organizations. They vary from a basic operational field patrol team to a complex assemblage of teams for administration and management purposes as well as operational purposes.

First, a look at the field operational team restricted to patrol. The agency is otherwise organized in the traditional manner. In the simplest approach, a team of officers is assigned to a given area with full responsibility to perform the normal duties of patrol and in addition develop a greater involvement in the community by participating in citizen-groups meetings and visiting people in their homes and businesses. The team provides twenty-four hour coverage for the area and officers are assigned on a reasonably permanent basis.

Another operational team approach is to assign a patrol sergeant and from six to ten officers or deputies to a given area during their tour of duty. The sergeant has flexibility in the assignment of personnel and methods of operation. An entire city or station/

precinct area will be policed by several teams, each team with the usual responsibilities for responding to calls and repressing crime. Responsibility for policing the entire area of assignment rests with the team as a whole rather than each unit having responsibility for a designated area or beat. This approach could be expanded by giving a lieutenant watch commander full authority to deploy the teams on a watch in varying area and operational patterns.

A third patrol operational team approach is to assign a lieutenant two or more sergeants and a sufficient number of officers to a designated area of a city or station/precinct with full responsibility for policing the area twenty-four hours a day, seven days a week, 365 days a year. The team has responsibility for determining deployment and hours of assignment as well as freedom in determining methods of operation.

Expansion of the patrol operational teams on either a shift or area basis to include detectives, juvenile officers, and/or identification technicians is feasible. In essence, this provides the team a capability to perform the total field policing functions with support from specialty units. For the most part, the team will have the capability to handle most situations and incidents which arise with flexibility for deployment and assignment of personnel and methods of operation.

Creation of the teams previously described requires a commitment from management that it is in fact delegating to the teams flexibility and full authority to operate as teams within the guidelines established. There is also the requirement that there be a reasonable permanency of assignment of personnel to teams. *Planning* and *training* are two additional components essential if the team approach is to be effective.

At the management level we note an increasing amount of team building taking place.

The administrative staff functions are removed from a level on the organizational chart and incorporated into the chief's office, a move calculated to make the staff an "extension" of the chief. They now, in escence, act as the chief's alter ego and are removed from a position of vying with the line units for "power." The chief's capability to *manage* the department has been considerably expanded.

Units usually incorporated into the office of chief of police include administration—budget and planning and research, personnel and training and community relations. Smaller agencies may combine these functional responsibilities into a position of adminstrative assistant, increasingly being filled by management trained individuals who do not become peace officers.

Thus, we have the genesis for building a comprehensive team policing organizational model in which there is a top management team and the necessary operational teams. There is a recognition that structure modifies behavior and behavior modifies structure. Organization becomes flexible and is constantly is a state of change. The inflexibility of the traditional organizations no longer hamper operations and circumscribe behavior.

Before discussing the model some "don'ts" and "do's" are in order. Do not introduce the model without careful planning and, insofar as feasible, involvement of all personnel in the planning process. Don't proceed by administrative *fiat*. Do have training sessions for all levels of personnel. Do use an outside consultant in the planning process to resolve interpersonal friction in the decision-making process. Do involve representatives from the Mayor's office, the City Managers or administrative officers office, the financial director or budget office, the personnel department and the governing body. Do not assume that every member of the agency is interested in change. Assume the opposite. Involve civilian as well as peace officer personnel. The team concept implies that all affected personnel *must* be involved in change as well as in operations. Allow for a sufficient period of time to get all personnel involved before implementation. (Plan roughly for a two-year planning period and from three to five years for full implementation.)

The team policing organizational model is based on a phenomena of accepting the premise that all personnel have a contribution to make to the policy and operational decision making process. It presupposes that all personnel have a major contribution to make to the operation of the department, are capable and willing to make that contribution, and that there will be acceptance of the contribution. It is a continuing process whereby involvement of all personnel takes place, a continuous input of ideas, suggestions and constructive criticism is fostered, and new ideas and concepts are ingested into the system. Management personnel assume responsibility for broad policy making, planning of operational activities, for dealing with substantive issues regarding personnel, operations, budget and unusual situations and community relations. Operational personnel become concerned with routine day to day operations and implementation of broad policy guidelines laid down by the management team.

Team policing suggests that all personnel will be removed from a nice neat "box" which prescribes jobs, tasks and duties. All personnel will be perceived as having a major contribution to make to the success of the organization, and capabilities will be maximized. It is recognition that each person in the organization has talent, develops an expertise, has a basic know-how and a quality of leadership and an ability to contribute to the total operation of the department. Capitalizing on the talents of all persons in the organization is the key to success.

Can a police department match these lofty concepts? A break from tradition is required. The structures of the traditional and orthodox concepts of organization are being challenged. The giving up of command authority—not responsibility—are the first of the hallowed myths of organizational relationships which follow. Collegial decision making replaces orders by administrative *fiat*. Operational activities involves the maximizing or the expertise of the individual members of the organization, the development of a system whereby the most capable (not necessarily the highest ranked) officer is called upon to decide and handle specific situations.

Superior-subordinate relationships dealing with concepts of command and supervision are questioned. Perhaps it is better to approach activities from a basis dealing with programs, that is, investigation, traffic repression, juvenile and community relations as a foundational basis of departmental goals and objectives. The emphasis is placed upon work to be done rather than upon the bureaucratic relationships of "who is responsible to whom." The dealing with specific situations replaces emphasis on command relationships. Elaborating on this issue, program goals become the principle focus of operations. Personnel work as a team in dealing with problems and situations rather than being concerned with the chain of command. Organizational relationships focus on adaptation of the capabilities of individuals to deal with specific situations, be they management or operational, rather than on the rank of individuals.

Movement toward the team policing model suggests changes in perception about line and staff and the introduction of new terminology which is more descriptive of line and staff concepts. The traditional view of staff as advisory and aiding and assisting without "authority," and line as having sole responsibility for operations, is too delimiting. Rather, an abandonment of the traditional line-staff relationships in favor of a distinction between "program" and "sustaining" units acting within a network of power is in order. Briefly, programs refer to the functions and policies related to what is to be done—whereas sustaining activities deal more with procedures and how something is to be done. A team approach to the meeting of agency goals and objectives is suggested, in which "programs" and "sustaining" activities are reflected in a grid or "network of authority."

Police programs have traditionally been perceived along functional lines. That is, the patrol unit performs the patrol function, the traffic unit the traffic function, the investigations unit the investigation function, and so on. Sustaining activities are reflected in separate functional units. The fact that in actual operation the sustaining and program functions are intertwined is often overlooked. The tendency is toward overspecialization. However, an analysis of informal relationships which pertain in many departments, suggests that team efforts do prevail but are not officially recognized.

Conceptualization of programs in discrete terms has been weak in most police departments. The emphasis has been on vertical program subdivisions rather than horizontal integration. More specifically, for example, an investigative program has been viewed in terms of the preliminary investigation considered a patrol unit program function, and follow-up investigation an investigation unit program function.

What appears to be needed is a statement of the police agency role followed by a clarification of goals and objectives. If the police role is perceived as maintenance of ordered liberty which requires performance of coercive and non-coercive activities and the protection of personal liberties and civil rights, a basis is provided for the establishment of goals and objectives. Coercive activities include criminal, traffic, regulatory and juvenile law enforcement; keeping the peace and intelligence gathering. Non-coercive activities include social service, crime prevention, participation in the development of an environment of order and stability and the provision of services.

With the establishment of goals and objectives departmental programs follow logically. For example, an investigation program is essential for achievement of law enforcement goals and objectives; a community relations program becomes essential for achievement of crime prevention and participation in the development of an environment of order and stability. A program may be perceived differently by different departments, but programs can be operationally defined, an important process in changing structure.

Clarification of sustaining activities and their relationship to programs is necessary. Orthodox organizational models reflect sustaining activities in terms of functional units. Thus, the personnel and training functions are perceived as a responsibility of a singular unit. Records management is perceived as a function of the records unit. Integration of the sustaining activities with program operations and goals suggests a broader perception. For example, a career development activity (a personnel function) is not a singular function of the personnel unit. All units, of necessity, must be involved. This is also true of training. Likewise, records management may be integrated into program operations to provide a more meaningful support for programs, a most feasible operation with the advent of automatic data processing.

Thus, in developing a team policing organizational model, it is in order to perceive program and sustaining activities as integrated and intertwined. However, this does not just happen. Continuous planning is the key to success, including a continuous monitoring of all activities for control and coordination purposes. Planning does not just take place as a function of a planning unit. To be successful, all segments of the organization should be involved. True, basic guidelines must be developed by sustaining personnel, and "staff" work is necessary for data processing and preparation of reports, directives and procedures. Basically, long range planning is a function of top management, and day to day operational planning is a responsibility of operational personnel. However, the most successful planning both of a long range and operational nature takes place when all concerned personnel are involved, or at least are aware of what is taking place. Long range plans, of necessity, must be resolved into operational plans, and often operation plans require or lead to the development of long range plans.

Although there may be no best way to structure for the team policing model, a circular arrangement for the top management team and the operational teams is suggested. The circular nature of the structure suggests that continuous bargaining relationships will prevail between the members of each team with all members intimately involved in the decision making process. The top management team will be primarily concerned with the policy decisions, and the operational teams with the operations decision.

First, a look at the top management team. Normally, the team will consist of the chief of police and the top command personnel of a department. The command personnel will not have the traditional command responsibility but will bring to the team the "expertise" of their function speciality which will be utilized in planning, directing, coordinating and controlling an integrated operation. In essence, the team will be constituted as a "board of directors" with a collegial responsibility for overall management of the affairs of the department.

The chief of police is the "chairman" of the top management team. He may be perceived as the hub of the circle surrounded by other team members. His function is to provide leadership for the team as well as the department as a whole. The "buck" stops with him, but more importantly he is responsible for creating a management environment in which the top management team members may tackle and resolve major departmental issues. It is up to him to provide a proper balance in the resolution of issues to assure that no one program is overemphasized, and that the operation of the department proceeds in a reasonably smooth fashion.

The other members of the top management team will be responsible for designated program areas for which their background of education and experience reasonably provide them with an expertise. Using the orthodox organization's command officers as examples; the commander of the investigation division would bring to the team, expertise in investigation and juvenile programs; the commander of the operations division, expertise in patrol and traffic programs; the commander of the services division, expertise in records and property management, communications and custody; and the head of the management services division, expertise in personnel, training, planning and budget. Each in a sense would become the "director" of their individual program area. One "director," more than likely the expert in operations in a medium sized department would become "director of operations" with responsibility for coordinating and directing the work of the operations teams. Departmental size would dictate the number of personnel for each team member. Necessary support personnel would be provided to perform "staff" work.

How would the team function? Each member of the team would be responsible for inputting into the planning and policy decision making process, program requirements for his area of "expertise." Each input would be evaluated by the team as a whole, vis-a-vis all other inputs. The ultimate team decision with respect to each input would assure that implementation provides for a balanced operation. It is presupposed that there is a continuous flow of information on all operations and activities to the top management team.

Although the "director of operations" would have primary responsibility for coordinating and directing operational teams, all other directors would engage in a continuous process of monitoring the operation of their program areas. This would be done by review of information and by inspections. In essence, the top management "circle" would revolve for the purposes of a continuous evaluation and integration of all programs to assure a proper balance in operations.

The responsibility of the basic operational team under one concept would consist of the performance of substantially all program and sustaining activities essential for fulfilling the police role in the community and achieving departmental goals and objectives. In effect each team will be self-contained with respect to performing all operational functions. The field operational personnel will consist of generalist patrolmen capable of performing the basic traditional field tasks including traffic enforcement and accident investigation. Personnel traditionally performing investigation duties, including juvenile, will become an integral part of the field team. Thus the field personnel will have the capability to handle all routine operations and major investigations. Preplanning calls for development of a case assignment system whereby the expertise of the most capable officer on duty will be assigned to handle each case. Routine cases will be handled by the generalist officer. The officer in charge of each case will, in effect, supervise the investigation of that case.

One may ask what about supervision? The team approach suggests that basic supervision will be accomplished through a system of peer control. However, the model suggests that a "lead officer" approach could be used to supplant the traditional first line supervisor, which calls for a clearly defined superior/subordinate relationship. A clear delineation of the superior/subordinate relationship does not appear necessary when all personnel become a collegial body equally responsible for the success or failure of operations. Furthermore, careful preplanning of relationships in which all personnel participate in the decision making process can eliminate confusion and tension situations. In addition, a system can be preplanned for meeting emergency situations requiring precise leadership and supervision.

Sustaining personnel performance records, communications, property, crime laboratory and custodial functions will be an integral part of the operations team responsible for supporting the field operations. They too, will function under the general direction and in cooperation with the team manager. Again reliance is on the most capable or qualified person to give direction on the performance of work related to his field of endeavor.

As indicated, the number of basic operational teams will depend on departmental size and needs. One approach could be to determine total departmental program workload, then divide the department into two teams, each responsible for a twelve-hour period, seven days a week throughout the year. Deployment of personnel and scheduling of assignments would be a responsibility of each team. Preplanning should take care of overlapping responsibilities.

Another approach might be to follow the traditional three shift pattern with appropriate distribution of personnel to meet shift loads. This appears to be somewhat cumbersome since there would be unequal teams, but it may be the best approach under certain circumstances.

Inevitably there will be certain activities which may be incompatible for the basic operation team structure. Narcotics, vice and intelligence activities are probably better performed by use of a specialist team. The same could be true of warrant service, forged document investigations, and other types of activities which are somewhat incompatible with the basic team operation.

In the larger department with substations or precincts, a

number of structural relationships may be feasible. One approach could be for headquarters to consist of the top managment team, and to provide specialist support through a number of specialist teams. The substations or precincts could use the approach of having a "minor" top management team and then constitute several basic operational teams to provide twenty-four hour coverage of designated areas, workload to be relatively equal. Sustaining activities could be performed for all basic operational teams by a station team. Careful preplanning for designation of responsibilities and interteam relationships would be required.

What about such sustaining activities as recruit and inservice training, budget preparation, and purchasing? Again, specialist teams under the general direction of one of the top management directors may be the answer. Each department will have to decide. However, it must be remembered that insofar as is feasible the sustaining functions should be performed as an integral part of the activities of each operational team.

The team policing organization model is a drastic breach with tradition. However, it does provide an "ideal" to be strived for which can reduce the rigidity and the tension building features of current approaches to organization. It provides a means for involving the "new breed" of police officer who brings to the department talents and potential capabilities superior to his predecessor. It makes possible a continuous input of ideas and observations about the community, the operations and agency relationships which have been severely restricted and which are becoming extremely vital for meeting the dynamic and changing demands of society on the police. Admittedly, many problems are posed in changing over to adoption of the model. Serious questioning of many current practices must take place and a number of "myths" challenged. However, the end result may be a viable, meaningful and dynamic organization.

Team Policing Models*

In recent years, due in part to changes in the social climate and in part to changes in police patrol techniques (more patrol cars, less foot patrol), many police agencies have become increasingly isolated from the community. This isolation makes crime control more difficult. The need to increase police-citizen cooperation is self-evident.

Team policing is a modern police attempt to reduce isolation and involve community support in the war on crime. Team policing can be administered in small or large amounts. Syracuse, N.Y., the first U.S. city to try team policing, uses teams of 10 officers to patrol, investigate, and control crime. Team leaders have considerable discretion and authority. In Los Angeles, nine-member units have patrol responsibility; investigation remains separate, however, Dayton, Ohio, has an extremely advanced concept of team policing in operation.

What is team policing? Essentially it is assigning police responsibility for a certain area to a team of police officers. The more responsibility this team has, the greater the degree of team policing. For instance, team policing that has investigative authority is more complete than team policing that does not. Teams that have authority to tailor programs and procedures to the needs of their areas go even further.

The basic idea is that the team learns its neighborhood, its people, and its problems. It is an extension of the "cop on the beat" concept, brought up to date with more men and modern police services. It lessens the danger of corruption of a single officer in a single area.

The first experiment in team policing took place in Aberdeen, Scotland, in 1948. It ended in 1963. Several other attempts were tried but also abandoned. In 1966, a new modification known as "unit beat policing" appeared, also in Great Britain. This concept combined patrol and investigation personnel and stressed public cooperation. It is this form of team policing that made the trip across the Atlantic. It was discussed, and certain aspects recommended, in

---

*Source: National Advisory Commission on Criminal Justice Standards and Goals, *Police,* Washington: U.S. Government Printing Office, 1973, pp. 154-161.

the 1967 President's Commission on Law Enforcement and Administration of Justice.

Problems

Team policing has certain inherent difficulties. Combining patrol and investigation personnel into teams can cause friction. Investigation personnel are usually higher paid and think of themselves as farther up in the police hierarchy. This may cause friction among team members.

Team policing requires considerable individual initiative and responsibility. Many patrolmen are reluctant to exercise such authority. There are other situations that require quick, military orders and obedience. How can this be reconciled with team member equality?

These are the problems that are being worked out and answered in pilot programs now in progress. One definite benefit is that team policing concepts are rich in fresh thoughts; they stir police agencies to reexamine many assumptions about police procedures. Even when team policing is not adopted, examining other agency programs can prove beneficial to overall police thinking.

So far, even where team policing has proved most beneficial, programs are still in the experimental stage. This chapter advises any agency considering team police work to plan carefully before going ahead.

Larger agencies have the advantage of being able to try programs in certain precincts while carrying on routine police work in others. Smaller agencies have no such option. For the smaller agency it is all or nothing.

The benefits of team policing, primarily greater police-public cooperation, are not automatic. Team policing only affords the opportunity for such benefits. It is up to the participants to go out into the community and foster the cooperation needed. The police agency must let the public know about its new program and what it hopes to achieve.

Just as importantly, in agencies where a team policing experiment is being prepared, all employees—not just those to take part—must understand and support the program.

Police chief executives often lack information for determining the advisability of changes in police operations. Professional publications constantly report new ideas and programs, but they generally stress aspects flattering to the program's creator and avoid the problems of implementation. Most police administrators recognize the need for change in law enforcement in a changing society. Knowing what changes to make is difficult without objective reports from other cities.

Team policing has become one of the most popular forms of police reorganization. Practiced different ways in different agencies, it has received considerable publicity. No definitive study has yet been made that gives police chief executives a thorough understanding of what team policing actually is and what benefit an agency can expect from adopting the concept.

Team Policing Systems

Total team policing can be defined as: (1) combining all line operations of patrol, traffic, and investigation into a single group under common supervision; (2) forming teams with a mixture of generalists and specialists; (3) permanently assigning the teams to geographic areas, and; (4) charging the teams with responsibility for all police services within their respective areas. Most team policing systems have not taken this total approach; they have limited operation to a small area within the agency or have concentrated on reorganizing only the patrol function without including investigative personnel or other specialists in the team.

Certain structures and goals are common to all team policing programs. Structurally, they all assemble officers who had previously functioned as individuals or two-man teams and assign them shared responsibility for policing a relatively small geographic area. The common goal is improved crime control through better community relations and more efficient organization of manpower.

Syracuse, N.Y., was the first police agency to combine the patrol and investigative function into one unit with a geographic responsibility for crime control. The crime control team was implemented in July 1968, and consisted of a team leader, deputy leader, and eight policemen. The team was relieved of many routine, noncriminal duties and given responsibility for controlling serious crime, apprehending offenders, and conducting investigations in a small area of the city.

The team leader, a lieutenant, was given considerable discretion in directing the activities and operations of the team. The program was decentralized and operated independently of the rest of the agency. The crime control team concept was later extended to other agency operations after the project report on the experiment indicated considerable success in reducing crime and increasing crime clearance rates.

The stated objectives of the basic car plan were to help society prevent crime by improving community attitudes toward the police; to provide stability of assignment for the street policeman; and to instill in each team of officers a proprietary interest in their assigned area and a better knowledge of the police role in the community. Beginning in November 1969, the plan was tested in two divisions; it was expanded citywide in April 1970.

Each police division has geographic areas of varying size based on workload and crime occurrence data. A team of nine officers is assigned to each basic car plan area and given responsibility for providing police service on a 24-hour basis. Each team is headed by a senior lead officer. Supervisory responsibilities of the patrol watch commanders and field sergeants remained unchanged.

Formal meetings between the team and citizens in each area are held monthly; informal meetings occur frequently throughout the month. Investigative, traffic, and other specialized personnel in each division are not assigned directly as members of the basic car teams at this time. A comprehensive experiment in total team policing begin in the agency's Venice Division in June 1972.

The beat commander pilot program began in April 1970, in two scout car areas in Detroit's 10th precinct. The beat commander, a sergeant, was given command of approximately 20 men, including three detectives who investigated only those cases originating in the beat command area. Two additional sergeants were later assigned to provide around the clock supervision.

The principal element of the system was stability of assignment of the beat commander and the team to a specified neighborhood. The goals were to improve police-community understanding, cooperation in crime control and police efficiency, and job satisfaction.

In January 1971, a neighborhood police team consisting of a sergeant and 18 policemen began operations in one radio motor patrol sector. As a result of this experiment, the system was later expanded throughout the agency. The structure of the N.P.T. is similar to the Detroit system. N.P.T. patrolmen, however, take greater investigative initiative; detectives are not directly involved in the program.

Crime control and community relations are two principal goals of the project. Additionally, improved supervision and motivation have resulted in increased productivity and efficiency. Substantial reductions in response time to calls is also attributed to the team program.

Dayton designed its team policing project to test the generalist approach to police work, to produce a community-based police structure, and to change the police organization from its traditional military structure to a neighborhood-oriented professional model. All specialized assignments in the test area were eliminated. Discretion was allowed in the wearing of uniforms, modes of operations, and program development.

The experiment began in October 1970, in a district covering about one-sixth of the city area. The personnel consisted of 35 to 40 officers, 12 community service officers, a lieutenant in charge, and four sergeants who acted as leaders for teams of 10 to 12 men. The lieutenant was selected by the chief and approved by neighborhood groups. The officers selected by vote their team leaders from a slate of sergeants.

The Dayton team project is probably the most fundamental attempt to change police field operations. Most internal matters are settled democratically among team members. The project decentralized authority and function, and concentrated upon community participation in achieving its goals.

Most agencies that have tried team policing have tested the concept on a limited basis. Many of the currently operating systems have not been expanded to include the entire agency but continue in a small area on a scale conducive to testing and evaluation.

Agencies that have attempted team policing innovations have usually approached the subject from the standpoint of demonstrating that the innovation will work rather than trying to prove or disprove team policing through extensive experimentation. A demonstration project attempts to prove that a particular idea is an improvement. An experiment tests alternative ways of solving a problem.

The overriding objective of most agencies—the avoidance of disruption of ongoing operations—precludes true experimentation. Police agencies should conduct extensive research and plan comprehensive testing prior to formal implementation of team policing agencywide.

For most medium size and large agencies, a limited test of a proposed team policing system can be accomplished by designating one geographic area as a test site. Smaller agencies with limited personnel may have to test the system agencywide. Agencies with less than 75 employees should insure that adequate planning and appropriate training is completed prior to agencywide testing of a team policing project.

In all cases, the research and planning stage must include the establishment of program objectives and goals. It must provide for evaluation of the effectiveness of the program in reducing crime, increasing arrests, increasing the general level of service rendered to the community, and enhancing police-community cooperation. The police chief executive and the agency's staff, command, and supervisory personnel should understand and support the proposed plan. Personnel charged with making the concept work must be given appropriate training, support, and authority to achieve the project's goals and objectives.

The team policing concept has been viewed by some as a return to a bygone golden age of police work typified by the friendly, well-known corner cop who helped community residents manage the problems of urban life and learned a great deal about their lives in the process. In the late 1940's and early 1950's, however, reformers found this friendly officer on the foot beat to be corrupted by his familiarity with local residents and slow to respond to the scene of emergencies. To solve the latter problem they put him in a radio car. To solve the former they transferred him so frequently that he would not have a chance to know people well enough to become corrupt.

The impersonal police officer created new problems. The President's Crime Commission and the President's Riot Commission both emphasized lack of community contact between the police and the citizens as a serious weakness in patrol operations. The Crime Commission, in fact, urged that patrolmen should be thought of as foot officers who use vehicles for transportation from one point to another. Many patrolmen resist the idea of getting out of their patrol cars to talk to citizens; some even view this activity as a degrading form of appeasement. The idea also runs contrary to the tactical principle that the continual moving presence of motorized patrol on the street is required to provide adequate preventive patrol. Many police agencies embrace this principle to the extent that unnecessary or unofficial conversations with the public are discouraged by agency regulations. Recent research on the effectiveness of preventive patrol, however, indicates that any crimes prevented by passing patrol vehicles can be and usually are, committed as soon as the police are out of sight. Police presence can only prevent street crime if the police are everywhere at once.

A more effective way to increase the risk involved in committing a crime is to raise the probability of apprehension after the crime has been committed. Without information supplied by the community, apprehension is quite difficult. The easiest way to obtain information about a crime is for the police to talk to community residents who may have knowledge of the crime or the offender.

Rapid response to serious or emergency calls for services, particularly those involving crimes in progress, is essential to crime control and criminal apprehension. Once a crime is committed, the police must switch their major tactical emphasis from prevention to interception. Again, police must depend to a great extent on information supplied by the public to increase their chances of intercepting a criminal while the crime is in progress or during his flight from the immediate scene.

Once prevention and interception have failed, the only tool left to the police is investigation. But like prevention and interception, investigation requires cooperation from the public in apprehending the suspect and providing testimony in any subsequent court proceeding.

Team policing in any of its various forms is an attempt to strengthen cooperation and mutual coordination of effort between the police and the public in preventing crime and maintaining order. The concept is rather simple but complex in its implementation. The degree of success of any agency's team policing program depends on the active participation of all agency personnel in cultivating the active and willing support of the public.

Personnnel Participation

The lack of involvement of agency personnel in the planning and implementation of programs has been a basic defect in many team policing experiments.

Involvement begins with a personal and continuing commitment on the part of the police chief executive. This does not mean that the idea must originate with the chief, or that he must personally direct all planning and implementation. However, the chief executive must assume leadership by supporting the project and identifying himself with its implementation and long range operational aspects. The same is true of the agency's high ranking staff and command personnel. If support from all or some of these key personnel is withheld, the project's chances for success are reduced appreciably.

Middle management support has been missing from several team policing programs. The lack of this support, in many cases, can be traced to top management failure to communicate to middle management the value of the team policing concept and middle management's role in making the concept work. This problem can be avoided by providing sufficient planning and implementation time to allow full participation by middle management to develop.

Planning input and participation also are required from the officers and the supervisors who will carry out the program at the operational level. Horizontal and vertical team building can be supported by a participative planning process that reduces the traditional distance between ranks.

The Kansas City, Mo., Police Department developed a planning process that involved representatives from all ranks. Task forces were assembled in each division. The task force members communicated with all other divisional personnel to solicit ideas, obtain reactions to early plans, and build support for the proposed organizational change. This type of participative planning increases support of management and supervisory and field personnel, and greatly enhances the potential for success of any project.

Training for Team Policing

The initiation of a team policing project offers an excellent opportunity to provide comprehensive training to police personnel at all ranks. With proper planning, there is great potential for obtaining widespread support for the project by acquainting all agency personnel with the project's objectives and goals, and with their roles in achieving those objectives and goals. Specific preparatory and inservice training should be provided to personnel according to their involvement in the team policing effort.

The training of team policing supervisors or team leaders should emphasize planning, managing, directing, and coordinating the activities of team members. Team leaders should also be taught the techniques for teaching and training team members on an ongoing basis. The role of the team leader requires knowledge for maintaining liaison between the team and all other involved agency entities and community organizations.

Placing the responsibility for decision making at the lowest possible organizational level is an integral part of the team policing concept. In many agencies first line supervisors and patrolmen have never made decisions to the extent required in team policing. Adequate training in decision making is essential for team leaders and, to a lesser extent, for other team members.

Patrol officers assigned to teams should be trained in the theory of team dynamics and provided with information to enable them to function effectively as team members. Communications, conference leadership, and interpersonal relations training will assist team officers in their work with the community. Traditional patrol, traffic, and investigative techniques and skills should also be emphasized.

The training of other agency personnel should emphasize the role of administrative, staff, and support personnel in attaining team policing objectives and goals.

Community Involvement

One of the most serious problems confronting police agencies today is isolation from the community. Several factors, including police organizational inflexibility and the attitudes of both the police and the public, have caused this isolation.

Team policing places the police officer in an environment that encourages cooperation with the public and thus reduces isolation. Team policing brings the police organization down to the community level. This enables individual officers to cultivate community support and build personal relationships essential to the goal of police-community partnership.

Effective police-community cooperation is critical to the success of a policing project. The public must be informed of the team policing concept, its objectives and goals; public assistance and participation must be solicited actively. Successful community involvement programs depend on direct participation of citizens in the planning stages. Ongoing public commitment is encouraged by continually seeking the opinions, ideas, and assistance of citizens in resolving problems of mutual concern.

Team Leadership*

Team policing is a relatively new police management theory which relies upon job enlargement to motivate individuals to the accomplishment of group goals. It emphasizes the development of generalists as opposed to specialists, and relies upon the self-worth and direction of individuals. It is an unstructured organizational theory which provides for collegial decision making in an atmosphere of participative management.

Organizing for Team Leadership

The *strict* application of a team policing theory, however, does not allow for the adverse influences, the exceptions to its theory, the individuals of the group who simply cannot rise to the level of team competence.

With appropriate modification to allow for team leadership, team policing *may be* the answer to providing the most police service for the least cost. Revision of the theoretical concept to allow for standard human character traits enables the theory to become operationally effective, and allow for a better program of individual generalist development. It is such a development, established within

---

*Source: George H. Savord, "Organizing for Team Leadership," *California Law Enforcement Journal,* Vol. 8, No. 1, January, 1974, pp. 22-26. Reprinted by permission of the author.

workable guidelines, which provides the foundation for the concept's fiscal desirability. The dynamics of the group must make way for the dynamics of a team leader.

Recent trends in management lead toward participative involvement of all organizational levels in the management process. The Human Behavior School is emphasized to motivate managers to work *with* people rather than through people to create an atmosphere of shared responsibility and self-imposed controls. The methodology of participative management is thus contrary to a centralized authority which is more regulatory and autocratic in nature. Responsibility is delegated and free discussion is encouraged at all levels, while quality of product is controlled by the workers themselves through suggestions and criticisms. By relying upon a feeling of interdependence among the workers, goals are achieved more through self-direction than established decision making levels within a rigid chain of command.

The difficulty with participative management in police administration is that it overlooks the Social System School of management theory. Peer group pressures through the greater involvement of police agencies on a regional basis can affect individual agency activity. Police officers are frequently trained on a regional basis through all operational and managerial levels. Thus, they have a natural tendency toward peer group influence on a wider scale. Their work experience is in an environment of such a wide ranging scale of operations that they identify with the police of a total area, rather than the police of a particular place. The resulting social or peer pressure upon involved personnel is considerably greater, a fact which results in more rigid attitudes toward change.

While it may be argued that participative management develops leadership, initiative, creativity, and self-worth, the job may not get done for want of that which is expected. The police, like their military counterparts, are trained in an atmosphere of expected leadership, direction, and discipline. The result of not providing these stimuli can be an inconsistent and ineffective operation which lacks uniformity simply because of the lack of formal controls and direction.

A managerial dilemma can also arise from the regulatory theory of management. In order to achieve uniformity and control of facets of the operation by pinpointing responsibility, initiative and creativity can be stifled resulting in personnel becoming followers unable to individually develop. The peer dependency of police personnel on leadership and direction lends itself to a continued regulatory approach, but the marked disadvantages of stifling the individual and perpetuating lesser but assured production, merits some consideration for change.

Compromise

A compromise must be developed to involve some principles of participative management and yet include some aspects of the more autocratic schools of police management. While the human behavior theorists may rely on self-motivation and self-fulfillment, all personnel are not self-directed or self-motivated, and find their self-fulfillment in a lesser level of achievement than might be theoretically desired. Their behavior and job performance can be equated with their individual desires. The management art is to recognize this fact without treating personnel as inferior and lacking motivation, to avoid the self-fulfilling prophesy that people will perform in accordance with that expected of them. The compromise must, therefore, include the theory that personnel will respond favorably if treated responsibly, emphathetically, and with provision for their individual development.

Team policing *can* provide the compromise between participative and regulatory management without sacrificing basic management principles. While total team policing cuts across organizational lines and delineated responsibilities and relies upon self-motivated personnel being responsive to job enlargement, a modified form of team policing can satisfy the need for getting the job done in compatibility with peer group standards, while providing many advantages to individual personnel by managerial involvement.

By maximizing the importance of the individual and his contribution to the organizational goals, the desire to be appreciated is satisfied and becomes an important motivator. As man enjoys a sense of his own importance, he contributes to the measure of that importance. Thus, the personnel of an organization must be influenced through the dynamics of leadership. That leadership must lend itself to the accomplishment of organizational goals by inspiring others to the same end.

Not everyone has the conceptual ability to visualize organizational goals or the intellectual ability to comprehend complete job knowledge. Everyone does not have the experience necessary to perform complex, technical skills or the ability to concentrate on a particular problem or a number of problems at the same time. These limitations must be recognized and dealt with in setting a leadership style.

Principles

There are some basic principles of organization to which most theorists and proponents of the various schools of management seem to agree:

Similar tasks and functions should be grouped together:

> Team policing, in its broadest, theoretical application, does not consider that police operations and support services are comprised of many complex, dissimilar tasks and functions. While team policing groups all tasks and functions of the organization into one or few areas of concern for the purpose of maximizing individual worth and job enlargement, many of the tasks and functions simply cannot be performed by the same individual during a singular tour of duty, for want of time or broad expertise.

Clearly defining tasks and functions:

> When defining the tasks and functions of the total team policing organization, it becomes apparent that they are too numerous, complex, dissimilar, and dysfunctional for individual performance. Thus, the definition of tasks and functions lends itself to better analysis of the police role and a modification of the team policing concept.

Fix responsibilities for tasks and functions:

> Broad application of the team policing concept does not provide for the fixing of individual responsibilities. All police

officers are considered generalists capable of performing any task and function. It is, therefore, difficult to determine administratively who is responsible for what task or function.

Delegate authority with responsibility:

It is always necessary to delegate authority commensurate with the responsibility for the performance of a given task. It is difficult to delegate broad authority to a group accountable for broad responsibilities in that the larger the number of those with authority, the more likely an abuse of that authority will result.

Those to whom authority is delegated must be held accountable:

It is more difficult to hold a large number of persons accountable for their responsibilities, if those responsibilities are delegated on a broad scale, with little control other than reporting systems.

The span of control must be reasonably limited:

The total team policing model has greater than a desirable number of persons reporting to individual supervisors, which tends to diffuse supervisory effectiveness and inhibit the communication process. Capabilities of individual supervisors vary, and the number of subordinates that can be adequately influenced should be kept within the bounds of efficiency to insure that organizational goals are met.

Communication channels must be clearly established:

Strict team policing can confuse communications channels by eliminating many of the communications routes within the organization. The team organization depends more upon informal and inter-personal relationships for adequate communications than upon formal communications channels. Thus, team policing seriously challenges the basic management tenet of clearly established communications channels.

Unity of command must be maintained:

The team concept can interpose a group to make decisions

rather than the individual. This diffuses the responsibility for decisions, causes delays in decision making, and leaves subordinates without a focal point of leadership.

There are several advantages and disadvantages to the generalist theory of team policing which contends that one person can be trained to competently perform a variety of tasks. The stated objective of the approach is to maximize individual worth and potential, motivate the individual by emphasizing his individual contribution to the organization, and to enlarge the scope of individual responsibilities to increase job performance.

The advantages are that the individual's usefulness does not become limited to one specific function. Individuals work as a part of the entire organization rather than as an autonomous unit, and most personnel develop the necessary skills for performance of specialized tasks. The disadvantages are that the responsibility for the performance of various tasks is not firmly fixed and accountability suffers. Optimum proficiency is seldom really achieved, and intensive training in an attempt to develop proficiency will develop special interests rather than a generalist attitude. It is noted that not all persons are either interested in or capable of performing all tasks and functions equally.

Job Enlargement

Job enlargement, as a team concept, increases the sphere of responsibility for each individual by providing a variety of tasks to be performed as a safeguard against boredom and apathy. It is believed that the wider scope of responsibility will present a needed challenge and sense of accomplishment among all personnel. Thus, productivity will increase as each individual exercises the freedom to set his own pace and discretion in the method of performing the various tasks.

However, since all persons are not interested in the same variety of tasks, properly accommodated interests can provide greater accomplishment with the same diminution of boredom and apathy. Additionally, not all persons enjoy greater responsibility. Many persons find additional responsibilities painful, and work better in an atmosphere of limited responsibility. Thus, individual evaluation prior to assignment of responsibilities is vital to provide for a more comfortable assignment of responsibilities among personnel.

Experience shows that not all persons are capable of self-direction and self-discipline in setting priorities and the pace of productivity. Whenever any group strives for a common objective, a leader will emerge from that group who assumes that responsibility, and marks progress. This natural process must be reconciled with the team theory that all persons are capable of common leadership.

Similarly, not everyone is capable of exercising discretion in selecting methods of performance, and many persons lack the conceptual ability to estimate subsequent events. What is individually expedient at the moment, may not be organizationally wise in view of later consequences. In police work a proper choice of methods can be crucial because every police act affects a human life. The exercise of authority demands discretion, and only those who have demonstrated sound judgment can be afforded the risk of exercising it for both the individual and common welfare.

The strength of the team concept lies with motivated individuals willingly working together toward a common achievement. Its weakness is in the dependency upon the group for leadership. Someone must best define the group objectives, reconcile differing methods, recognize and develop individual capabilities, and inspire the others to action. The advantages of team effort can best be realized by modifying the team theory to allow for designated leadership. This theory lends itself to a newly emerging Leadership School of management; an application of the dynamics of leadership to the team model. Although it preserves some of the more traditional regulatory management modes, it allows for reciprocal respect, confidence, and loyalty between management and employees who share a mutual concern for the achievement of mutual goals.

Figure 8-1
Strict Team and Modified Team Policing Theory
Team Resonsibilities

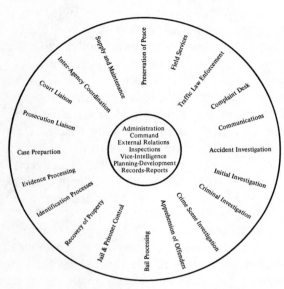

Strict Team Policing Theory
Team responsibilities are all encompassing of the entire organizational functions. Individuals are generalists capable of any task at any given time, and are self-motivated and self-directed.

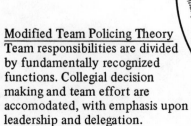

Modified Team Policing Theory
Team responsibilities are divided by fundamentally recognized functions. Collegial decision making and team effort are accomodated, with emphasis upon leadership and delegation.

Topics for Discussion

1. Discuss the key features of the colleague model.
2. Discuss program implementation under the team policing concept.
3. Discuss the value of team policing.
4. Discuss the need for the involvement of agency personnel in the planning and implementation of team policing.
5. Discuss the compromise between some principles of team policing and traditional police management.
6. Discuss the modified team policing theory.

FOOTNOTES

Introduction

[1] Elliott, J.F. & Sardino, Thomas J.; *CRIME CONTROL TEAM: An Experiment in Municipal Police Department Management and Operations,* Springfield, Illinois, Charles C. Thomas, Publisher 1971.

[2] Etzioni, Amitai: *MODERN ORGANIZATION,* Englewood Cliffs, N.J., Prentice Hall, 1964.

[3] Golembiewski, Robert T.: *ORGANIZING MEN AND POWER: PATTERNS OF BEHAVIOR AND LINE AND STAFF MODELS,* Chicago, Rand McNalley, 1967.

[4] International City Managers Association: *MUNICIPAL POLICE ADMINISTRATION,* Chicago, C.M.A., all editions.

[5] Kenney, John P.: *POLICE ADMINISTRATION,* Springfield, Ill.: Charles C. Thomas, Publisher, 1972.

[6] _____: "Team Policing Organization: A Theoretical Model," *POLICE,* Vol. 16, No. 12, August 1972.

[7] Pfiffner, John M. and Frank Sherwood: *ADMINISTRATIVE ORGANIZATION,* Englewood Cliffs, Prentice Hall, 1960.

[8] Wilson, O.W.: *POLICE ADMINISTRATION,* New York, McGraw Hill, 1972.

[9] It is suggested that the reader check with Syracuse, New York; Los Angeles, Palo Alto, Culver City, Richmond, Cypress, and Menlo Park, California Police Departments; Los Angeles County Sheriff's Department; and Dayton and Kettering, Ohio relative to their implementation of team policing programs. Also check with St. Petersburg, Florida.

Team Policing Models

[1] Cann, W. "4/40-Basic Team Concept," *Journal of California Law Enforcement,* October 1972.
[2] Cincinnati, Department of Safety, Division of Police, "Community Sector Team Policing Program, Com-Sec." Proposal for Action Funds Submitted to the Police Foundation, April 1972.
[3] Cordrey, J., and G. Pence, "An Analysis of Team Policing in Dayton, Ohio," *Police Chief,* August 1972.
[4] Elliot, J.F., and Thomas J. Sardino. *Crime Control Team.* Springfield, Ill.: Charles C. Thomas, 1971.
[5] Goldstein, Herman. "Police Response to Urban Crisis, *Public Administration Review,* September-October 1968.
[6] Holyoke, Massachusetts, Model Cities Program. "Evaluation Report on the Model Cities Team Policing Unit of the Holyoke Police Department." April 1972.
[7] Leek, C.C. "Management by Objectives," *Police Journal,* July-September 1971.
[8] Likert, Rensis. *The Human Organization: Its Management and Value.* New York: McGraw Hill, 1967.
[9] Police Foundation, "Project Strategy Recommendations." January 29, 1971.
[10] Police Foundation. "The Simi Valley Community Safety Agency." Report by Police Foundation Staff.
[11] President's Commission on Law Enforcement and Administration of Justice. *The Challenge of Crime in a Free Society.* Washington, D.C.: Government Printing Office, 1967.
[12] Syracuse Police Department, and General Electric Company Electronics Laboratory, Syracuse, New York. "Crime Control Team, Final Report, January 1, 1970-June 30, 1970." National Institute of Law Enforcement and Criminal Justice, Award No. NI-0046.
[13] Zurcher, James A. "Team Management/Team Policing," *Police Chief,* April 1971.

SELECTED READINGS

Bloch, Peter B. and David Specht, *Neighborhood Team Policing.* Washington: U.S. Government Printing Office, 1973. A practical manual and theoretical guide to neighborhood team policing. Summarizes current knowledge and describes an ideal neighborhood team policing system.

Cordrey, John and Gary K. Pence, "An Analysis of Team Policing in Dayton, Ohio. *The Police Chief,* Vol. 39, No. 8, August, 1972, pp. 44-49. Describes the analysis of a team policing experiment in Dayton, Ohio. Reviews the achievements and problems that occurred. Includes specific considerations that agencies should analyze if they implement a similar program.

McArdle, Edward C., and William N. Betjemann, "Return to Neighborhood Police," *FBI Law Enforcement Bulletin,* Vol. 41, No. 7, July, 1972, pp. 8-11 and 28. Analyzes the Albany Police Department, New York experiment with a neighborhood police unit. This model was designed to reduce crime, provide expanded police and social services, and build rapport with the populace.

Phelps, Lourn and Lorne Harmon, "Team Policing," *FBI Law Enforcement Bulletin,* Vol. 41, No. 12, December, 1972, pp. 2-5 and 28. Examines the results of four years of team policing in the Richmond Police Department. Reviews positive results, negative results and directions for the future.

Sherman, Lawrence, Catherine H. Milton and Thomas V. Kelly, *Team Policing,* Washington: Police Foundation, 1973. Presents the elements of team policing and an analysis of team policing in seven communities. Includes an analysis of obstacles to team policing and the problems of evaluation.

Zurcher, James, "Team Management/Team Policing," *The Police Chief,* Vol. 38, No. 4, April, 1971, pp. 54-56. Reviews the application of team management in the Palo Alto Police Department. This concept is based upon the premise that all personnel have a contribution to make to the policy and operational decision-making process.

 **The study of this chapter will enable you to:**

1. Identify the limitations of the Uniform Crime Reports as a management tool.
2. List five types of data the police generally use to evaluate performance.
3. Differentiate between input and output.
4. Define productivity.
5. Differentiate between productivity and effectiveness.
6. Identify the five stage process of productivity improvement.
7. List three substitutes that might be used in the absence of a direct measure of deterrence.
8. Identify the three segments of response time.

9

Police Productivity

Introduction*

Quantitative measures are nothing new in police services; in fact, they may be more familiar to police managers than to many other State and local government officials. Nevertheless, many of the measures currently being applied to police services do not provide managers with the information they need to help them improve operations. This chapter discusses some of the problems experienced with current measures and the potential and limitations of better measurement in helping police managers to improve productivity.

Problems With Current Police Measures

The most common data used for judging overall police performance are crime rates—such as compiled in the *Uniform Crime Reports (UCR)* published annually by the Federal Bureau of Investigation.[1] However, because the incidence of crime is a function of many factors unrelated to police activity, crime rates alone are

---

*Source: National Commission on Productivity, *Opportunities for Improving Productivity in Police Services,* Washington: National Commission on Productivity, 1973, pp. 7-12.

insufficient measures of police performance. Police managers need other measures that more directly reflect the significance of police activities.

The *UCR* has additional limitations as a management tool, many of which are cited in the *UCR* itself. The most critical of these is the fact that the *UCR* is not an altogether accurate reflection of crime. The *UCR* documents only *reported* crime, which, as several surveys have shown, is only a fraction of crime actually committed. In fact, recent surveys of "victimization" have indicated that reported crimes may represent, in some jurisdictions, as little as 25 percent, and rarely more than 75 percent, of the actual incidence of crime. One reason many types of crime go unreported is the victims' fear of embarrassment, of family or personal involvement, or of retaliation on the part of the offender. In some cases victims failed to report crimes because of lack of confidence in the police.[2]

Furthermore, in the FBI's Crime Reporting Program, data are not published on both offenses and arrests for all categories of crime. The *UCR* identifies "index crimes," which include the major crimes of murder, forcible rape, robbery, aggravated assault, burglary, larceny $50 and over in value, and auto theft. It then groups offenses into two categories: "Part I" offenses, which include the Index crimes, and "Part II" offenses, which, while they are lesser crimes, consume much of every police department's time and effort. The *UCR* reports both offense and arrest information for "Part I" offenses, but only arrest information for "Part II" offenses. The distinction between major and minor offenses is all the more a problem since many police agencies do not adhere strictly to *UCR* definitions. Some, for example, classify a burglary attempt as a malicious destruction of property, thus demoting it to a "Part II" offense.

Some hope is offered for getting more accurate crime data through victimization surveys—confidential and detailed surveys of scientifically selected samples representative of the population as a whole to detect the true number of crime victims. Scientifically and consistently administered, victimization surveys may provide new measures for crime-control and crime-prevention programs.[3] They may reveal not only the true incidence of crime but also the reasons why crimes were not reported and the victims' attitudes toward the police and police service.

Victimization surveys, however, are expensive if conducted properly, primarily because a large sample is needed before the data are valid. The Law Enforcement Assistance Administration, in collaboration with the Bureau of the Census, is gathering victimization data on a national and citywide basis. The National Crime Panel, under technical development for about 3 years, is a nationwide survey of individuals and businesses which regularly provide statistical data on the incidence of common crime, its costs, the characteristics of victims, and the characteristics of criminal events.[4]

In addition to crime statistics (e.g., the *Uniform Crime Reports*), there are several other types of data upon which police generally rely to help them monitor their workload and evaluate their performance. These include:

- Numbers of arrests by crime category.
- The clearance rate. (As used in the *UCR*, police "clear" a crime when they have identified the offender, have sufficient evidence to charge him, and actually take him into custody. The arrest of one person can clear several crimes, or several persons may be arrested in the process of clearing one crime.)
- The exceptional clearance rate. (Once again using the *UCR* definition, crime solutions are recorded in exceptional instances when some element beyond police control precludes the placing of formal charges against the offender, such as the victim's refusal to prosecute after the offender is identified, or local prosecution is delined because the subject is being prosecuted elsewhere for a crime committed in another jurisdiction.)
- Complaints received from the public about the department or about specific actions by officers.
- Activity measures of field operations and other services.
- Workload measures of clerical functions (e.g., number of additions per month to the fingerprint files.).

All of the above data are useful, but they are limited in the amount and quality of information they supply. Activity and workload measures can be usefully integrated into an information system to help managers estimate the demand for additional manpower resources or to identify concentration of clerical or

Police Productivity 395

administrative activity. Arrest data provide crude estimates of the activity of the patrol force, and the clearance rate is thought to provide some indication of a department's ability to solve reported crimes.

However, the majority of these data are not sufficiently refined to provide police managers with dependable and useful information which can lead to better performance. Clearance rates, for example, frequently do not correspond to police arrests made in the same time period; for example, crimes committed in one year may not be solved with the offender apprehended and court action taken until the following year. Similarly, a simple measure of arrests may tell little about how effective the police were in arresting the "right" person. A study for the 1967 President's Crime Commission estimated that only 24.1 percent of arrests for Index crimes survived a "formal accusation and detention" stage, and only 22 percent of all arrests resulted in conviction.[5]

In another example, measures of simple "workload" may reflect the amount of activity in a given function or field operation, but do not indicate how well the job is done or the amount of resources devoted to that activity.

Complaints received by the police department regarding general police performance or specific police activities potentially can give insight into their effectiveness and the quality of both the crime and noncrime services they provide. Frequently, however, police departments fail to distinguish the sources of such complaints (e.g., individuals vs. community groups vs. political pressure), and respond equally to all types of pressure for improved service. If tabulated carefully, complaints from the general public could provide useful guidelines for police as well as "feedback" regarding the public's perception of the relative importance of various police activities.

Many police departments presently have a solid base for gathering information regarding police activities. However, these data as presently aggregated can be misleading. One reason that existing data are not put to better use is that the police mission is complex. The specific objectives of the force are not always clear. Nor is it always certain what some police activities are contributing, or how they relate to higher department goals. In short, it is often difficult to know what to measure.

Consequently, a first step to improved measurement is to

understand how the various functions of police work relate to the broader mission of the department and the goals of State and local government.

Improving Measurement of Police Services

Measurement of police activities, as is true for most government organizations, is complicated by the absence of goals and objectives that are easily quantifiable. While there may be agreement on the broad goals of the police force, the specific activities that "rightly" fall under the jurisdiction of a police department are subject to debate.[6]

The Realm of Police Management

A police department is part of several public service systems, each of which may include a variety of public agencies. Although overall police performance may be judged by the general public on the basis of crime prevention or some perceived level of public security, the police are also responsible for non-crime-related and nonemergency services. Among the categories of service in which the police department plays a role are the following:

- Criminal justice, which includes (depending upon one's definition) the courts, correctional institutions, probation, parole, and many other public and private agencies concerned with crime and the criminal offender.
- Maintenance of public order.
- Emergency response, for fire, accidents, natural disaster, medical emergencies, etc.
- Community relations, which affects the community's feeling of confidence in or alienation from the government.
- Nonemergency general services. Police are called upon for a variety of non-crime-related tasks which do not fall under the responsibility of any other public agency, or which, because of the 24-hour nature of police duty, befall them when other agencies are closed. This may range from directing a stranger to a historical landmark, to registering bicycles, or to stoking the townhall furnace.

The relative amount of time and resources a police department devotes to meeting responsibilities within each of these systems naturally varies from community to community. But more important is the fact that the police both affect and are affected by other elements of the several systems of which they are a part. Effectiveness in preventing crime, for instance, depends in part on how well the corrections agency performs in rehabilitating felons. Or, whenever patrolmen spend time in court beyond the minimum required for efficient assistance in the judicial process, the less time they will be available for crime-related work. On the other hand, police skill in investigation and apprehension increases the likelihood that district attorneys can obtain convictions of arrested persons.

Consequently, measures of police performance must take into account the other system components that affect the outputs of police work.

Despite these problems, measuring police activity need not await, nor depend upon, a final resolution of the "proper" police role and responsibilities. Certain goals can be agreed upon, and certain activities clearly are important enough that measurement of them can proceed. Indeed, it may be that careful measurement and analysis of specific police activities will gradually produce a clearer understanding of their relationship to broader police and community goals.

Measurement to Assist Management

The principal purpose of measurement is to provide sufficiently precise information to enable police managers to: (1) evaluate their department's performance; (2) identify and diagnose problem areas; and (3) design solutions.

But measurement can provide other advantages as well. For example, measures frequently stimulate constructive thinking—e.g., measurement of crime deterrence requires in-depth analysis of the nature of deterrence—thus increasing the understanding of police activity. Measurement also may provide a means for linking one activity to another, or one part of the management process to another—e.g., relating resources to output.

Measurement, of course, also has its limitations. It is not a substitute for sound professional judgment; it is meant to assist the manager, not dictate action to him. Furthermore, care must be taken to guard against measures that provoke negative activity. To use a familiar example, measuring a patrol unit solely on the basis of arrests made, without considering whether or not the arrests are valid, can reward the apprehension of innocent people. Nor should uncritical enthusiasm for measures result in meaningless and costly measurement activity. Some measures may require data gathering that is more expensive than their value would justify, and consequently should be avoided.

But the most difficult problem in measurement is to assure that measures provide information that genuinely assesses how well the department is doing its job. Where the agreed-upon objectives of the department can be quantified, this should not cause problems. The difficulty lies in constructing indirect or proxy measures for objectives which defy quantification. In such cases professional judgment must be used to determine what activities contribute toward the accomplishment of the objective, with quantitative measures then established for the intermediate objectives of these activities. The identification of intermediate activities and objectives, however, requires great care and constant evaluation to assure that they do, in fact, contribute to higher departmental goals. Otherwise a department may erroneously judge its performance by measuring an activity that bears little or no relation to the real role of the force.

There are many types of measures, and as many names to describe them as analysts have time to devise. Given the state of the art of measuring public services, we need only be concerned with a few basic distinctions.

There are two fundamental types of measures which may be used separately or in combination to provide useful information. They are measures of results (or output) and measures of resources used (or input).

Police departments, as is true of most public services, traditionally have been more concerned with measures of resources than with measures of results. The budget gives the most basic measures of what resources are being used for what activities. Resource use might also be measured in terms of man-time, or units of equipment. Variations of resource measures include simple percentages of total resources devoted to a particular activity or subactivity.

Results, as noted earlier, are generally more difficult to define and measure. Traditionally, police departments have relied upon easy-to-quantify results such as miles driven by a patrol vehicle. Such measures (often referred to as workload) have some use as indicators of intermediate results, but they clearly do not provide an adequate assessment of whether higher level objectives are being met. A comparison between results achieved and results intended gives a simple measure of effectiveness.

Result and resource measures can be compared to indicate productivity. A productivity measure indicates the cost (in money, men, and/or equipment) of accomplishing a given result. Such a measure can apply to a whole police department, a division of the department, or a unit within a division. It should be noted that where the results of a smaller division are meaningless to overall departmental goals, the productivity of that division may increase (it gets more results per unit of resource) but with no improvement in the department's overall productivity. Thus, it is essential that the measurement of individual activities or organizational components always be understood in the context of overall departmental goals and performance. This suggests the need for budgeting systems that integrate the diverse operations of the police department into a coherent package.

Quantitative measures can take an endless variety of forms. At some future date it will no doubt be useful to establish a more precise system and language of public service measurement. For the time being, these simple distinctions should suffice. The important thing is to devise quantitative measures that provide better information to management, and to constantly be alert to what those measures are and are not revealing.

Productivity and Police Services*

State and local governments are challenged to provide more effective police services at a time when the growing desire for public safety is surpassed only by the increase in police costs. For a police department to create one more round-the-clock post actually requires adding five officers to the force at a cost that may exceed $80,000 a year. To place an officer in a police car with a partner 24 hours a day may exceed $175,000 in annual costs to the community.

These costs are reflected in the growing nationwide expenditures for police services. In response to the mounting fear of personal harm, loss of property, and public disorder in recent years, municipal police expenditures increased 70 percent from $2.1 billion in 1967 to $3.5 billion in 1971. Total Federal, State, and local expenditures for police services reached $6.2 billion in 1971, a 20 percent increase over the previous year. Those funds went principally to cover the compensation for over 530,000 law-enforcement officers employed full-time in over 10,000 public police agencies at all levels of government.[1]

These fiscal facts of life have forced many communities to recognize that the demand for more police services cannot be met simply by expanding the police force. Rather, police departments must learn to use more effectively the personnel and other resources currently available to them. That means increasing their productivity.

What is Productivity?

Productivity means the return received for a given unit of input. To increase productivity means to get a greater return for a given investment. The concept most often is used in reference to the production of goods, e.g., more agricultural products, automobiles, or tons of steel per man-hour. Specialists argue over the precise definition of the term "productivity," but it is generally assumed to

---

*Source: National Commission on Productivity, *Opportunities for Improving Productivity in Police Services,* Washington: National Commission on Productivity, 1973, pp. 1-5.

be a ratio of "output" (or what results from an activity) to "input" (or the resources committed to the activity).

Police services are not as easily defined as the process of producing a television set or an eggbeater. In general, higher police productivity means keeping the police department's budget constant and improving performance, or keeping performance constant and reducing the size of the budget. Productivity gain can also mean increasing the budget but improving performance at an even higher rate. But the concept of productivity cannot simply be transferred in its raw form from the economics of production to the operations of a State or local police department. Rather, increasing productivity in police services might be considered in the following four ways.

First, *increasing police productivity means improving current police practices to the best level known, to get better performance without a proportionate increase in cost.* In its simplest form, this means doing the things that are considered to be a necessary part of good police work, but doing them as well or efficiently as the best current practices permit. For example, officers assigned to patrol spend a great deal of time on such activities as filling out unnecessarily long reports, or on activities that are important but that would require less time if better coordinated, such as the long hours spent waiting to testify at a trial. These activities could be minimized through better administrative procedures, thus increasing the time available for more important assignments.

Of course, freeing up more police officer time—or improving upon other practices—will not guarantee that the force will be more effective in deterring crime, apprehending criminal offenders, or providing noncrime services. But it is a first step in making the force more effective, and can be accomplished at little or no cost to the department.

Second, *increasing police productivity means allocating resources to activities which give the highest return for each additional dollar spent.* A police department carries out a range of activities, many of which are non-crime-related and most of which are necessary to its overall capability and its responsibility to the public. Beyond a given scale, however, expanding certain activities will give the force less value than initiating or expanding others. For example, experiments already in progress tend to support the contention of some criminal-justice analysts that random patrol has a limited effect in deterring criminals. Thus, it may be possible to take, say, 10 percent or more of the patrol force off random patrol[2] without any significant negative effect and shift those officers to activites that focus on anticipating crime or "hardening" likely crime targets (e.g., improved building security), which may provide a higher payoff.

Or, to give another example, would a 500-man force get more value from adding a few more officers than from providing the existing 500 men with mobile radios? These are the kinds of decisions—rarely so simple in reality—that continually confront police managers, but that are often made with insufficient understanding of the options available or of their true costs and potential values. They require asking not just whether the force is doing things right, but also whether it is doing the right things.

Third, *given the uncertainties of police work, increasing productivity means increasing the* probability *that a given objective will be met.* The professional police officer—from the chief to the patrolman—must deal constantly with many unknown or ambiguous factors. He is continually assessing the likelihood that this or that may happen, and consequently the more skillful he becomes at increasing the probability that each activity will result in useful accomplishment, the more productive the overall operation will be.

The clearest example of increasing the probability of achieving intended impact is having personnel assigned when and where crime is highest or calls for service are heaviest. Simple observation can indicate the "when and where" in general terms; careful analysis of available data can more accurately pinpoint likely times and places of crime occurrence, thereby significantly increasing the probability of putting officers where they are needed.

Fourth, *increasing productivity in police work means making the most of the talents of police personnel.* Sworn officers are better trained and more expensive than ever before. This means that they are capable of higher performance, that economy requires they be used more effectively, and that they expect to be treated with greater respect and intelligence. Too often the individual talents of sworn officers are overlooked or suffocated by rigid organizational procedures. This represents both a squandering of public resources and a stifling of human potential. Our system should not—and increasingly will not—tolerate either.

Examples of better human resource development and management abound and can be expected to become increasingly important to police managers. They may include making patrolmen responsible for following through on investigations; permitting senior patrolmen to refuse promotion but receive a higher salary and prestige as a patrolman; and developing alternative career paths for professional police officers.

Productivity and Effectiveness

For any police activity, productivity must be considered in relation to effectiveness. The two concepts are closely related and at times may be difficult to differentiate.

In simple terms, effectiveness is a measure of the extent to which a goal is achieved. In this sense, it does not include any notion of resources committed to the activity.

Productivity includes not just what was accomplished but what resources were required to accomplish it.[3] It is important to recognize that productivity does not necessarily indicate the extent to which the result actually accomplished a given goal. Productive performance—was the job done efficiently?—must also be seen in terms of effective performance—how well was the job done and how significant was the activity in contributing to departmental objectives?[4]

One thing that is always common to both productivity and effectiveness is "output," or results. Under the pressure of growing demands for service and spiraling costs, government is being compelled to identify more precisely what it is trying to accomplish, and what the real results are from its activities. The former requires a clearer identification of objectives, and the latter a more precise way of assessing the results of activities that contribute to those objectives. Better measurement of results leads to better productivity assessment. And better productivity assessment, in turn, is an important step in the process of productivity improvement.

The Process of Productivity Improvement

Getting a greater return for the dollar spent is not a "one shot" activity. It is an ongoing, long-term process that should be an integral part of police management. The Advisory Group has identified the following five-stage process as one approach to productivity improvement in police agencies.

Establishment of Objectives

Ideally each police department establishes its goals in concert with the political and professional leadership of its government and the people they serve. It then proceeds to identify intermediate objectives, the achievement of which will contribute to the attainment of the broader goals.

In practice, the process of setting objectives is often reversed. Instead of determining the mission of the department and then organizing to accomplish it, more often the apparent aims of ongoing activities are described and are built into departmental goals.

What is important is that the different levels of objectives be clearly related and understood.

For example, police agencies usually have a broadly defined goal of reducing the amount of crime to tolerable levels. An intermediate objective would be to reduce the incidence of a specific crime during a specific time period. A police department can then choose various strategies, such as reducing the opportunity for burglary by a citywide campaign on building security or increasing the visibility of the police in areas of high burglary rates.

The key is that activities at rank-and-file level should be contributing to higher level departmental goals. Simple as this sounds if often is not the case.

Systematic Assessment of Progress

Police management needs to know how well it is doing in meeting its objectives. Most police chiefs, mayors, or managers have some judgment on how their police force is doing; for example, good, better than before, about the same, not quite up to par, or, it appears that we have a problem. Often these are "gut reactions" based on little more than intuition and informal evaluation.

In contemporary police work, effective assessment requires more precise measurement. This is not to suggest that all assessment must be based on quantified information, but without more precise measures it is difficult to determine how much better or worse a particular unit, strategy, or piece of equipment works. Even scant quantifiable data can be used as a limited aid to assessment.

Search for Improved Operating Methods

The Advisory Group agreed that many improved operating methods, types of equipment, and ideas being used in certain police agencies could and should be made known to and be applied in other jurisdictions. While numerous journals, conferences, and other communications media provide information on innovative and improved methods, some are not presented fully or clearly enough to be usable by busy police managers. Nevertheless, police managers ought to play an active role in searching for new and better methods.

Efforts to learn about developments in other jurisdictions are frequently passive, at best. Often the most valuable ideas come from within an agency, but people familiar with staff and line problems either are not asked for suggestions for improvement, or are asked to address themselves to the wrong questions. Similarly, nonpolice agencies within the same government, such as management analysis staffs, are too often ignored. If police managers sought their cooperation, such agencies could become sources for ideas or assistance for better police performance.

Experimentation

Most police chiefs are understandably cautious when it comes to doing something out of the ordinary. "Innovation" is a luxury many police departments feel they cannot afford. However, neither can they afford to hold to the *status quo* while conditions around them change.

Clearly a prescription for a balanced approach to risk-taking is needed. It is important to recognize that useful information often comes from the idea that did not work as expected. Managers must also learn how to take reasonable and controlled risks, that is, to try things out, and establish a consistent approach to risk-taking. Experiments should be designed in such a form as to make evaluation possible, to determine whether or not they are successful, as well as why and by what margin of quality or cost they are inferior or superior to existing methods, techniques, equipment, and ideas. In addition, those who ultimately are to use a new idea should be involved with the development and testing process.

Implementation

A new method that has been tested and proven feasible remains to be implemented. At this point, the sense of caution and resistance to change that it might have met from department leadership extend throughout the department, the government, and the citizenry as well. Overcoming this resistance requires involvement of those people at the experimentation stage, as well as thorough preparation, patience, cooperation, close monitoring of the innovation, and clear accountability.

Productivity in Police Patrol*

The patrol force—the men and women who know and "work" the streets—is the front line of every police department. These men and women are responsible for carrying out the wide variety of crime and noncrime services that a police department provides.[1] During a single tour, a patrol unit may respond to a bank robbery, assist a resident who forgot his housekey, rush a coronary victim to the hospital, and quell an argument between irate neighbors. Thus, the patrol force is more than a single man walking a beat or a lone patrol car; it is the principal operational arm of the police department.

The Advisory Group chose the following three objectives of police patrol for consideration in this report:[2]

- deterrence of crime;
- apprehension of criminal offenders; and
- satisfaction of public demands for noncrime services.

These three objectives are closely related. Apprehension of criminal offenders, an end in itself, also has some effect—to what extent is uncertain—in deterring crime. Deterrence of crime, of course, reduces the need for apprehension. And better noncrime services enhance the image and public support of the police department, thereby strengthening crime deterrence and apprehension efforts.

*Source: National Commission on Productivity, *Opportunities for Improving Productivity in Police Services*, Washington: National Commission on Productivity, 1973, pp. 13-29.

To meet these objectives, the patrol force carries out a variety of activities, any one of which may contribute simultaneously to one, two, or all three of the objectives. The activities include observation; response to calls for service; enforcement of the law; investigation; maintaining order; and various administrative and postarrest activities (e.g., report writing, court duty). Since any one activity may contribute to all three objectives, and since the objectives themselves are interrelated, the measurement and analysis of the patrol force can be a complex undertaking.

In an attempt to cope with this complexity in a practical way, the Advisory Group has identified three problem areas which begin to sort out the easier patrol problems from the more difficult ones These areas are:

- Making a greater proportion of the existing patrol division (up to a reasonable limit) available for street assignment,
- Increasing the real patrol time of those who are given street assignments; and
- Utilizing patrol time to best advantage (i.e., to achieve the greatest impact in accomplishing patrol objectives).

It is important that the relationship among these three areas be clearly understood: The first two—the easier ones—are *preparatory* to increasing patrol productivity, a mobilization of resources with which to do the job. The maximum number of personnel is made available, and then the time of those people is unfettered by relieving them of useless or marginally useful activities. Thus, the pool of real manpower available to do the job is increased without any additions to the patrol force.

Neither step one nor two alone, however, will guarantee increased productivity. The payoff comes in using that manpower to the greatest advantage. That is the concern of the third problem area where the more difficult questions arise and the activities of the patrol force are related directly to patrol objectives.[3]

This chapter discusses, first, measurement; and, second, actual means of improvement, in each of these three problem areas. Many of the measures suggested are not "productivity measures" as some strict definitions may hold. But taken together they do suggest a set of quantitative measures that should prove helpful to police departments in assessing the performance of their patrol force.

Measuring Patrol Activity

In order to determine the type of data now collected and used for patrol evaluation, and to assess the range of performance among departments for specific activities (use of patrol time, response time, misdemeanor vs. felony arrests, etc.), the Advisory Group surveyed several police departments throughout the United States. The results of the survey show that:

- many police departments keep statistics needed to compute productivity measures that are adaptable for widespread use; and
- the range of performance for a variety of measures (allowing for probable differences in definition) suggests a potential for productivity improvement in most departments.

Other, more specific, results of the survey are quoted, where appropriate, throughout the report.[4]

Making More Patrolmen Available for Street Assignment

Assigning more of the patrol force to street work and increasing their effective time on patrol are important steps toward expanding the use of the existing resources of the patrol force. Although these efforts may not insure better performance, they are important to maximizing useful patrol time, and to minimizing or even eliminating the need for increasing the overall size of the force.

At any given time, only a small percentage of a patrol force is on the street. In great part this is due to the need to provide 7-day-a-week, 24-hour coverage; at least five men must be on the roster in order to fill any one patrol position around the clock. However, not all of the positions themselves are given to street assignments. Many are assigned to police headquarters, precinct stations, and other facilities in jobs that may not contribute directly to the crime-control and service-delivery objectives of the patrol function and that may not require the skills of a sworn officer.

A simple measure to help determine the ability of management to make manpower available for patrol is:

$$\frac{\text{Patrolmen Assigned to Street Patrol Work}[5]}{\text{Total Patrolmen}}$$

The percentage of the patrol force assigned to patrol appears to vary considerably among departments, indicating that there may be opportunity for improvement. Figure 9-1 presents percentages of the patrol force with street patrol assignments reported by six of the police departments that responded to this question in the Advisory Group's survey. At least two departments were able to maintain almost 90 percent of the patrol division on patrol duty.

It is impossible and unwise, of course, to put the entire patrol force out on the street. Some experienced patrolmen are needed for supervisory and other essential assignments at the station. The range of values for this measure, as shown in Figure 9-1, however, does indicate a potential in some departments for increasing the proportion of men on the street without adding any more sworn personnel.

Increasing the Real Patrol Time of Those Assigned

Once personnel are assigned to patrol duties, their time should be devoted to activities that may potentially meet patrol objectives. A simple measure to indicate the extent to which patrol time out in the field is being committed to patrol activities is:

$$\frac{\text{Man-Hours of Patrol Time Spent on Activities Contributing to Patrol Objectives}^6}{\text{Total Patrol Man-Hours}}$$

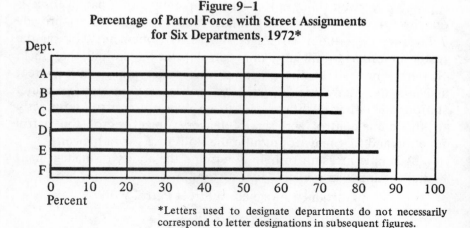

**Figure 9–1
Percentage of Patrol Force with Street Assignments
for Six Departments, 1972***

*Letters used to designate departments do not necessarily correspond to letter designations in subsequent figures.

410 *Police Productivity*

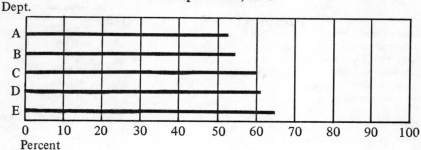

Figure 9–2
Percentage of Patrol Time Spent on Activities
Contributing to Patrol Objectives for Five
Police Departments, 1972

Time can be "lost" by performing nonpatrol tasks during duty hours. Examples are filling out unnecessary forms, servicing vehicles, running errands, and spending unnecessarily long hours waiting for court appearances. An analysis of the percentage of time spent performing such activities would be a preliminary step in diagnosing how patrol man-hours are really used.

The percentages of time spent by a patrol force on activities germane to their function (that is, random patrol, directed patrol, and responding to crime-related and noncrime calls for service) are shown in Figure 9-2. It is, of course, unlikely that close to 100 percent of time would be spent directly on patrol-related activities. Some time must be allocated to meal breaks, vehicle servicing, clerical tasks, court appearances, and the like. But the statistics indicate room for some improvement in several of the departments surveyed.

Given the rapidly rising costs of sworn personnel, even a small increase in the percentage of time spent on patrol activities can lead to a significant savings and, potentially, to increased effectiveness. Breaking down these statistics into portions of time spent on specific kinds of activity shows significant ranges in time allocations to different activities performed on a normal tour of duty. (See Table 9-1). The spread of time allocations is no doubt due to differing styles of operation and differing responsibilities of the police in their communities. But these ranges also suggest that some patrolmen may be spending too much time on nonpatrol activities.

**Table 9–1
Ranges in the Percentage of Time Allocated to
Different Patrol Activities, 1972 Survey Data**

Activity	Range (%) Low	Range (%) High	Time Allocation Should Be*
Random and other preventive patrol	16	36	High**
Crime-related calls for service	6	38	High
Noncrime calls for service	2	30	High
Training (on duty)	0	20	Medium
Report writing	2	6	Low
Arrestee processing	4	10	Low
"On duty" time in court	2	8	Low
Meal breaks	2	16	Low
Other	5	36	Low

* This column is included as a guide to help interpret the ranges in the table.
** Assuming a variety of patrol strategies and methods of deployment.

Maximizing the Impact of Patrol

So much for the relatively easy problems that focus on increasing the availability of existing resources. The more difficult problem is how to increase the effectiveness of those available resources. This requires relating patrol activities more directly to the three patrol objectives selected for consideration by the Advisory Group: crime deterrence, apprehension, and noncrime service.

Deterrence of Crime.[7] A principal objective of most police departments is to deter crime. The patrol force bears an important responsibility in this effort. Through its "preparedness" to respond to calls for service and its patrol activities, apprehensions, and investigations, the patrol force is expected to reduce the amount of crime actually committed by impressing would-be criminals with its ability to detect, to respond, to apprehend, and to marshal the support of the community.[8]

There are no altogether satisfactory measures of the success or failure of a patrol force's efforts to deter crime. Whereas apprehension, for example, can be measured directly from the number of arrests made, the number of crimes not committed—except for those few stopped in the act—is impossible to measure directly and can only be inferred. In fact, no persuasive relationship between overall patrol activities and crime deterrence has been established, as yet.[9]

In the absence of a direct measure of deterrence, three types of substitutes might be used:

- Existing reported crime indices used with discretion.
- Victimization surveys.
- Quantitative measurement of activities which professional judgment suggests contribute to deterrence.

For all its problems, reported crime is still one of the few measures that police managers have to provide some check—however general and unreliable—on their activities. Used judiciously, for specific types of crime, in specific districts, over specific periods of time, and with specific knowledge of what other factors (such as employement or age of population) may be affecting the result, reported crime can be a useful tool in evaluating the effects of patrol activities in deterring crime. There is hope that victimization studies will provide more reliable information; perhaps such information will permit a more accurate relationship to be established between the amounts of actual and reported crime, thereby increasing the usefulness of reported crime data.

This discussion focuses on the third method of assessing crime deterrence: the more precise measurement of patrol activities which are thought to contribute to deterrence.

Among the patrol activities thought to contribute directly to deterrence are apprehension and the ability to respond quickly to calls for assistance. Theoretically, a high likelihood of arrest undermines the confidence of would-be criminals and deters them from future crime. The extent to which this assumption is valid remains open to question. In any case, apprehension is considered to be an appropriate objective of the patrol force in and of itself, and is treated as such in the following section.

The remainder of this discussion will address the question of how to measure response time.

Assessing Patrol Response Time. Rapid response time may contribute to deterrence in at least three ways. First, there may be some deterrent effect in the knowledge that police can respond quickly to crimes in progress, although no indisputable correlation has been so established. Second, there is evidence that suggests that below certain time levels quicker response to crimes in progress does result in higher apprehension rates;[10] higher apprehension rates, in turn, may have some deterrent effect, although with qualifications as mentioned above. Third, rapid response probably does or could increase citizen confidence in the police, which in turn could encourage greater citizen involvement in the observation, reporting, and prevention of crime; such public involvement may, in turn, have some effect in deterring crime.

In short, there is no definitive relationship between response time and deterrence, but professional judgment and logic do suggest that the two are related in a strong enough manner to make more rapid response important. Moreover, response time is an important factor in achieving other police objectives, especially apprehension, which in turn contribute to increased deterrence.

Given these assumptions, we can turn our attention to the measurement of response time, bearing in mind that more rapid response is not an end in itself but a means of achieving patrol objectives.

Several factors must be considered in establishing a measure for response time. In the first place, different kinds of calls require different speeds of response. Nonemergency calls, for example, need not be answered as quickly as reports of crimes in progress. Some crime calls, in turn, may warrant a quicker response time than others, depending upon community priorities. Clearly, there are trade-offs in using existing men and equipment to respond to different kinds of calls. A low *average* response time for *all* calls, emergency and nonemergency, may mean sacrificing a quicker response capability for emergency calls. A decision to focus on emergency crime-related calls by deferring or "stacking" nonemergency calls may lengthen the overall average response time but significantly speed up responses to crimes in progress.

The desired response time to emergency calls must also be based upon some knowledge of the relationship between quicker responses and higher rates of apprehension. If reducing response time from 14 to 10 minutes produces little or no apparent increase in apprehensions or prevention of crime, its value is doubtful, except insofar as it may increase citizen confidence. Reductions from 5 to inside 3 minutes, on the other hand, may prove to produce significantly higher apprehension rates.

Still another important factor to be considered is the cost of reducing response time by given increments. If response time is already low, shaving off additional seconds may require heavy additional inputs of men and equipment or shifts of resources from other activities. The desired response time cannot be established on the basis of the projected result alone, but must also include consideration of the cost in new resources or men and equipment diverted from other activities.

The Advisory Group attempted to consider all of these factors in developing possible measures for patrol response time. No one measure adequately accounts for all of these factors, and consequently at least two measures should be used in concert with the kind of knowledge and judgment discussed above. Those two measures are:

$$\frac{\text{Number of Calls of a Given Type Responded to in Under ``X'' Minutes}}{\text{Total Calls of That Type}}$$

And

$$\frac{\text{Number of Calls Responded to in Under ``X'' Minutes}}{\text{Resources Devoted to the Response Function}}$$

"Resources devoted to the response function" is used in the denominator instead of patrol man-hours because of the potential for introducing capital-intensive technologies for improving the efficiency of response activities. Resources include patrol force salaries and benefits (still the major component) as well as the cost of computer-assisted dispatching systems and the salaries of nonsworn dispatchers.

"X" minutes is used in the numerator to indicate that different response times are appropriate for different types of calls. The value of "X" would depend on whether the call was an emergency or nonemergency call, or whether the call was about a crime-in-progress, suspicious activity, or previously committed crime. Additional breakouts by type of crime may also prove helpful. A report of a bank robbery, for example, may require a more rapid response than a larceny in progress. In each case, the department must determine for itself what is a desirable response time ("X") for a particular kind of call, based upon the considerations noted above.

To the extent that the measures reveal inefficient resource use, it would help, in diagnosing the problem, to divide response time into three segments:

- Dispatching delay
- Queue delay
- Travel delay

Dispatching delay is the time from receipt of a call to the time the dispatcher is ready to assign a unit. The queue delay is the time a dispatcher must wait before a unit is available for dispatch, and thus is calculated as zero if a car is available. Travel delay covers the time from dispatch of a unit to its arrival at the scene of an incident. To the extent that these components of response time can be recorded separately, they can be quite helpful in diagnosing the cause of an inefficient activity.

If a police manager is using his resources in the most effective manner, however, but response time still has not been reduced to an "acceptable" level, then the only available alternative is to seek an increase in resources that will be sufficient to enable him to obtain the desired response time. This assumes that emergency calls are being given priority, that nonemergency and service calls are being "stacked," that maintenance of patrol cars is managed in a way that keeps the maximum number available for responding to calls, and that shifts and positioning of cars place them where they are most likely to be required at the times of greatest need. When a police manager can show that all these things are being managed with maximum efficiency, but the response time still is not meeting his and the community's needs, then he can present a justified request for the resources that are needed.

Table 9-2 displays the range of response times obtained in the survey of departments. It demonstrates that at least some departments are able to furnish the kinds of response-time statistics useful in troubleshooting the causes of poor productivity. Of course, the data cannot point directly to inefficiencies in the use and deployment of patrol resources, but they may assist in searching them out. Unfortunately, the survey did not yield sufficient data on the extent of patrol resources devoted to the response function, although most

Table 9-2
Response Time Component Delays, 1972 Survey Data

Response Time Component	Category	RANGE (Minutes)	
		High	Low
Dispatch Delay	Emergency	3	1
	Nonemergency	10	2
Queue Delay	Emergency	1.5	0
	Nonemergency	10	0
Travel Delay	Emergency	5	3
	Nonemergency	14	3

departments have the data available to do this. Consequently, ranges in the response productivity measure cannot be presented.

No matter how quickly a department can respond to a call for service, productivity is sacrificed if the quality of the response is not up to par.[11] Thus, a further indication of this quality is a necessary adjunct to the principal response measure, and in some cases can be provided by a followup recipient-of-service survey.

Most of the departments responding to the Advisory Group's survey indicated that they are already carrying out some such form of survey. Telephone surveys could reveal what percentage of recipients were satisfied with the service. Questions asked should cover the effectiveness of the officers in performing the particular service as well as their courtesy and general helpfulness at the time. Criteria could be developed for these surveys which differentiate between satisfactory and unsatisfactory police responses to calls for service, and only a sample of recipients need be surveyed to obtain valid results. This information can be useful in uncovering persistent problems (or new ones) which may require some retraining of patrol officers.

Since one of the benefits of lowering response times is the opportunity to make more quality arrests by arriving at the scene of a crime in progress or by intercepting a fleeing suspect, departments might use the following measure of response effectiveness in leading to arrests:

$$\frac{\text{Arrests Surviving the First Judicial Screening}^{12} \text{ Resulting From a Response to a Crime Call}}{\text{Crime-Related Calls for Service}}$$

Again, this measure should be applied to appropriate categories of arrest (felony, etc.) and be calcualted separately for each major type of call. Suspicious activity and past-crime calls may not result in many arrests, but they are important for maintaining public confidence in the police and a feeling of security. As noted above, rapid responses to calls, especially to crimes in progress, can result in a higher rate of apprehension.[13]

Apprehension of Criminal Offenders. Traditionally, number of arrests has been used as an output measure of apprehension. Occasionally, other outputs, such as clearances and convictions, also are used. Although these may be useful "workload" measures for some police activities, as measures of output or results for productivity measurement, the Advisory Group found them subject to the following qualifications:

- Arrests themselves may be too easily subject to inflation; e.g., by making arrests of dubious merit (or increasing arrests for minor public nuisance offenses).
- Clearances (i.e., crimes for which police identify an offender, have sufficient evidence to charge him, and actually take him into custody) may be unsuitable because crime frequently cannot be attributed accurately to offenders. This figure can be adjusted according to particular department incentives—for example, over-attribution—if it becomes important to keep clearance rates high. In addition, clearance rates are not due solely to patrol activities, but also reflect investigative and prosecutorial activity.
- Although convictions reflect the quality as well as the quantity of patrol work, conviction rates usually are subject to many forces outside the control of the police (actions of courts, prosecutors, etc.).

Because of these difficulties, the use of arrests *surviving the first judicial screening* is a more appropriate "output." Although the process of judicial screening differs from one jurisdiction to another, thereby making interjurisdictional comparisons difficult, it usually involves an appearance before a judge or magistrate to assess whether or not a case has enough substance to merit a trial (probable cause). Survival of the screening process implies some measure of quality which arrests by themselves do not reflect (although some prosecutors and judges refuse to accept certain charges for various reasons).[14] Furthermore, the survival of arrests past the first judicial screening is less susceptible than convictions to forces outside police control. A poorly prosecuted case, for instance, can mean that an otherwise "valid" arrest will not result in conviction.

A suggested measure for apprehension productivity, then, is:

$$\frac{\text{Arrests Resulting From Patrol Surviving the First Judicial Screening}}{\text{Total Patrol Man-Years}}$$

Because patrol is a labor-intensive activity (80 to 90 percent of the costs of deploying a patrol force are salaries and benefits), patrol man-years is probably a more appropriate measure of resources than dollar costs.[15]

According to this measure, patrol productivity, in terms of the apprehension objective, would increase:

- if the number of arrests surviving first judicial screening per patrol man-year increased (e.g., through change in methods, deployment, etc.); or
- if the number of arrests per patrol man-year remained the same but the effective patrol time (number of patrol man-years actually employed in patrol activities) of the existing patrol force increased sufficiently to permit fewer sworn officers to be assigned to the patrol force.

To illustrate briefly, consider a patrol division having 500 sworn personnel which only manages to maintain 50 positions in the field round-the-clock. If each position accounts for, say 30 quality arrests per year, the above productivity measure computes to (50 X 30)/500 = 3 quality arrests per man-year. If the department fielded more than 50 round-the-clock positions from its force of 500 (it is generally recognized that 5 men are required for every round-the-clock street position), then any number of additional quality arrests made as a result would increase patrol productivity so long as patrol strength remained at 500. An extra 10 fielded positions also averaging 30 quality arrests per year would yield an overall result of (60 X 30)/500 = 3.6 quality arrests per man-year. If an extra 18 positions were fielded with a lower overall arrest rate of 25, patrol productivity would still be increased from 3 to 3.4 quality arrests per man-year.

In addition to the productivity measure suggested above, police departments can develop other, more detailed, measures to provide useful information. Among the most important of these is an apprehension productivity measure for each major arrest category For example:

$$\frac{\text{Felony Arrests Resutling From Patrol Activities Surviving First Judicial Screening}}{\text{Total Patrol Man-Years}}$$

This measure can be modified for consideration of different kinds of arrests, including:[16]

- Felonies.
- Misdemeanors that involve a particular victim.
- Consensual crime misdemeanors as determined by local jurisdiction.
- Other violations.

Different types of arrests have different values, which can in turn vary from community to community. An armed robbery arrest and an arrest for public drunkenness clearly are not equivalent. Arrest totals may be inflated by legitimate arrests for petty and often so-called "victimless" crimes, which do not reflect police goals for more serious crimes.

Moreover, using the measure to calculate separately one arrest productivity measure for each of these categories may tell managers where their arrest emphases lie, and will allow them to assess whether or not the results are in accord with their particular priorities.

For example, the ratio of felony to misdemeanor arrests in Long Beach, California, is 0.22, while in nearby Compton the ratio is 0.77. For drug offenses, the ratios for the same two cities are even more extreme—0.66 in Long Beach and 114.3 in Compton. As the study reporting these statistics explains:

> They (the range of these ratios) . . . cannot be accounted for by differences in crime patterns; they flow mainly from differences in police arrest policies.[17]

Schemes that give weights to different types of arrest and that compute a combined arrest index are generally undesirable because the weights often must be arbitrarily chosen and are therefore unrelated to public concerns about certain crimes.

A variant of the preceding measure can be applied when evaluating the success of a specific, directed patrol strategy—for example, the concentration of uniformed or plainclothes preventive patrols in areas with particular crime problems. The measure would read:

$$\frac{\text{Arrests (Resulting From a Directed Patrol Strategy)}}{\text{Total Directed Patrol Man-Years}}$$

This measure could be adjusted, as above, by specifying the type of arrest to evaluate the impact of the strategy on a particular crime category.

It is important to remember that the amount of crime in a jurisdiction may bias apprehension productivity data. Because different crime rates represent different opportunities for making arrests, it is often easier for a patrol force to make more arrests in the presence of a higher crime rate. By close examination of the productivity improvements, those due merely to an increase in the crime rate could be distinguished from those due to better use of the patrol force. The clearance rate also may be of some use as an adjunct because it reflects to some extent how well a department is matching apprehensions to crimes that actually occur.

A second important source of information for police managers is the ultimate disposition of arrests, which provides an additional check on the quality of apprehensions and postarrest activities. Even though judicial screening, used in the original arrest productivity measure, imposes some quality standards for arrests, such screening occasionally can be perfunctory. Two additional measures of the quality of arrests are:

$$\frac{\text{Convictions}}{\text{All Arrests Made by Patrol Force}}$$

Convictions

Arrests Resulting From Patrol That
Survive the First Judicial Screening

These measures also may be calculated separately for each arrest category to provide more detailed information. Although these two measures are determined by factors beyond police control, they do reflect somewhat the quality of police discretion in making arrests and the effectiveness of postarrest activities (e.g., preparation for testimony).

Thus, police managers may determine—by breaking down the apprehension productivity measure according to crime category and comparing the results to crime statistics—what relative importance the department places, implicitly or explicitly, on various types of crimes. Police managers can evaluate, further, the quality of police arrests by examining the portion of all arrests surviving first judicial screening made in a given crime category and resulting in conviction of the offender.

Responses to an Advisory Group survey question on the number of felonies and misdemeanors surviving first judicial screening per patrol man-year showed a wide variation among police departments.[18] Whereas one department reported a combined number of felony and misdemeanor arrests surviving judicial screening of 61.5 per man-year, another reported only 9 per man-year.[19]

Figure 9-3 shows not only a range of arrests per patrol man-year among seven police departments, but also reflects the relative emphasis placed by these departments on felony vs. misdemeanor arrests. The department which reported the highest number of arrests surviving judicial screening also reported over 80 percent of these as misdemeanors.

Four other departments, though with lower overall arrest productivity levels, also reported a much higher percentage of misdemeanor arrests than felony arrests. Two departments reported more felony than misdemeanor arrests.

Figure 9-4 shows the number of felony and misdemeanor arrests surviving first judicial screening for the portion of the patrol force actually engaged solely in patrol activities (for the four departments that could provide such information). Although arrest rates are higher for all four departments because the patrol force resource base

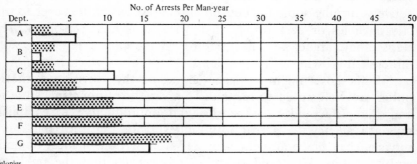

Figure 9–3
Felony and Misdemeanor Arrests Surviving the
First Judicial Screening (resulting from patrol)
Per Patrol Man-year, 1972 Data for Seven Police
Departments

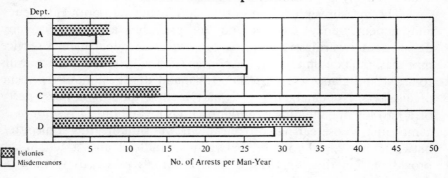

Figure 9–4
Felony and Misdemeanor Arrests Surviving the
First Judicial Screening (resulting from patrol)
per Man-Year Actually Spent on Patrol, 1972 Data
for Four Police Departments

is lower, those departments that have the largest percentage of the patrol force on the street and expend the largest percentage of time in patrol activities have the highest number of arrests per patrol man-year.

Police Productivity 423

Provision of Noncrime Services. Services provided by the patrol force that do not relate to incidents of crime or suspicious activities make up the large majority of calls for service, often 70 percent or more. Despite their predominance in the patrol workload, police departments put the most emphasis on crime-control activities and stress crime control in their training programs. Many departments are actively turning some of their noncrime responsibilities over to other city agencies or performing them with nonsworn personnel. These steps, they argue, are essential to release police resources to be directed at growing crime problems.

Some government managers argue, on the other hand, that the police are well suited to respond to a variety of noncrime situations and that it would be expensive and unproductive to establish a separate agency to perform those tasks. In the end, the mix of services that the police do provide is a function of local objectives and priorities. Whatever the mix, it is certain that the public will continue to expect the police to provide a 24-hour response capability for a variety of emergency and nonemergency needs. Even if some of these needs are met by nonsworn personnel, they still remain the concern of police managers. To the extent that police continue to provide noncrime services, with both sworn and nonsworn personnel, these services should be provided as efficiently and effectively as possible.

The measures presented in this section apply to both emergency and nonemergency services and are probably applicable to any department regardless of its mix of services. A department's service mix may include emergency responses such as ambulance runs, calls to assist ambulance crews, rescue runs, and deployment at the scene of disasters. Nonemergency calls may include quieting a noisy party or a barking dog, helping a resident who is locked out of his house or a motorist whose car has broken down, or "adjudicating a dispute" between two neighbors or between a landlord and a tenant. In providing noncrime services, a patrol force's productivity may be determined by the following measure:

$$\frac{\text{Number of Noncrime Calls for Service Satisfactorily Responded To}}{\text{Man-Hours Devoted to Noncrime Service Calls}}$$

Here the number of calls includes both emergency and nonemergency situations. The quality of the response should be sufficient to satisfy the recipient of the service. Therefore, supplementary information provided by a followup survey should be used in conjunction with this measure.

Calculating the measure separately for emergency and nonemergency noncrime calls may be useful, but not as useful from a management point of view as the calculating separately of measures for major categories of noncrime service calls. This more detailed kind of information is likely to indicate where action can be taken to improve noncrime service productivity.

For example, the effectiveness of the force in responding to medical emergencies (accidents, impending births, etc.) could be assessed by a measure of this type:

$$\frac{\text{Medical Emergency Calls That Emergency Room Personnel Evaluate as Having Received Appropriate First Aid}}{\text{Total Medical Emergencies}}$$

An evaluation procedure can be developed to provide data for this measure in cooperation with emergency room staffs at local hospitals. A sample, rather than an evaluation of every case, may be sufficient for determining effectiveness in this area. Too low a value for this measure may indicate inadequate first-aid training or equipment.

Another measure may be developed for calls regarding noisy disturbances in the community:

$$\frac{\text{Noisy Disturbance Calls for Which No Further Attention Is Required (For the Remainder of the Patrol Tour)}}{\text{Total Noisy Disturbance Calls}}$$

Too low a value of this measure may indicate a lack of respect for the police in the community or a lack of tact on the part of officers handling such incidents. A low value may also indicate a need for additional training of some officers and/or a better community relations program.

Regardless of the service mix provided by a particular department, these measures may be helpful to the police manager in diagnosing a productivity problem in the patrol force's provision of noncrime services.

Several of the departments responding to the data survey were able to provide data on numbers of calls for noncrime services and the man-hours devoted to answering those calls during 1972. Figure 9-5 presents the ratios of noncrime calls for service to man-hours spent in answering such calls.

Some of the reasons for the wide range of these statistics may merely be such factors as larger patrol sectors and the longer travel times they imply. Other factors, however, may be excessive service times or excessive amounts of paperwork associated with each incident.

Police departments can be more productive in meeting noncrime service objectives if they carefully analyze what is required to provide these services. For example, is a sworn officer always needed? Can a report be taken by phone? How much time is really needed for this type of call? By answering such questions, police managers can apply the necessary resources and accomplish noncrime service objectives more productively.

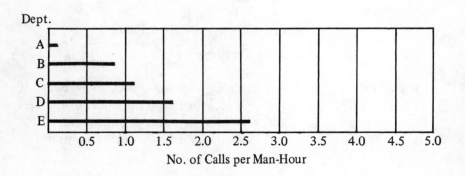

Figure 9-5
Number of Noncrime Calls for Service per Man-Hour Spent Answering those Calls, 1972 Data for Five Police Departments

Topics for Discussion

1. Discuss the problems experienced with current police measures.
2. Discuss the use of victimization surveys.
3. Discuss ways of increasing productivity in police services.
4. Discuss the process of experimentation.
5. Discuss the objectives of police patrol.
6. Discuss the means of assessing patrol response time.

FOOTNOTES

Introduction

[1] The full title of the *UCR* dated August 8, 1973, is *Crime in the United States, 1972–Uniform Crime Reports.* Since it was first published in 1930, the *UCR* has been the only nationwide data source on crimes committed throughout the United States. The report has been improved over the years by refinements in data collection, comparison, and dissemination, and it will undoubtedly continue to be used in the future as an indication of crime rates and police performance.

[2] The rape victim fears the reaction of relatives or suitors. Parents fear sex crimes involving their children may bring unwanted publicity. Burglary victims fear higher insurance rates, loss of coverage, or simply that publicity will make them a target for other burglars. Fear of retaliation by the perpetrator and a reluctance to involve spouses or friends are important factors in underreporting personal assault or robbery crimes.

[3] See Athony G. Turner, "Victimization Surveying–Its History, Uses, and Limitations," Appendix A in *Report on the Criminal Justice System*, Uncorrected Proof Copy, National Advisory Commission on Criminal Justice Standards and Goals, Washington, D.C., soon to be published.

[4] A brief though more detailed description of the National Crime Panel can be found in the forthcoming *Report on the Criminal Justice System*, ("The National Crime Panel," part of Appendix A, 110-112 of the uncorrected proof copy), National Advisory Commission on Criminal Justice Standards & Goals, 1973. It may be that after several years of victimization survey data collection, stable correlations may be demonstrated by crime category, type of population, or by some other breakdown of the data, between reported

crime and total crime. In this event, police agencies would have a relatively cheap tool of very great utility for productivity management. However, it cannot yet be said whether or to what extent this may happen.

[5] *Task Force Report: Science and Technology,* Report of the President's Commission for Law Enforcement and the Administration of Justice, U.S. Government Printing Office, Washington, D.C., 1967, pp. 60-61.

[6] One list of police goals provided by the American Bar Association's Project on Standards for Criminal Justice (which was also approved by the executive committee of the International Association of Chiefs of Police) ranges from such general goals as safeguarding freedom and developing a reputation for fairness, civility, and integrity, to more specific goals such as identifying and apprehending offenders and facilitating the movement of people and vehicles. The Advisory Group did not attempt to define the overall police responsibility or to develop an authoritative list of police goals and functions. Rather, it decided to focus its attention on selected activities which are of top concern to police chiefs. The debate over proper police goals no doubt will, and should, continue. And efforts to measure police activity must be attentive to the changing perception of police responsibilities.

Productivity and Police Services

[1] *The Municipal Yearbook 1973,* International City Management Association.

[2] Subject to acquiescence of citizens in the affected areas, for they have been made to believe, over the years, that random patrol has real deterrent value.

[3] By the definition employed here, productivity is akin to, if not indistinguishable from, efficiency. However, the Advisory Group has used the term in a way that implies a greater concern for effectiveness and quality or value of service than is usually associated with the term "efficiency."

Others concerned with this broader concept may define productivity as effectiveness over input, or a cost-effectiveness ratio. There are no hard and fast rules for such measurements. Different kinds of problems lend themselves to different kinds of analytical formulations. At this stage in the process of police measurement, it is probably less important to quibble about these conceptual differences—although they no doubt will become more important as the art advances—and more worthwhile to get on with trying to establish and test measures that seem useful to police managers.

[4] While conceptually it may be useful to assume that effectiveness does not concern itself with resource input or cost, in practice the term "effective" often is used in a way that assumes a reasonable economy of resource use. This consideration, and the relationship between effectiveness and productivity, may be clarified in the following example.

Suppose that one measure of police productivity is the number of *valid* arrests per patrolman per year. If in a year a force of 100 men made 500 such arrests, then its productivity for that activity would be 5.0 arrests per patrolman.

However, if the force were reduced by 20 men, and if the remaining 80 men made 480 arrests the following year, then the productivity of the force would increase to 6.0 arrests per patrolman. But while productivity increased, the actual number of valid arrests made decreased. In that sense, the effectiveness of the force in apprehending criminals (assuming the same level of criminal activity) has declined.

Whether effectiveness actually declined, of course, depends upon how the goal of the police department is defined. If its goal is to apprehend as many criminals as possible commensurate with a reasonable degree of public order *and* cost, then perhaps the slight decline in arrests is acceptable, especially in light of the apparently large cost savings that resulted. In this type of case the goals of increasing productivity and increasing effectiveness become intimately related.

The ideal productivity gain, of course, is where the same force of 100 men is able to increase its number of valid arrests from 500 to, say, 600. In this case, both the productivity and effectiveness of the force—by this measure—are increased: more valid arrests are made with no increase in manpower.

Productivity in Police Patrol

[1] For examples, see Richard C. Larson, *Urban Police Patrol Analysis,* Cambridge, Massachusetts: The MIT Press, 1972, p. 27 ff.

[2] Two other important objectives should also be mentioned. The first is the recovery of stolen property, which is relatively easily measured either by the value of stolen goods or units of stolen items recovered. The second objective is to provide the community with a sense of security and a feeling of confidence in its police force, a more difficult objective to measure. The public's attitudes toward crime and the police nevertheless can be assessed through a variety of means, including opinion surveys. And while the Advisory Group did not have the time to examine this question, its importance should not go unrecognized.

[3] Clearly, discretion must be used in determining the division of labor within any police department or patrol division. Some sworn officers will always be needed for other than street assignment, and some of their time may be required for what are considered to be non-patrol-related activities. The Advisory Group's approach focuses on the situation found in many departments across the country: within the patrol division, too many sworn officers are performing jobs they should not be performing, and of those officers with street assignments, too much time is spent on activities unrelated to patrol objectives. Improvement in these two areas could, in most

departments, contribute significantly toward making more people available for the real role of the patrol force, regardless of how departments wish to define that role. This does not necessarily mean that additional men should be assigned to traditional patrol activities, as will be discussed further in this chapter.

[4] The survey was conducted by distribution of three questionnaires (one each on Patrol, Crime Prevention, and Human Resources) to the following 11 law-enforcement departments and agencies: Police Departments of Boston, Massachusetts; Charlottesville, North Carolina; Cincinnati, Ohio; Kansas City, Missouri; Miami Beach, Florida; New York, New York; Oakland, California; and Washington, D.C.; the Los Angeles (California) Sheriff's Department; the Michigan State Police; and the St. Petersburg (Florida) Department of Public Safety.

This sample was not intended to be statistically valid on a nationwide basis. It was considered sufficient, however, for the limited objectives of the survey, which were:

- To check whether data required for use in measures being developed are normally available.
- To test whether the measures are feasible when actual data are applied to them.
- To obtain some idea of the ranges (or disparities) existing so that the potential for improvement could be assessed.

Time available for conducting the survey did not permit a pretesting of the questionnaires, which would have enabled the Advisory Group's staff to refine definitions of the terminology used. Therefore, not all respondents provided comparable data in response to all questions. For that reason, the survey results, as depicted in figures and tables throughout this report, provide approximate data for those departments (8 in some cases, only 4 in others, etc.) whose responses seemed to reflect a common understanding of the categories of data solicited.

It should be noted in particular that, to preserve the anonymity of respondents, identification of departments is by letters only; moreover, Department "A" in Figure 9-1 does not necessarily correspond with Department "A" in Figure 9-2, etc.

[5] This measure does not, of course, indicate whether the patrolmen thus assigned are accomplishing anything useful. It is an indication of the department's success in making sworn officers available for more directly patrol-related activity.

[6] As noted for the previous measure, this measure does not indicate whether the time made available is put to good use. It does measure success in making more time available which can be turned to good use.

[7] This section deals with the crime deterrrence activities generally associated with the work of patrol.

[8] Richard C. Larson, *op. cit.*, p. 32.

[9] For a thorough discussion on this subject, see Franklin E. Zimring, *Perspectives on Deterrence,* prepared for the National Institute of Mental Health Center for Studies of Crime and Delinquency, Public Health Service, Publication No. 2056, January 1971, U.S. Government Printing Office, Washington, D.C.

[10] See Appendix B, "A Study of Communications, Crimes, and Arrests in a Metropolitan Police Department," Herbert H. Isaacs, *Task Force Report: Science and Technology,* President's Commission on Law Enforcement and Administration of Justice, 1967. U.S. Government Printing Office, Washington, D.C.

[11] The arrival of an officer at the scene of an incident assumes a certain "quality" factor—i.e., a police officer is on the scene. That has some intrinsic value, but more discriminating information is needed to determine what the officer does when he gets there, i.e., the "quality" of the response.

[12] See the following section on "Apprehension of Criminal Offenders" for explanation of this language.

[13] Herbert H. Isaacs, *op. cit.*

[14] In Los Angeles County, District Attorney rejection rates in cases involving possession of dangerous drugs vary from 26 percent for the Whittier Police Department to 69 percent for the Long Beach Police Department. For robbery, the rejection rates vary from 6 percent in Compton to 53 percent for the Los Angeles Sheriff's Department. A major reason cited in the report from which figures are taken, though by no means the sole one, for these rejection rates is that police departments vary greatly in their own screening of felony cases. See Peter W. Greenwood et al, *Prosecution of Adult Felony Defendants in Los Angeles County: A Policy Perspective,* Report R-1127-DOJ. The RAND Corporation, Santa Monica, California, March 1973, page ix.

[15] "Total Patrol Man-Years" refers to all sworn officers in the patrol division whether or not they are assigned to street work. Note that, instead of years, such other units of time as months, days, or hours can be used, depending upon which makes more sense to the user.

[16] Individual police departments should develop and use crime categories which reflect as accurately as possible the true mix of arrests in the community and which provide them with the specific information that they require.

[17] Greenwood, *op. cit.,* pp. viii and ix. It should be mentioned that the report does not make clear whether these arrest ratios are for arrests that survive the first judicial screening. The ranges indicated, however, would not change very much if the data were qualified by judicial screening.

[18] The man-years include those expended by members of the patrol force in supervisory, clerical, and other nonpatrol assignments to give a clear picture of the total resources expended.

[19] Although these data have not been adjusted for definitional differences among departments, differences in the definition of misdemeanor and felony alone could not account entirely for the range in performance.

SELECTED READINGS

Gylys, Julius A., "Application of a Production Function to Police Patrol Activity," *The Police Chief,* Vol. XLI, No. 7, July, 1974, pp. 70-71. Describes the application of production theory to police patrol activities. Discusses the lack of knowledge about deterrence and the disparity in the quality of police labor and capital inputs.

Jones, E. Terrence, "Evaluating Everyday Policies—Police Activity and Crime Incidence," *Urban Affairs Quarterly,* Vol. 8, No. 3, March, 1973, pp. 267-279. Discusses a mode of analysis to study the impact that year-to-year changes in police protection expenditures, uniformed police manpower, and civilian police manpower have on the rates of eight types of crime in 155 American cities.

Kelling, George L., Tony Pate, Duane Dieckman and Charles E. Brown, *The Kansas City Preventive Patrol Experiment,* Washington: Police Foundation, 1974. A summary report of the findings of an experiment to test the effectiveness of the traditional police strategy of routine preventive patrol.

Lind, Robert C. and John P. Lipsky, "The Measurement of Police Output: Conceptual Issues and Alternative Approaches," *Law and Contemporary Problems,* Vol. 36, No. 4, Autumn, 1971, pp. 566-588. This paper discusses the conceptual and practical difficulties of defining and measuring police output. Emphasizes the need to measure and evaluate the police as part of the criminal justice system and as part of the social environment.

Morgan, J.M., Jr., "Police Productivity," *The Police Chief,* Vol. XLI, No. 7, July, 1974, pp. 28-30. Reviews the activities of the advisory group on productivity in law enforcement in terms of the concept of productivity and a consideration of means of measurement.

National Commission on Productivity, *The Challenge of Productivity Diversity,* Part IV, Washington: National Commission on Productivity, June, 1972. Includes three case studies in police crime control with an emphasis on the explanatory work of the procedures for locating and evaluating innovations.

Name Index

Angel, J.E., 163, 171, 218
Argyris, Chris, 47, 226
Ashby, John, 180
Bard, M., 177
Barnard, Chester, 170
Bell, Cecil H. Jr., 225
Bendix, R., 168
Bennis, Warren, 165, 229
Bidwell, Alvin C., 88fn
Blake, Robert T., 88fn
Columbus, Eugene C., 113 fn
Coppock, J. Laverne, 54fn
Cribbin, James J., 334
Davis, Richard M., 149fn
Enthoven, Alain C., 140
Etzioni, Amitai, 47
Fielding, Henry, 4, 5
Fielding, John, 4, 5
Fosdick, Raymond B., 16, 20
French, Wendall, L., 225
Fuld, Leonhard F., 12, 13, 14, 15
Galvin, Raymond T., 180
Gibb, Jack, 71
Goodall, Robert, 304
Gourley, G. Douglas, 45fn
Graper, Elmer D., 20, 21, 22, 23, 24, 25
Hughes, Charles L., 302
Katz, Robert L., 92
Kenney, John P., 361
Lawrie, J.W., 321
LeBaron, Melvin J., 244
LeGrande, James L., 180

Leonard, V.A., 34, 35
Likert, Rensis, 75
Maslow, Abraham, 68
McAllister, John A., 97
McGregor, Douglas, 70, 77fn, 302
Moore, Sir John, 7
Mouton, Jane S., 88fn
Millsson, Ernst K., 119fr
Peel, Robert, 6, 11
Pence, Gary, 205
Presthus, Robert, 46
Quade, E.S., 143
Randall, Lyman K., 230
Reith, Charles, 7
Rhodes, W.R., 68fn
Robinson, Revis O., 259
Rowan, Charles, 7, 11
Savord, George H., 381
Schmidt, Warren H., 343
Scott, William, 50
Shell, Richard L., 140fn
Simon, Herbert, 46
Smith, Bruce, 25, 26, 27, 28
Stelzer, David F., 140fn
Szanton, Peter, 140
Tannenbaum, Robert, 53, 343
Thibault, Fred D., 244
Tullock, G., 170
Unsinger, Peter C., 192
Vollmer, August, 35, 362
Wasserman, R., 177
Weber, Max, 164, 168
Wilson, O.W., 31, 32, 33

Subject Index

Administration
    status quo, 55
Apprehension, 418
Arrests, 417
Authority, 167

Basic car plan, 97, 375
Behavioral sciences, 45, 68
British, 8, 9, 10, 11
Budget
    calendar, 201
    guidance, 210
    involvement, 204
    priorities, 203
    structure, 211
Budget document, 199, 202
Bureaucracy, 163, 165, 167

Centralization
    detectives, 24
Centralization of authority, 5
Change
    environmental, 166
    management, 55
Collaborative management, 227
Community relations, 166
Complaint processing system, 150
    case study, 151
Coordination, 29
    tasks, 174
Coordination and Information Section, 174
Communication, 170
Computer
    decision making, 148
Constraints, 142
    philosophy, 142
Control, 170, 172

Corrections, 127
Courts, 127
Criminal Justice System, 126, 127, 133
Culture, 165

Data processing, 118
Decentralization, 105, 176
Decision, 213
Decision making
    involvement, 169
Democracy, 167
Detectives, 23
    administration, 13
    organization, 24
Diagnostic leadership, 330
    environment, 331
    motives, 331

Ethics, 178
Executive
    strong leadership, 13
Experimentation, 190, 406

Fiscal control, 200
Fiscal management, 198
Fledging police organization, 4
Formal organization, 45
Forces
    formal, 47
    informal, 47
Force Field Analysis, 60, 61, 62
    negative pressures, 61
    status quo, 61
Force field, 232

General Instructions, 8
General Services Section, 172

Goals, 286, 288
Group leadership, 387

Hierarchy of needs, 68
   self-fulfillment, 69
Human relations, 231
Human Side of Enterprise, 87

I.C.M.A., 28, 29, 30
Informal organization
   communications, 57
Information dissemination, 5
Inspection
   functions, 15
Interviewing, 189

Job enlargement, 386

Leader, 327
   democratic, 334
   forces in the manager, 351
   forces in the situation, 353
   forces in the subordinate, 352
   free-rein, 334
   organization, 329, 341
Leadership, 5, 21, 34
   administrative aids, 27
   behavior, 345, 346, 347
   focus, 344
   personality, 336
   process, 335
   situation, 340
Leadership styles
   charismatic, 322
   followers, 324, 325, 326
   magical, 322, 324
Leveling, 234
Los Angeles Police Department, 99

Man
   ego needs, 81
   physiological needs, 79

safety needs, 80
   self-fulfillment needs, 82
   social needs, 81
Managerial Grid, 88
   country club, 88, 89, 90
   dampened pendulum, 88, 92
   impoverished, 88, 91
   task, 88, 89
   team, 88, 92
Management
   control, 20, 23
   democratic, 51
   functions, 35
   paternalistic, 93
   principles, 28
   trends, 53
Management climate
   data collector, 75
Management theory
   behavioral school, 48
   classical school, 48
   systems approach, 49
Managing for results, 278, 305
   appraisal, 305
   attributes, 279
   audit, 282
   design, 282
   elements, 278, 280
   goals, 286, 288
   key results analysis, 308
   key tasks, 309
   standards, 310
Manpower
   distribution, 22
Measurement, 398
   output, 399
   quantitative, 400
Metropolitan Police Act, 6
Military model, 7
Model, 128, 154, 164, 171, 214, 215, 364, 367
   decision phase, 131

democratic, 171
military, 179
population, 129
results, 157
simulation, 149
team, 364
Morale, 75, 95, 168, 169
conflict, 169
low, 169
Motivation, 82
capacity, 84
carrot-and-stick, 83

Nickel auction, 244
Noncrime services, 424

Objectives, 119, 120, 139, 405, 40
criteria, 121
defined, 142
order maintenance, 120
public service, 120
system, 139
writing, 300
Observation, 190
Opportunity, 128
Organization
administrative man, 46
characteristics, 18, 164
definition, 28
distribution, 21
function, 21
informal, 30
line and staff, 34
mechanical, 17
principles, 28, 29, 31, 32
span of control, 26
structure, 21, 26
system, 50
unity of command, 34
Organization development, 225
change agent, 228
culture, 226
definition, 225, 230, 245

diagnosis, 262
feedback, 235
leveling, 234
objectives, 259
risk-taking, 235
self-renewal, 226
stages, 261, 262

Participative management, 57, 75, 382
contribution, 364
controller/moderator, 58
group meetings, 58
problem solving, 59
staff meetings, 58
Patrol, 408
deterrence, 412
measuring, 409
noncrime services, 426
response time, 413
Patrolman
supervision, 14
Peel's guidelines, 11
Personnel
alienative, 47
behavior, 46
calculative, 47
management, 19
normative, 47
types, 46, 47
Personality traits, 165
PERT
concept, 193
event table, 195
likely time, 193
optimistic time, 193
pessimistic time, 193
Planning, 180, 367
definition, 181
goals, 183
management, 181
operational, 181
procedural, 181

Plans
  intermediate, 182
  long-range, 182
  short-term, 182
Police, 127
  activities, 26
  auxiliary functions, 30
  detail, 22, 23
  objectives, 31
  problems, 166
  records bureau, 25
  women, 24
Police duties
  extraneous, 15, 19
Policeman
  morality, 13
Police service, 397
Police system, 125
  follow-up force, 135
  outputs, 136, 137
  preventive force, 135
  response force, 132, 134
Politics, 14, 167
PPBS
  components, 206, 207
  methodology, 207
  objectives, 208
  priorities, 209
  problems, 215
  purpose, 212
  requirements, 206
Principles, 384
Problem solving
  management problems, 66
  policy making, 67
  seven steps, 63, 64, 65, 66
  spiral model, 144
Process, 233
Productivity
  definition, 401
  effectiveness, 404
  objectives, 405

Questionnaires, 190

Reorganization, 99
Response time, 416
Responsibility, 21, 167

Salary, 178
Satisfaction, 90
Selection of personnel, 5
Specialization, 27, 29
Specialized Services Section, 175
Supervision
  first line supervisor, 14
  involvement, 73
  variations, 72
  view points, 71
    participative, 72
    traditional, 71
System
  analysis, 113, 116, 115, 119
  application, 149
  documentation, 117
  implementation, 117
  organization, 114
  survey, 115
  synthesis, 116
Systems analysis, 113, 119
  alternatives, 124
  basic acts, 141
  decision making, 140
  definition, 122
  needed, 145
  steps, 123, 141, 150
  sub-systems, 122

Tactical planning, 182
Task, 233
Team building, 246
  agenda, 248, 249
  nature, 247
  purpose, 249
  readiness, 257, 258
  results, 250

Team management, 96
Team policing
  community, 380
  definition, 372
  model, 364, 367, 371
  organizing, 381
  personnel, 365, 379
  prevention, 378
  principles, 384
  problems, 373
  programs, 366
  total, 374
Teams, 172, 362
  valuation, 173
Territorial imperative, 97
Theory, 173

Theory X, 70, 77, 236
  averageman, 78
  propositions, 77
Theory Y, 70, 77, 84, 236
  decentralization, 86
  job enlargement, 86
  motivation, 84
  self-control, 85
  self-direction, 85
Top management team, 368
Types of police work, 22

Uniformed crime reports, 394

Vice, 27

Work group, 337